278 COMBINATION
MACRO 57
CLASS 29
CAC

DIGITAL
COMPUTER
DESIGN

Prentice-Hall Computer Applications in Electrical Engineering Series

FRANKLIN E. KUO, *Editor*

ABRAMSON AND KUO *Computer-Communication Networks*

BOWERS AND SEDORE *SCEPTRE:*
A Computer Program for Circuit and Systems Analysis

CADZOW *Discrete-Time Systems: An Introduction with Interdisciplinary Applications*

CADZOW AND MARTENS *Discrete-Time and Computer Control Systems*

DAVIS *Computer Data Displays*

FRIEDMAN AND MENON *Fault Detection in Digital Circuits*

HUELSMAN *Basic Circuit Theory*

JENSEN AND LIEBERMAN *IBM Circuit Analysis Program:*
Techniques and Applications

JENSEN AND WATKINS *Network Analysis: Theory and Computer Methods*

KLINE *Digital Computer Design*

KOCHENBURGER *Computer Simulation of Dynamic Systems*

KUO AND MAGNUSON *Computer Oriented Circuit Design*

LIN *An Introduction to Error-Correcting Codes*

NAGLE, CARROLL, AND IRWIN *An Introduction to Computer Logic*

RHYNE *Fundamentals of Digital Systems Design*

SIFFERLEN AND VARTANIAN *Digital Electronics with Engineering Applications*

STAUDHAMMER *Circuit Analysis by Digital Computer*

STOUTEMYER *PL/1 Programming for Engineering and Science*

DIGITAL COMPUTER DESIGN

RAYMOND M. KLINE

Washington University
St. Louis, Missouri

Prentice-Hall, Inc., Englewood Cliffs, New Jersey

Library of Congress Cataloging in Publication Data

KLINE, RAYMOND M.
    Digital computer design.

    Includes bibliographies and index.
    1. Electronic digital computers—Design and construc-
tion. I. Title.
TK7888.3.K56   621.3819′58′2   76-9865
ISBN 0-13-214205-8

10   9   8   7   6   5   4   3   2   1

Printed in the United States of America

PRENTICE-HALL INTERNATIONAL, INC., *London*
PRENTICE-HALL OF AUSTRALIA PTY, LTD., *Sydney*
PRENTICE-HALL OF CANADA, LTD., *Toronto*
PRENTICE-HALL OF INDIA PRIVATE LIMITED, *New Delhi*
PRENTICE-HALL OF JAPAN, INC., *Tokyo*
PRENTICE-HALL OF SOUTHEAST ASIA PTE. LTD., *Singapore*

To *Marian*
Dave
Kathy
Sue
and John

CONTENTS

PREFACE   *xi*

1 | *AN OVERVIEW*   *1*

1.1	*The Evolution of Digital Computers*	*4*
1.2	*The Growth of Computer Technology*	*8*
1.3	*Classification of Computers*	*11*
1.4	*Important Digital Computer Characteristics*	*13*
1.5	*Prologue to the Remainder of the Text*	*16*
1.6	*Bibliography*	*17*
1.7	*Topics of Further Study and Student Reports*	*18*

2 | *NUMBER SYSTEMS*   *19*

2.1	*Arabic Notation*	*19*
2.2	*Conversion of Integers*	*20*
2.3	*Conversion of Fractions*	*23*
2.4	*Addition and Subtraction*	*24*
2.5	*Multiplication and Division*	*27*
2.6	*Codes*	*29*
2.7	*Bibliography*	*39*
2.8	*Exercises*	*40*

3 | *SOFTWARE*   *44*

3.1	*Basic Machine Organization*	*45*
3.2	*Machine Language Programming*	*47*
3.3	*Programming Languages*	*54*
3.4	*Bibliography*	*58*
3.5	*Exercises*	*59*

4 | **LOGICAL DESIGN** 63

4.1   Basic Logical Operations   63
4.2   Theorems   66
4.3   The Canonical Expansion of Functions   71
4.4   The Karnaugh Map   75
4.5   Logic Diagrams   83
4.6   NAND-NOR Logic   85
4.7   EXCLUSIVE-OR Logic   91
4.8   Implementation of Computer Circuits   95
4.9   Special Logic Blocks   100
4.10 Bibliography   106
4.11 Exercises   106

5 | **SEQUENTIAL CIRCUITS** 110

5.1   Description of Sequential Circuits   111
5.2   Circuit Synthesis   115
5.3   Synthesis of Counters   122
5.4   Heuristic Design Methods   133
5.5   Cost and Speed as Design Criteria   142
5.6   Bibliography   146
5.7   Exercises   147

6 | **DIGITAL ELECTRONICS** 152

6.1   Diode Gates   152
6.2   Transistor Gates   165
6.3   Flip-Flop Realization Methods   187
6.4   MOSFET Circuits   195
6.5   Integrated Circuits   203
6.6   Bibliography   208
6.7   Exercises   209

7 | **COMPUTER ARCHITECTURE: Part I**
Register Transfer Logic, Organization,
and Simulation 213

7.1   Register Transfer Logic   214
7.2   Applications of RTL   220
7.3   Computer Organization   232
7.4   Further Development of RTL   238
7.5   Simulation of Digital Systems   241
7.6   Bibliography   253
7.7   Exercises   254

8 | THE ARITHMETIC AND LOGIC UNIT *258*

8.1	*Basic Hardware for Addition and Subtraction*	*258*
8.2	*Fast Adders*	*263*
8.3	*Multiplication*	*269*
8.4	*Fast Multiplication*	*277*
8.5	*Division*	*290*
8.6	*Overflow*	*299*
8.7	*Bibliography*	*301*
8.8	*Exercises*	*301*

9 | SEMICONDUCTOR AND CORE MEMORY SYSTEMS *303*

9.1	*Magnetic Core Storage*	*304*
9.2	*Memory Logic Circuits and Electronics*	*312*
9.3	*Semiconductor Memories*	*320*
9.4	*Read-Only Memories*	*333*
9.5	*Control Memories*	*340*
9.6	*Bibliography*	*345*
9.7	*Exercises*	*346*

10 | COMPUTER ARCHITECTURE: Part II Organization, Microprocessors, and Large-Scale Systems *348*

10.1	*Organization of Instruction Words*	*348*
10.2	*Bus Structures*	*355*
10.3	*Introduction to Microprocessors*	*358*
10.4	*Microprocessor Systems*	*368*
10.5	*Basic Architectural Features of Modern Computers*	*378*
10.6	*Bibliography*	*392*
10.7	*Exercises*	*393*

11 | COMPUTER INTERFACE DESIGN *395*

11.1	*Data Transfer Techniques*	*395*
11.2	*Programmed Transfers*	*396*
11.3	*Interrupt Transfers*	*400*
11.4	*Direct Memory Access*	*403*
11.5	*Input/Output System Variations*	*407*
11.6	*Bibliography*	*414*
11.7	*Exercises*	*414*

APPENDIX

A   **FORTRAN SUBROUTINES**
          FOR THE LOGICAL MODEL   416

APPENDIX

B   **SUMMARY OF**
          PSEUDO-LINC I/O INSTRUCTIONS   421

          INDEX   423

PREFACE

This text presents a broad, though substantive, introduction to the three major levels of computer hardware design: architecture, logical design, and digital electronics. In addition it develops sufficient software concepts to establish the necessary background for making hardware design decisions. A major goal of the text is to provide a modern, quantitative overview of hardware design for people in all computer areas. Among the new topics presented are: an introduction to microprocessors and control memories, development and application of a register transfer language, use of heuristic-design techniques, investigation of various processor structures through simulation models based on FORTRAN, and application of MOS and bipolar integrated circuits. The text stresses general design tools and hardware principles rather than current specialized devices which may soon be discarded. (The treatment of NAND/NOR circuit design methods in Chapter 4 is an example of this approach.) On the other hand, the text includes practical details and well-recognized contemporary designs.

I assume that the *average readers* will be second- or third-year college students of electrical engineering or computer science. Many will have completed a basic course in a procedure-oriented language, like FORTRAN or PL/I, and a one-semester course in electrical networks. However, such courses are not *strict requirements,* for I have indicated those specialized sections which the reader may omit or cover less thoroughly without jeopardizing his understanding of later chapters. Moreover, I am assuming no advanced mathematical background, and I have presented sufficient details so that the book is suitable for individual study. The text contains the depth required by those readers majoring in computer technology, yet it is flexible enough for those who are in other areas of engineering or science and require only a limited introduction to digital systems. The text includes sufficient material for a three-hour course of one semester or two quarters in length; but it may be employed for a two-semester course with a limited amount of supplementing from the bibliography. In addition, it is an appropriate book for introductory computer courses such as those recommended by the COSINE Committee (Commission on Education of the National Academy of Engineering) and by the Association for Computing Machinery.

Undoubtedly, many students will later take specialized courses in the various topics covered by this text. However, I believe that a broad introduction to digital computers, with depth in some areas, gives the student the perspective and background required to assimilate more advanced courses. My text, therefore, is intended to foster the maturity and experience these students require to make the advanced course meaningful. As a specific example, most of our computer engineering students at Washington University follow their introduction to computer hardware by a three-hour course devoted to switching theory. Chapters 4 and 5 on logical design, while not intended to be exhaustive treatments, are complete enough that our switching course is able to progress at a more rapid pace and on a more advanced level.

I have observed that introductory computer courses often do not present the same challenge to college students that many of their other courses do; thus, students frequently do not learn as much and are not as highly motivated. Certainly, there is no need to create artificial challenges, but it is one of my objectives to present the material in sufficient depth to be useful in practical design situations as well as to command the interest of the most promising students. For several years I have taught undergraduate digital computer courses following this objective, and my text has been compiled from that experience. Preliminary versions have been extensively tested in a three-hour introductory course on computer organization and logical design.

Students and colleagues, too numerous to list, have been my teachers during the preparation of this book; their questions, suggestions, and ideas were indispensable. I am indebted to Russell Pfeiffer, Chairman of our Department of Electrical Engineering until his tragic death, for the encouragement he provided me during the writing process. Debbie Hawley did a praiseworthy job of typing the very early portions of the manuscript. Debbie Fivecoat did all the other typing and drafted the figures, and she deserves a special note of appreciation for accurately translating my very rough pencil notes and patiently completing many manuscript revisions. Finally I wish to acknowledge the unique climate for computer research, development, and teaching present in the Department of Electrical Engineering, the Computer Research Laboratory, and the Biomedical Computer Laboratory at Washington University for strongly influencing my thinking.

R. M. K.

DIGITAL
COMPUTER
DESIGN

1 | AN OVERVIEW

It is always very helpful to gain perspective about a new subject before design techniques and mathematical details are considered. Thus, this chapter emphasizes the role of computers in our society, their practical importance, their evolution, and other nontechnical aspects, while in all the following chapters, the emphasis is on engineering analysis and design. In addition to the traditional computer hardware topics, several innovative features will be found in later chapters. Typical members of this group are the heuristic design methods presented in Chapter 5, the register transfer logic developed in Chapter 7, and the microprocessors treated in Chapter 10.

Digital computers have assumed a prominent place in our society, and their value in business and industry is without question.[6,7,11] However, many people still have a distorted view of them. Some picture computers exclusively as big adding machines, when in reality they currently perform extremely complex logic; general nonnumerical tasks are also becoming increasingly important.

Typical of this trend is the application of computers in medical research and treatment. An interesting case in point is the development of "artificial hearing" for tens of thousands of deaf people who cannot be helped by conventional hearing aids. In an experimental program, scientists from the Ear Research Institute in Los Angeles and the University of Utah in Salt Lake City reported that a deaf subject was able to sense "sounds" produced by a PDP-8/F minicomputer. Figure 1.1 shows a 62-year-old engineer named Joe, deaf since infancy, who is participating in experiments to develop an artificial ear. Touching a key or combination of keys on the keypad in his lap, Joe causes the PDP-8/F, left side of photo, to send signals at selected frequencies to electrodes implanted in his inner ear. The computer provides a flexible system for conducting a wide variety of tests, eliminating the need for building special equipment for each experiment. When the experiments are completed, a portable artificial ear will likely be constructed using a subminiature computer, a *microprocessor*, not much larger than a conventional hearing aid.

Another example of an important nonnumerical application is the computerized knitting machine shown in Figure 1.2(a). One of the problems in the textile industry is the long time required for development of new fabric

Figure 1.1 Patient and Computer-Controlled Artificial Hearing System
(Courtesy of Digital Equipment Corporation).

patterns on conventional knitting machines. By using a PDP-11/20 mini-computer [background at the right of Figure 1.2(a)] to control one of its knitting machines (left foreground), the Uxbridge Knitting Mills of Massachusetts have been able to reduce the fabric development time to a matter of hours, or even minutes for less complex tasks. The computer-aided pattern design begins with sketches made, at the operator's console, on an electronic tablet coupled to a color television monitor by the computer. [See Figure 1.2(b).] As the work progresses, the computer transforms the information displayed on the monitor into stitch data recorded in the core memory of the PDP-11/20. When the designer wants to examine a piece of the new fabric, a brief set of instructions is given through the console, and the stitch data in the memory provides control of the knitting machine to produce a sample. Should corrections be required, the designer can recall the pattern to the television monitor and indicate the adjustments with the electronic tablet. When the designer is satisfied with the modified pattern, he allows the computer to resume production. A much more complicated procedure is required with older methods.

Other representative nonnumerical tasks are automated typesetting,

(a)

Figure 1.2 A Computer-Aided Fabric Pattern Design System: (a) Knitting Machine Controlled by a PDP-11/20; (b) Operator's Console and Television Display (Courtesy of Digital Equipment Corporation).

(b)

computer-aided drafting, derivation of theorems, and weather prediction. The reader should consult References 2, 5, 6, and 7 for details concerning these applications.

Popular magazine and newspaper articles frequently imply that computers are some kind of magical device. In reality, they are based on well-known engineering principles, and the quality of the results is critically dependent on the quality of the computer's logic design, the input data, and the skill of the programmer. (There is a well-known motto in computer engineering which aptly expresses this: Garbage in . . . garbage out!)

Another false notion concerning computers is that they are found exclusively in multimillion-dollar computer centers; in reality, a small general-purpose machine, a *minicomputer*, with a teletypewriter, 12,000 words of memory, and a FORTRAN compiler is currently available for less than $5,000, and surprisingly the prices are coming down. So-called microprocessors, and pocket calculators constructed from large-scale integrated circuits and costing two orders of magnitude less than a minicomputer, are available; but limitations on the ability of these devices to independently handle a complete spectrum of computational tasks prevents them from being classified as truly general-purpose machines. Minicomputers are being applied to such tasks as automatic monitoring of industrial processes, rapid testing of complex electronic systems, and controlling the sequence of programs at television stations. (In 1975, more than 30,000 minicomputers were operating in the United States.)

1.1 THE EVOLUTION OF DIGITAL COMPUTERS[1]

Although the digital computer itself is a quite recent development, it has had a complex evolution that goes far back into the history of technology. One of the earliest sources of digital concepts occurred thousands of years ago in Hindu-Arabic arithmetic notation. Somewhat later, yet before the birth of Christ, the odometer was invented, and this eventually lead to the development of the mechanical adder and multiplier in the 17th century. Then, improvements in machine tools and other developments finally lead to the first mass-produced desk calculator in 1911.

Meanwhile, in approximately 1800, Joseph Jacquard perfected a punched-card process* for automatically programming the location of threads in cloth-weaving looms. The punched-card method was so successful in automating the weaving process that Herman Hollerith decided to consider a variation of it to solve the massive data-processing problem (for that time) of tabulating statistics for the 1890 census. The result was a great improvement in tabulating speed over the previous manual methods. (In 1896,

*Actually, it was a sequence of cards jointed together in a long belt.

Hollerith formed his own company to manufacture punched-card equipment, and this later became part of IBM.)

Of all the contributions to the computer, by far the most remarkable was the contribution of Charles Babbage[4,8] (1791–1871), who was a professor of mathematics at Cambridge University. He conceived many of the basic ideas of the modern digital computer over 100 years before they were perfected. His *difference engine*, shown in Figure 1.3, was intended to prepare mathematical tables of various types by accumulating differences through mechanical means. The apparatus shown in the picture is capable of producing tables to 20-place accuracy. In the process of his work, however, Babbage conceived of a much more general machine, the *analytical engine*, having several of the properties of a modern computer. Although his theory was sound, the mechanical technology of his day was not capable of supporting his very ambitious ideas, and his analytical engine was never completed.

Figure 1.3 Babbage's Difference Engine (Photo Courtesy of IBM).

Although there were improvements in desk calculators, in punch card tabulating machines, and in other calculating devices between 1900 and 1937, no real breakthroughs occurred. Then, from 1937 to 1950, revolutionary changes occurred in such rapid succession, by so many different people, and in so many different places that it is impossible, in this short chapter, to give credit to all those concerned. Some of the major contributors were:

(a) John Atanasoff and a colleague, Clifford Berry, at Iowa State College during the period 1937–1942 demonstrated an experimental special-purpose electronic digital computer for solving simultaneous linear equations.

(b) During the early 1940s George Stibitz, at Bell Telephone Laboratories, built several special-purpose digital computers based on relay technology. He was one of the first to implement floating-point arithmetic.

(c) The Mark I relay computer was completed in 1944 by Howard Aiken at Harvard. It is considered to be the first general-purpose machine.

(d) The ENIAC, Electronic Numerical Integrator And Calculator, was completed in 1946 by J. Presper Eckert and John W. Mauchly of the Moore School of Electrical Engineering of the University of Pennsylvania. Although somewhat similar in structure to the Mark I, it employed vacuum tubes instead of relays; thus, it became the first operational electronic computer.

(e) The EDSAC, Electronic Delay Storage Automatic Computer, the first stored-program computer, began operating in 1949 at the University of Manchester in England.

An abbreviated family tree for the computer is shown in Figure 1.4. It includes the history just described, some important events since 1950, and certain related facts not previously mentioned. Most of the individual entries are self-explanatory; a few, such as core memory and assembly language, for which the reader may only have a vague definition, will be explained in considerable detail later. More historical details may be found by consulting the Bibliography.[3,4,8,10] (Because of the multiplicity of commercial computers, only a few of those which represent landmarks in the development of the field appear.)

From the density of dots in Figure 1.4, it may appear that the activity peak in the development of computers was reached around 1960. This is misleading because a vantage point in the 1970s does not allow one the proper historical perspective to evaluate the long-range importance of most of the techniques which were originated since the early 1960s; hence, only such obvious developments as Large-Scale Integration, LSI, have been added to the latter portion of the chart. An example of a development which may be later inserted in the chart is the highly parallel machines such as Illiac IV. (It was proposed in 1968 and completed in 1972.)

A number of items, such as the slide rule and Babbage's difference engine, are not directly connected into the mainstream. Although they had a role in the development of computer science, their contribution to the digital computer can not be identified through a specific single path.

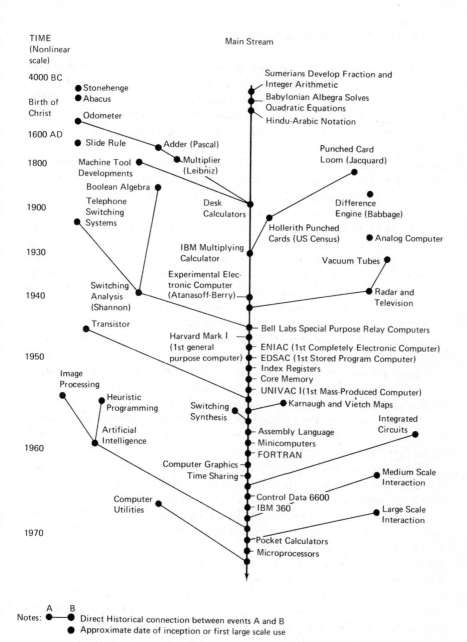

TIME
(Nonlinear
scale)

Main Stream

4000 BC

Sumerians Develop Fraction and
Integer Arithmetic

● Stonehenge
● Abacus

Babylonian Albegra Solves
Quadratic Equations

Birth of
Christ

Odometer

Hindu-Arabic Notation

1600 AD

● Slide Rule   Adder (Pascal)

Punched Card
Loom (Jacquard)

1800

Machine Tool ●
Developments

Multiplier
(Leibniz)

Boolean Algebra ●

1900

Telephone
Switching
Systems

Desk
Calculators

Difference
Engine (Babbage) ●

Hollerith Punched
Cards (US Census)   ● Analog Computer

1930

IBM Multiplying
Calculator

Vacuum Tubes ●

1940

Switching
Analysis
(Shannon)

Experimental Elec-
tronic Computer
(Atanasoff-Berry)

Radar and
Television

● Transistor

Bell Labs Special Purpose Relay Computers

Harvard Mark I
(1st general
purpose computer)

ENIAC (1st Completely Electronic Computer)

1950

EDSAC (1st Stored Program Computer)

Index Registers

Image
Processing

Core Memory

UNIVAC I(1st Mass-Produced Computer)

● Heuristic
Programming

Switching
Synthesis

● Karnaugh and Vietch Maps

Integrated
Circuits ●

Artificial
Intelligence

Assembly Language

1960

Minicomputers

FORTRAN

Computer Graphics

Medium Scale
Interaction

Time Sharing

Computer ●
Utilities

Control Data 6600

IBM 360

Large Scale
Interaction

1970

Pocket Calculators

Microprocessors

Figure 1.4 A Brief Chronology for Digital Computers.

7

1.2   THE GROWTH OF COMPUTER TECHNOLOGY

Having just considered the general evolution of computers, we shall now emphasize recent developments. We do not intend to present a lot of potentially boring statistics, but at this point it does seem that quantitative information is one of the best ways to give the reader a perspective concerning the rapid changes that have recently occurred in the industry.

Table 1.1 presents a contemporary and estimated near-future overview of computer technology as measured by various economic and engineering parameters. The first parameter, computer generations, is a popular means for indicating the major periods of development. To a large extent, these intervals are determined by the available electronic components. (See the second line in the table.) There are other factors, however, which are unique to the various generations. One of these is programming languages. During the first generation, only elementary languages were available, but in later generations very sophisticated programming systems have been employed. (Some of these will be considered in Chapter 3.) An extensive discussion of the various computer generations is found in the literature.[3]

Observe from the table that during the decade from 1960 to 1970 there was approximately an order of magnitude increase in both the economic importance of computers (e.g., note the increased value of computers shipped) and their performance (e.g., note the decrease in logic delay). Moreover, these were years of generally rising manufacturing costs; yet the prices for multiplications, minicomputers, and memory all showed very sharp decreases.

Another aspect of the progress in computer hardware can be obtained from the photograph in Figure 1.5, which shows comparable circuitry in each of the first three computer generations. The original vacuum-tube circuitry, a transistor version, and an integrated-circuit revsion are all easily identified. Actually, the reduction in size is compounded in that there is a corresponding change in power supplies, cooling, and other support functions.

Another facet of computer technology is concerned with the corporations which constitute the industry. Four companies, IBM, Honeywell, Sperry Rand, and Control Data, with sale volume in that order, do over 85% of the business. Moreover, IBM itself has close to 75% of the industry's income. This company employs approximately 350,000 people (for comparison, General Electric employs about 400,000 in all divisions), and it has had a 390% increase in gross income since 1959 (the gross national product increased 88% in the same period). Perhaps even more important, IBM is the top company of the world from the standpoint of the market value of its stock (approximately $40 billion as compared to about $25 billion for American Telephone and Telegraph, which is in second place). One of IBM's largest computers, the System 370/168, is shown in Figure 1.6.

Table 1.1 The Growth of Computer Technology*

Parameter	1950	1960	1965	1970	1975	1980
Computer generations	First	Second	Early third	Late third	Fourth	Fifth
Electronic components	Vacuum tubes	Transistors	Integrated circuits	Some medium-scale integrated circuits	Medium- and large-scale integrated circuits	Large-scale integrated circuits
Value of computers shipped (billions of dollars)	—	0.8	2.5	7.2	14.6	—
General-purpose digital computers in the world (thousands)	≪1	5	35	70	200	500
Dollar cost per 10,000 ten-digit multiplications	300 (Desk calculator)	1.40	0.12	0.01	0.008	0.0006
Cost of a minicomputer with a 4000-word, number, memory (thousands of dollars)	Not in production	80	25	7	2.2	—
Logic delay—delay of one stage of digital logic (nanoseconds)	1000	30	10	3	1	0.3
Cost of main memory (dollars per binary digit)	—	1.00	0.1	0.025	0.005	0.003

*The tabular entries are approximate state-of-the-art values. The data are intended to give the reader a feeling for the magnitude of computer technology. Anyone desiring the most up-to-date information and data for which the parameters have been completely qualified should consult statistical abstracts or other specialized sources.

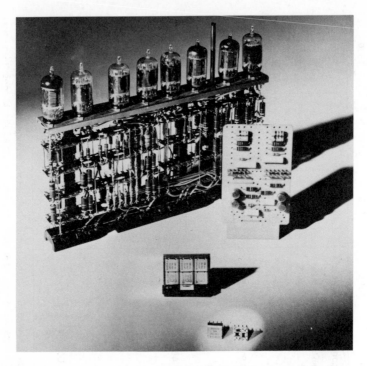

Figure 1.5   Typical Circuitry for the First Three Computer Generations (Photo Courtesy of IBM).

Figure 1.6   The IBM System 370/168 Computer (Photo Courtesy of IBM).

1.3 CLASSIFICATION OF COMPUTERS

In order to understand computers, it is necessary to be aware of their two main classes—analog and digital. Typically, the magnitude of an electric voltage is used to represent each variable in an analog computer, but the key to the classification is that the quantity representing the variable must be continuous in both time and magnitude. This should be contrasted with a digital signal, where the voltage (or other quantity) representing the variable is discrete (quantized) in both time and magnitude.

A slide rule is one of the most common types of analog calculators. Here, numbers are represented as continuous distances along the scales, and multiplication (division) is achieved by the addition (subtraction) of logarithmic scale lengths.

Table 1.2 presents a summary of the basic properties of digital and analog computers. Note that digital methods have the advantage of accuracy, ease of programming, and versatility, while analog methods may be cheaper if only a simple calculation is required. The cost parameter is difficult to generalize, particularly since the birth of large-scale integrated circuits. It is still true, however, that there is a lower practical limit on the cost of a digital computer, even to perform a simple task—a type of overhead charge must be paid. Once the basic digital system is available, however, a variety of other tasks can be carried out with little or no additional cost. (For more information concerning analog computers, consult Reference 12 in the Bibliography.)

The field of digital systems can be further divided into *general-purpose* and *special-purpose* machines. The general-purpose computers are those such as the IBM System 370, which is completely programmable and can be used to perform an almost unlimited variety of numerical calculations in the engineering, scientific, and business areas. A special-purpose machine is designed for a single type of application and frequently has most of its program prewired. The navigation computers aboard space vehicles are examples of this class of machine. Another very interesting example, shown in Figure 1.7, is the Electronic Switching System, ESS, being employed by the Bell Telephone Company to replace its older electromechanical systems in telephone exchanges. ESS allows a considerable improvement in the performance of the exchange and permits a number of new features to be sold to telephone customers. A few of these features are two-digit dialing for frequently called numbers, a simpler method for implementing conference calls, and a call-back option for busy numbers.

While not actual computers, there are a large number of special-purpose digital devices which use many of the same principles and require the same design techniques as an actual computer. Examples are digital voltmeters, counters, and various other types of automated digital instruments.

Table 1.2 Comparison of Typical Digital and Analog Computational Methods

	Form of Number	Accuracy	Minimum System Cost	Time to Obtain Result	Ease in Programming	Major Strengths
Digital	Two-level signal (e.g., 0 and +5 volts)	Very high (almost unlimited)	Practical limit for a general-purpose system is a few thousand dollars; but special-purpose units, microprocessors, are now available for under $100.	Finite—nanoseconds to hours	Excellent (Special problem- and procedure-oriented languages are available.)	*Versatility* in the solution of a wide variety of problems; *fast, accurate* (see Section 1.4).
Analog	Continuous signal (e.g., 0 to +5 volts)	Approximately 1%	Proportional to size of computation; may be only a few dollars.	Output is continuously available, but transients may require one second or longer to die out.	Fair (Extensive manual electrical and/or mechanical connections are usually required.)	*Simulation* of electrical, mechanical and chemical systems; solutions to *differential equations.*

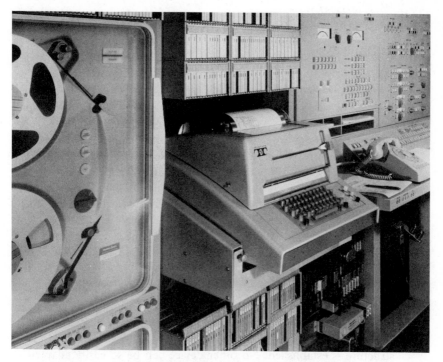

Figure 1.7 The Master Control Center in an ESS Telephone Exchange (Control and lamps on the panels at the far right are employed for various maintenance and administrative functions. The teletypewriter supplements the panel controls. The tape drive on the left is used for automatic message accounting recordings. Courtesy of Bell Laboratories).

1.4 IMPORTANT DIGITAL COMPUTER CHARACTERISTICS

In addition to the fact that analog processing of data is sometimes more appropriate, the digital computer, like any other tool, has its disadvantages. Perhaps the chief disadvantage is the number of decimal places printed and the format employed sometimes give an apparent accuracy to results which is entirely erroneous. Thus, the printed output may show 10 place numbers when in reality, due to inaccurate input data, the results may only be accurate to 1 or 2 places. The consumer, computer designer, and programmer must constantly be on the alert to avoid this trap. Also one must avoid becoming a lazy thinker by letting the computer produce a volume of numerical results for a particular problem when a little physical reasoning and an equation or two would provide a more complete and concise solution.

However, digital information processing does possess a number of characteristics which, when properly applied, can be very powerful. Some of the most important of these will now be considered:

1. *Problem-Oriented Languages.* Problem-oriented languages allow instructions to be given to the computer in the language of the problem, not that of the machine. This permits people in all fields (from theoretical physicists to high school students) to have direct access to the computer without working through an intermediary (programmer). Two of the most well-known languages are FORTRAN and BASIC; however, there are many other languages which are even more useful and efficient in special applications. The language ECAP (Electronic Circuit Analysis Program), for example, permits the engineer to easily obtain solutions to very complicated electrical network problems without having to go to the trouble of writing mesh or loop equations. (Additional details concerning this subject are presented at the end of Chapter 3.)

2. *Versatility.* The general-purpose computer is truly a versatile device in that a program can rapidly create an information processing structure to solve virtually any problem. Moreover, new programs can be inserted in a very short time so that an entirely different processing structure is established. No other device can be so easily adapted to such a wide variety of tasks.

3. *Accuracy.* Results to 10 or more significant decimal places is an achievement not even approached by most other systems. When this is compared with the two significant figures obtainable with most analog devices and the time required to manipulate large numbers manually, the achievement is all the more remarkable.

4. *Automation.* The ability to specify a list of tasks (arithmetic, logical, or other operations) and have them completed without human intervention is a key to the automation of a wide variety of processes, including the automated milling machine shown in Figure 1.8. Actually more can be accomplished than just slavishly producing a fixed sequence of commands. There is no trouble in accepting information from the process and using this to change program parameters or even to switch to an entirely new program. Mechanical and electromechanical systems previously used for such tasks were limited in their ability to accept parameter changes. They required an expensive and time-consuming hardware change whenever the process was modified, and they lacked the arithmetic and logical capabilities of digital systems.

Figure 1.8 A PDP-8 Computer Controlling a Large Milling Machine for Manufacture of Precision Aircraft Components (Courtesy of Digital Equipment Corporation).

5. *Speed.* Individual steps of an information-processing job are accomplished in a few microseconds or less. This means that extensive computational tasks can be performed concurrently with the operations of other systems (in real time), e.g., during the reaction time of a machine operator.

6. *Large Memory.* Another aspect of computer power is storage capacity for millions of words (numbers). Semiconductor, core, disk, and magnetic tape are all popular storage media in the current generation of machines. The fastest of these, semiconductor, allows any word in the memory to be selected in less than 100 nanoseconds. However, it is also the most expensive memory type, being orders of magnitude more costly than magnetic tape.

7. *Ease of Communication.* Programs (data too) are easily and inexpensively communicated. This is accomplished via telephone channels,

magnetic tape, decks of cards, etc. In most cases, a FORTRAN or similar program from one installation will run on another computer with little or no modifications. When this is compared with the difficulty of duplicating electrical or mechanical hardware using drawings produced by another organization, the real beauty of digital systems becomes evident.

8. *Simulation.* A digital computer is capable of *simulating* (or *modeling*) any process that can be described in quantitative terms. In recent years, great progress has been made in biology, economics, engineering, and other fields through the use of computer programs to model situations which would be impossible to study through ordinary physical experiments because of cost, the time required, or other limitations.

An interesting facet of computer simulation was recently brought out by Dr. Ruth M. Davis of the National Bureau of Standards. She made the following observation at an International Federation of Information Processing Societies meeting:

World War I was fought with chemistry. World War II was fought with physics. World War III is being fought with Computer Science. The first battles of World War III may well have occurred when mathematical formulations of strategies and counterstrategies of realistic proportions were able to be tried out as war games on computers.

With realistic wars being able to be fought in 20 minutes or 20 hours on computers, decisions to engage in such encounters have been nil. No statistical correlations are needed to validate the fact that no major encounters have occurred between "large computer possessing" nations.

1.5 PROLOGUE TO THE REMAINDER OF THE TEXT

Having very briefly examined some of the historical, statistical, and general background for computer technology, we should now be prepared to "dive" into the engineering details. The first topic that must be considered is number systems. Part of this material will be a review for some readers, but the information is a prerequisite for the machine language programming in Chapter 3, the study of arithmetic unit design in Chapter 8, and the tools to be employed in several other chapters.

In Chapter 3, computer software will be considered. There is a very intimate relationship between the areas of *hardware* (the equipment) and *software* (the step-by-step directions for solving a problem); thus, a person cannot be a good designer in one area without some knowledge of the other. This chapter also serves as an introduction to machine organization and to basic operating characteristics.

In Chapters 4 and 5, Logic Design and Sequential Circuits, we begin to reach the heart of our subject, since it is here that the basic tools and physical understanding employed for the remainder of the text are developed. Chapter 6 is a slight digression, but an important one, in that the electronic components and circuits considered as blocks in other chapters are analyzed in detail. The arithmetic unit and the memory unit, considered in Chapters 8 and 9, are two of the major subsystems from which computers are constructed In many respects, the climax of the text is reached in Chapters 7 and 10 since their subject, architecture, is the highest-level structure employed by the designer in planning any digital system.

Finally, Chapter 11 rounds out some of the practical applications and provides information concerning the possible interfaces between a computer and other systems.

1.6 BIBLIOGRAPHY

1 BELL, GORDON C., and ALLEN NEWELL, *Computer Structures: Readings and Examples.* New York: McGraw-Hill Book Company, 1971. (Especially the beginning of Chapter 3.)

2 *Computers and Computation.* (Readings from *Scientific American*). San Francisco: W. H. Freeman and Company, 1971.

3 DENNING, PETER J., "Third Generation Computer Systems." *ACM Computing Surveys,* 1971, pp. 175–216.

4 GOLDSTINE, HERMAN H., *The Computer from Pascal to von Neumann.* Princeton, N. J.: Princeton University Press, 1972.

5 JACKSON, PHILIP C., Jr., *Introduction to Artificial Intelligence.* New York: Petrocelli Books, 1974.

6 MYERS, CHARLES A., *Computers in Knowledge-Based Fields.* Cambridge, Mass.: M.I.T. Press, 1970.

7 PYLYSHYN, ZENON W. (ed.), *Perspectives on the Computer Revolution.* Englewood Cliffs, N.J.: Prentice-Hall, Inc., 1970.

8 RANDELL, BRIAN, *The Origins of Digital Computers—Selected Papers.* Berlin: Springer-Verlag, 1973.

9 RICHARDS, R. K., *Electronic Digital Systems.* New York: John Wiley & Sons, Inc., 1966. (Especially Chapter 1.)

10 ROSEN, SAUL, "Electronic Computers: A Historical Survey." *ACM Computer Surveys,* Vol. 1, 1969, pp. 7–36.

11 TAVISS, IRENE, *The Computer Impact.* Englewood Cliffs, N.J.: Prentice-Hall, Inc., 1970.

12 WEYRICK, ROBERT C., *Fundamentals of Analog Computers.* Englewood Cliffs, N.J.: Prentice-Hall, Inc., 1969.

13 WILSON, IRA G., and MARTHANN E. WILSON, *What Computers Cannot Do.*
New York: Vertex Books, 1970.

1.7 TOPICS FOR FURTHER STUDY AND
STUDENT REPORTS

1.1 Consult recent issues of *Scientific American* for articles concerning various facets of computers, and write a critical review of one of them.

1.2 Report on some of the nonnumerical applications of computers found in References 5 and 7 of the Bibliography.

1.3 Using References 4 and 8, prepare a brief history of digital computers.

1.4 Consult *Fortune* or a similar magazine, and report on important current industrial and business applications of computers.

1.5 Using References 7 and 11, write a paper on the impact of computers on our world.

1.6 Discuss the strengths and limitations of computers; References 6 and 13 may be helpful.

1.7 Read completely and report on Reference 2 in the Bibliography.

1.8 Report on the development of the IBM 360 by using the article: "Rough Road to the Marketplace," in *Fortune* magazine, October 1966, page 138.

1.9 There has been much discussion concerning whether computers can think.
(a) Make a list of pros and cons on the subject. (Reference 5 may be helpful.)
(b) Propose an objective way to decide whether a particular system can think.

1.10 Find information on the analog computer and write a two-page summary on its principles and major applications.

1.11 Repeat Exercise 1.10 for the field of Computer Graphics. (Reference 2 is a possible source of information.)

1.12 Repeat Exercise 1.10 for the field of pattern recognition.

2

NUMBER SYSTEMS

Knowledge of number systems forms a basic tool which is useful in all phases of computer technology. It is important to the understanding of all subsequent chapters of this text, and it is crucial for our work on machine language programming in Chapter 3.

Have you ever tried to do mathematical calculations with Roman numerals? A comparison of the capabilities of the Roman with the Arabic (or positional) number system emphasizes the necessity of a well-developed notation for satisfactory progress in mathematics and engineering. We frequently take for granted such innovations as negative numbers, the concept of zero, and the ease with which fractions can be expressed in the positional system. However, these were not trivial or accidental advancements but required brilliant insight and centuries of evolutionary development.[3] Moreover, it is only our provincialism that makes the decimal system seem so unique. Other civilizations have successfully applied alternate number systems, and there is even a group today, the Duodecimal Society of America, which advocates changing from base 10 to base 12. Although it is highly unlikely that the revolution advocated will ever occur, there are several advantages that would accrue from such a modification.

2.1 ARABIC NOTATION[2]

Consider a positional number such 591.3 which is a concise representation for $5 \times 10^2 + 9 \times 10^1 + 1 \times 10^0 + 3 \times 10^{-1}$. The beauty of this notation is that it may be extended without limit in both directions from the decimal point to represent both larger numbers and ones of greater precision. Moreover, it holds for all other bases, not just the base ten. In general terms then, a number, N_b, in any base,* b, may be represented by the equation

$$N_b = a_n b^n + a_{n-1} b^{n-1} + \ldots + a_1 b^1 + a_0 b^0$$
$$+ a_{-1} b^{-1} + a_{-2} b^{-2} + \cdots + a_{-m} b^{-m}$$
$$= \sum_{-m}^{n} a_i b^i$$

Radix is another common term which is used interchangeably with *base*.

where a_i is any of the b digits of the notation; the integers n and m control the size of the number and its precision. Thus, in representing $N_{10} = 591.3$, $b = 10$, $n = 2$, $m = 1$, $a_2 = 5$, $a_1 = 9$, $a_0 = 1$, and $a_{-1} = 3$. Special note should be taken concerning the use of the subscript following a number to represent the base used, e.g., 473.6_8 indicates that the given number is in base 8. (Of course, the number 591.3_8 is meaningless since a number in the base 8 system cannot possess the digit 9; it is only allowed 8 digits, 0 through 7.)

The three bases most frequently used in computer technology are described in Table 2.1.

Table 2.1 Comparison of Bases Most Frequently Used

Name	Base	Digits	Main Use
Binary	2	0, 1	Internal represen- tation of numbers
Octal	8	0 through 7	Convenient form for
Hexadecimal	16	0 through 9, A, B, C, D, E, F	input and output

2.2 CONVERSION OF INTEGERS

Since most data found in engineering, science, and business are in the decimal system and since most computers operate in other systems, it is important for the computer designer to be skilled in converting between various bases.

Example: Binary-to-Decimal Conversion

Convert 1101_2 to decimal.   From the definition in Section 2.1:

$$1101_2 = 1 \times 2^3 + 1 \times 2^2 + 0 \times 2^1 + 1 \times 2^0 = 8 + 4 + 0 + 1 = 13_{10}$$

The above method is satisfactory as long as there are only a few places in the binary number. But computers frequently use numbers with more than 36 places; thus, the faster technique shown below is recommended. This method is described recursively by the equation

$$S_{i-1} = 2S_i + a_{i-1}$$

where a_{i-1} is defined in Section 2.1, S represents the sum, $S_n = a_n$, and S_0 (the last sum term) yields the decimal equivalent of the original binary number.

A convenient three-line form for the conversion is produced by writing the binary number on the first line, calculating the $2S_i$ terms on the second line, and calculating the S_{i-1} terms on the third line:

Binary number	1	1	0	1
Product terms ($\times 2$)	—	2	6	12
Sum terms	1	3	6	13

Note how the calculation proceeds from left to right and how the sum in one column, when multiplied by 2, yields the product term in the next column to the right, which when added to the binary digit* in that column yields the new sum.

Example: Octal-to-Decimal Conversion

Convert 3767_8 to decimal. Here the recursive equation becomes $S_{i-1} = 8S_i + a_{i-1}$, and the calculation is completed as follows:

Octal number	3	7	6	7
Product terms ($\times 8$)	—	24	248	2032
Sum	3	31	254	2039

Thus $3767_8 = 2039_{10}$.

It now appears that the general recursive equation to convert an integer in any base to base 10 is:

$$S_{i-1} = bS_i + a_{i-1}$$

Proof: Given $S_n = a_n$. Using the recursive equation:

$$S_{n-1} = bS_n + a_{n-1} = ba_n + a_{n-1}$$

Continuing,

$$S_{n-2} = bS_{n-1} + a_{n-2} = b(ba_n + a_{n-1}) + a_{n-2} = b^2a_n + ba_{n-1} + a_{n-2}$$
$$S_{n-3} = bS_{n-2} + a_{n-3} = b(b^2a_n + ba_{n-1} + a_{n-2}) + a_{n-3}$$
$$= b^3a_n + b^2a_{n-1} + ba_{n-2} + a_{n-3}$$

$$\cdot \quad \cdot \quad \cdot$$

Finally,

$$S_0 = b^na_n + b^{n-1}a_{n-1} + \ldots + b^{n-n}a_{n-n} = \sum_{i=0}^{n} a_ib^i$$

which is the integer definition for the position notation ($m = 0$ for integers) which we had in Section 2.1.

There is also an easy way to convert from decimal to any other base, but first let us consider conversion to binary. The principle of the method involves successive division by 2, with the remainders forming the binary number and the final remainder becoming the *most significant digit*, MSD. Table 2.2 demonstrates the process.

---

*The contracted form *bit* will also be employed.

Table 2.2 Decimal-to-Binary Conversion

Quotients	Remainders
13	—
6	1
3	0
1	1
0	1 (MSD)

Thus $13_{10} = 1101_2$.

The only change required in the method to go from decimal to any base, b, is division by b rather than division by 2. (Proof of the above and several other proofs are left as exercises at the end of the chapter.)

Let us demonstrate by converting 3809_{10} to octal (Table 2.3).

Table 2.3 Decimal-to-Octal Conversion

Quotients	Remainders
3809	—
476	1
59	4
7	3
0	7 (MSD)

Thus $3809_{10} = 7341_8$.

Usually in converting between two bases, neither of which are decimal, it is easiest to convert the original number first to decimal and then change from decimal to the desired final base. However, conversion between binary and bases which are a power of 2 is particularly simple as the following illustration shows.

Due to the fact that $8 = 2^3$, there is a simple relation between octal and binary which permits conversion between the two systems by inspection. In converting from octal to binary, we merely change the octal digits individually to binary. For example, convert 7341_8 to binary:

$$7341_8 = 111\ 011\ 100\ 001 = 111011100001_2$$

To accomplish the reverse conversion, we break the binary number into groups of three digits and then convert each group to the equivalent octal digit. For example, convert 110101010111_2 to octal:

$$110101010111_2 = 110\,|\,101\,|\,010\,|\,111 = 6527_8$$

We may take advantage of the simple relation between binary and octal in converting between binary and decimal. First we convert the decimal number to octal by the division method, and then we use inspection to go to binary. For example, from the results of Table 2.3, $3809_{10} = 7341_8 =$

$111\ 011\ 100\ 001_2$. To accomplish the direct conversion between decimal and binary, 12 divisions by 2 would have been required; in this case, however, 4 divisions by 8 and inspection produced the same results.

2.3 CONVERSION OF FRACTIONS

Converting a decimal fraction to another base follows the same general principles as for integers except that it requires a multiplication process, while for integers it requires division. The following example (Table 2.4) shows how 0.6875_{10} is converted to binary by successive multiplication of the fractional part by 2, with the integer part after each step becoming the next lower digit in the binary fraction.

Table 2.4   Decimal-to-Binary Conversion

Multiplication Process	Integer Part
$0.6875 \times 2 = 1.3750$	1 (MSD)
$0.375 \times 2\ \ = 0.750$	0
$0.75 \times 2\ \ \ = 1.5$	1
$0.5 \times 2\ \ \ \ = 1.0$	1

The result is $0.6875_{10} = 0.1011_2$.

Of course, the conversion from decimal to octal follows the same pattern, as the example in Table 2.5 shows.

Table 2.5   Decimal-to-Octal Conversion

Multiplication Process	Integer Part
$0.8632 \times 8 = 6.9056$	6 (MSD)
$0.9056 \times 8 = 7.2448$	7
$0.2448 \times 8 = 1.9584$	1
$0.9584 \times 8 = 7.6672$	7

The result is $0.8632_{10} = 0.6717_8 \cdots$.

Since conversion of fractions from decimal to other bases requires multiplication, it is not surprising that going from other bases to decimal requires a division process. The exact procedure for converting a binary fraction to decimal starts with a division by 2 of the least significant binary digit. Then the next binary digit is added to the left side of the decimal point of the previous quotient, and the result is again divided by 2. The method terminates in the desired decimal equivalent after the most significant digit has been added to the quotient and the result is divided by 2. In the following example (Table 2.6), the decimal equivalent of 0.1011_2 is determined.

Table 2.6 Binary-to-Decimal Conversion

Binary Fraction	Division Process
1 →	2⌊①.0
1 →	2⌊①.5
0 →	2⌊⓪.75
(MSD) 1 →	2⌊①.375
	0.6875$_{10}$ Decimal result

2.4 ADDITION AND SUBTRACTION

The four arithmetic operations (addition, subtraction, multiplication, and division) follow the same basic rules of carry, borrow, etc., whether they are accomplished in decimal or in some other base. The major changes are the actual entries in the addition and multiplication tables, e.g., one plus one in binary must yield 10 since a separate symbol for 2 does not exist. Similarly, $7_8 + 2_8 = 11_8$. Binary addition is concisely summarized by Table 2.7. (Table entries are for $A + B$.)

Table 2.7 Binary Addition Table

$_A$\\$^B$	0	1
0	0	1
1	1	10

Use of Table 2.7 is demonstrated by the addition shown below.

$$
\begin{array}{r}
1011.01 \\
+\ 0111.10 \\
\hline
10010.11
\end{array}
$$

In this example, note that carries are handled in a way analogous to that done in decimal arithmetic. (As a check, the reader should convert all of the above numbers to decimal.)

An easy way to accomplish octal addition (a similar procedure can be employed in any base) without committing a full addition table to memory is to observe that the sum of two numbers is the same as in decimal unless the result is greater than or equal to 8. In that case, the sum is *modulo* 8 and a carry is produced. Modulo 8 means that at 8 the sum starts counting over from zero; for example, $7_8 + 4_8 = 13_8$. Here

$$7 + (1 + 3) = (7 + 1) + 3 = 10 + 3 = 13_8$$

There are two main ways to perform subtraction: with a table, as we all learned in elementary school, or through the *complement* method. Particularly in minicomputers, the complement method is preferred because it requires less circuitry and it is the method which will be emphasized here.

To get a better physical understanding of the complement method, examples using the decimal system will be presented before binary is discussed. The 10's complement of a number, N, is defined the following way:

$$\text{10's complement} \equiv 10^n - N$$

where n is the number of digits contained in the computer's register. (A register is an electrical circuit which temporarily stores numbers in a computer.) If $N = 252$ and $n = 8$, the 10's complement becomes:

$$
\begin{array}{r}
100000000 \\
- \quad\quad 252 \\
\hline
99999748
\end{array}
$$

To perform the operation $621 - 252$, we add the 10's complement of 252 to 621 as follows:

$$
\begin{array}{r}
00000621 \\
+ \; 99999748 \\
\hline
\cancel{1}00000369
\end{array}
$$

which is the correct result. Now you should be objecting to the fact that the leftmost 1 was crossed off. However, recall that the 10's complement is really $10^n = 10^8$ too large; thus, 10^8 should be subtracted from the final answer, which is exactly what we did when the 1 was dropped.

Another form of complement, called the *9's complement*, is defined in the following way:

$$\text{9's complement} \equiv (10^n - 1) - N$$

If $n = 6$ and $N = 252$, the 9's complement becomes:

$$
\begin{array}{r}
999999 \\
- \quad 252 \\
\hline
999747
\end{array}
$$

To perform the operation $621 - 252$, we add the 9's complement of 252 to 621 as follows:

$$
\begin{array}{r}
000621 \\
+ \; 999747 \\
\hline
\text{①}000368 \\
\hookrightarrow +1 \\
\hline
000369
\end{array}
$$

Using reasoning similar to that for the 10's complement problem, we should not have been surprised that the leading 1 in the answer was removed. But why was it added to the least significant digit? The answer, of course, is found in the fact that the 9's complement is one less than the 10's complement. Incidentally, addition of the leading 1 to the least significant digit is known as *end-around carry*.

As mentioned previously, our real purpose here is to learn how to perform the complement method for binary numbers since this method must be finally implemented electronically in the computer. The *2's complement* is analogous to the 10's complement and is defined in the following way:

$$\text{2's complement} \equiv 2^n - N$$

Rather than presenting an example for this method, let us immediately consider the *1's complement*, which is analogous to the 9's complement:

$$\text{1's complement} \equiv (2^n - 1) - N$$

If $n = 6$ and $N = 4_{10} = 100_2$, the 1's complement becomes:

$$
\begin{array}{r}
111111 \\
-000100 \\
\hline
111011
\end{array}
$$

Here is where the real savings in the 1's complement method occurs. Note that the result of the above subtraction could have been produced by merely complementing the individual digits, i.e., making the exchange $0 \leftrightarrow 1$. Later we will see that this is very easily accomplished electronically.

Let us complete our example by using 1's complements to subtract $4_{10} = 100_2$ from $13_{10} = 1101_2$:

$$
\begin{array}{r}
001101 \\
+\ 111011 \\
\hline
①001000 \\
\hookrightarrow +1 \\
\hline
001001
\end{array}
$$
(The reader should check
this result by converting
final answer to decimal.)

Note how the end-around carry was used in the above example much as it was in the 9's complement method.

If we subtract a large number from a smaller one, we will obtain a negative result. But with the complement method, how is this detected just by examining the result? The conventional way to solve the problem is to employ the MSD as a sign bit. (For emphasis, we also employ the symbol ⌋ to separate the sign bit from the others.) If the sign bit is 1, the number is negative and the magnitude bits are in complemented form. Thus, the 1's complement number $1\rfloor0110$ is a negative number since the MSD $= 1$. To convert it to a number having a positive magnitude with an algebraic sign, we affix the sign and complement all bits, e.g.,

$$1\rfloor0110_2 = -(\overline{1\rfloor0110})_2 = -(0\rfloor1001)_2 = -9_{10}$$

where the bar over the bits is a mathematical notation for complement.

Now let us employ our complement method and a six-bit number (five bits for magnitude and one for sign) to determine $4_{10} - 13_{10}$. Now,

$$-13_{10} = -(\underline{0|}01101) = \overline{\underline{0|}01101} = \underline{1|}10010$$

Continuing with our problem:

$$\begin{array}{r} \underline{0|}\,00100 \\ +\underline{1|}\,10010 \\ \hline \underline{1|}\,10110 \end{array}$$

Note that the MSD of both the augend and the result indicates that they are negative numbers. (The fact that there is no end-around carry is also an indication that the final answer is negative.) To get the result into the form of a magnitude plus its sign, we merely complement the original result and affix the sign:

$$-(\overline{\underline{1|}\,10110}) = -(\underline{0|}\,01001) = -9$$

as expected. (Some additional facts concerning the complement method are brought out in Exercise 2.15 of Section 2.8.)

2.5 MULTIPLICATION AND DIVISION

Since multiply and divide operations will be considered in detail in Chapter 8, only a brief presentation will be given here with emphasis on the fact that the main difference between decimal operations and those in other bases is in the multiplication tables. Binary multiplication is concisely summarized by Table 2.8. (Table entries are for $A \times B$.)

Table 2.8   Binary Multiplication Table

A \\ B	0	1
0	0	0
1	0	1

Use of Table 2.8 is demonstrated by the multiplication shown below.

$$\begin{array}{r} 1011.01 \\ \times \quad 110.1 \\ \hline 101101 \\ 1011010 \\ 101101 \\ \hline 1001001.001 \end{array}$$

(The reader should check the result by converting both the problem and final answer to decimal, performing the decimal multiplication, and comparing the answers.)

Division is also a simple process as the following example shows:

$$
\begin{array}{r}
10.1 \\
1100\,\overline{\big)\,11110.0} \\
1100 \\
\hline
1100 \\
1100 \\
\hline
0000
\end{array}
$$

It is really not necessary to memorize the complete multiplication table to solve an octal division problem. This is demonstrated by the next example.

Perform the division $5362_8 \div 65_8$ and round the answer to three places to the right of the octal point. The key to the method is to first find the 2 to 7 multiples of 65_8 by adding 65_8 to itself (see Table 2.9).

Table 2.9   Multiples of 65_8

Multiplier	Results
2	152
3	237
4	324
5	411
6	476
7	563

Since the needed multiples of 65_8 are now handy, it is a fairly simple matter to complete the division process as shown below. Of course, we must be careful to carry out octal rather than decimal subtraction, but this is fairly simple if the previously mentioned modulo concept is kept in mind.

$$
\begin{array}{r}
64.674_8 \\
65_8\,\overline{\big)\,5362.000_8} \\
476 \\
\hline
402 \\
324 \\
\hline
560 \\
476 \\
\hline
620 \\
563 \\
\hline
350 \\
324 \\
\hline
24
\end{array}
$$

Therefore $\dfrac{5362_8}{65_8} = 64.674_8$.

2.6 CODES[4, 5, 6]

Suppose we want to place the number 14_{10} into a computer. We could write 1110_2, or the individual digits of the decimal number could be coded separately as 0001 0100. The second method is called *binary-coded decimal,* BCD, and has been used in some commercial computers because of its similarity to the decimal system.

One should carefully distinguish between the terms *conversion* and *encoding.* As is obvious from the above example, conversion is the more drastic process in that the structure of the number itself is changed, e.g., a different base is used. Encoding leaves the basic structure of the number the same, but the individual digits are represented by different symbols.

One of the first questions that must be answered is "How many bits are needed to code the ten symbols 0 through 9?" A solution is desired for the equation $2^n = 10$, where n is the required number of bits. Using a calculator, we find that $n = 3.32$; but since n must be an integer, four bits are needed. However, $2^4 = 16$; thus, there are $16 - 10 = 6$ excess symbols. The next question that may be asked is "By using 10 of 16 symbols, how many different codes can we construct?" We are thus asking for the number of permutations of 16 things taken 10 at a time, $_{16}P_{10}$. Now,

$$_{16}P_{10} = \frac{16!}{6!} \approx 3 \times 10^{10}$$

which certainly is a staggering number. Fortunately, these codes can be broken down into a few classes and, except for very special applications, the computer engineer need only become familiar with a few members of each class. (Actually, we are speaking here of only a small part of the subject of coding, which has become a separate discipline of its own within the field of communication theory.)

Some typical BCD codes are shown in Table 2.10. The (8, 4, 2, 1) code is the common binary numbering system employed at the beginning of the chapter. The name "excess-3" is an apt description of the next code since it is formed by the addition of three (0011) to the corresponding common binary entries. The remaining codes, Gray and code X, will be considered shortly.

Weighted Codes

Consider the common binary code:

$$1001 = 1 \times 2^3 + 0 \times 2^2 + 0 \times 2^1 + 1 \times 2^0$$
$$= 1 \times 8 + 0 \times 4 + 0 \times 2 + 1 \times 1$$

Note that, starting at the MSD, the places of the number are weighted in decreasing powers of 2. Because of this, it is called an 8, 4, 2, 1 *weighted* code. (Codes may also have negative weights, e.g., 2, 3, 8, -4.) Weighted

Table 2.10 Typical BCD Codes

Decimal Digit	Common Binary (8, 4, 2, 1)	Excess-3	Gray*	Code X
0	0000	0011	0000	0100
1	0001	0100	0001	0010
2	0010	0101	0011	0111
3	0011	0110	0010	1110
4	0100	0111	0110	0011
5	0101	1000	1110	1100
6	0110	1001	1010	0001
7	0111	1010	1011	1000
8	1000	1011	1001	1101
9	1001	1100	1000	1011

*The most common form of Gray code containing 16 members is shown in Table 5.18; but for convenience, variations containing fewer members are considered here and in other chapters.

codes are useful because of the concise way they can be described, the ease of converting them to decimal numbers, and the circuit economy realized when they are employed for certain tasks, e.g., analog-to-digital conversion.

For a weighted code, the relation between the code and the corresponding decimal number is given by the equation

$$N = \sum_{i=0}^{n-1} a_i w_i + B \qquad (2.1)$$

where N is the decimal equivalent, a_i is the code coefficient, w_i is the weight, n is the number of bits in the code, and B is the positive (or negative) bias expressed in decimal.

Some authors omit the bias term in Equation (2.1). This results in a less general definition and considerably reduces the number of possible weighted codes, e.g., under these conditions, excess-3 is not a weighted code. (We will always employ the more general definition, except where specifically indicated.)

Now let us determine whether code Y, shown in Table 2.11, is a weighted code by using Equation (2.1) to generate the following five equations:

$$0 = 0 \times w_2 + 0 \times w_1 + 1 \times w_0 + B \qquad (2.2)$$

$$1 = 0 \times w_2 + 0 \times w_1 + 0 \times w_0 + B \qquad (2.3)$$

$$2 = 0 \times w_2 + 1 \times w_1 + 1 \times w_0 + B \qquad (2.4)$$

$$3 = 0 \times w_2 + 1 \times w_1 + 0 \times w_0 + B \qquad (2.5)$$

$$4 = 1 \times w_2 + 0 \times w_1 + 0 \times w_0 + B \qquad (2.6)$$

In order to solve a set of equations such as these, it is helpful to note that many of the a_i coefficients are zeros and the remainder are ones; thus, certain unknowns may be solved either by inspection or with very little effort, as the following equations show:

By inspection of Equation (2.3), $B = 1$

From Equation (2.2), $w_0 = -B = -1$

From Equation (2.5), $w_1 = 3 - B = 2$

From Equation (2.6), $w_2 = 4 - B = 3$

Checking the remaining equation, (2.4), with the above values, we get $2 = 0 + 2 - 1 + 1$, which is consistent and proves that code Y is the weighted code 3, 2, -1; 1. Note that the bias is separated from the weights by means of a semicolon. When the semicolon is absent, the bias is assumed to be zero.

Table 2.11 Elementary Codes to be Tested for the Weighted Property

Decimal Digit	Code Y	Code Z
0	001	011
1	000	110
2	011	111
3	010	000
4	100	101

As a second example, consider code Z in Table 2.11. Again, use of Equation (2.1) generates five equations.

$$0 = 0 \times w_2 + 1 \times w_1 + 1 \times w_0 + B \qquad (2.7)$$

$$1 = 1 \times w_2 + 1 \times w_1 + 0 \times w_0 + B \qquad (2.8)$$

$$2 = 1 \times w_2 + 1 \times w_1 + 1 \times w_0 + B \qquad (2.9)$$

$$3 = 0 \times w_2 + 0 \times w_1 + 0 \times w_0 + B \qquad (2.10)$$

$$4 = 1 \times w_2 + 0 \times w_1 + 1 \times w_0 + B \qquad (2.11)$$

By inspection of Equation (2.10), $B = 3$. Only three of the four remaining equations are required to solve for the three unknown weights; this leaves one equation to check for consistency. Using Equations (2.7) through (2.9) and the known bias value, we can write the following new equations:

$$w_1 + w_0 = -3 \qquad (2.12)$$

$$w_2 + w_1 = -2 \qquad (2.13)$$

$$w_2 + w_1 + w_0 = -1 \qquad (2.14)$$

These may be solved to give $w_2 = 2$, $w_1 = -4$, and $w_0 = 1$. ($B = 3$ was previously established.) When these are substituted into the remaining equation, (2.11), the result is $4 = 2 + 0 + 1 + 3$, $4 \neq 6$; therefore, Equation (2.11) is not consistent with the other equations, which, of course, means that code Z is not a weighted code.

Although any convenient method can be employed to solve equations such as the above, it is helpful to note that frequently two equations contain exactly the same variables, with one exception. Then, subtraction will immediately yield the value of the odd variable. This process can be applied between Equations (2.12) and (2.14), yielding $w_2 = 2$, and between Equations (2.12) and (2.14), yielding $w_0 = 1$.

It is left as an exercise for the reader to show that, in Table 2.10, excess-3 is a weighted code while the Gray code and code X are not.

Self-Complementing Codes

Another useful way of classifying a code concerns whether or not it is *self-complementing*. This means that the complement of the binary code leads to a number which is the 9's complement of the original number. Table 2.12 illustrates this situation for the excess-3 code. For example, decimal 4 has the excess-3 code 0111; the complement $\overline{0111} = 1000$, which is the excess-3 code for 5, which in turn is the 9's complement of 4. Self-complementing codes are very useful in BCD computers because of the simple electronic circuits needed to perform subtraction. It is left as an exercise for the reader to show that, for the other codes of Table 2.10, code X is self-complementing while the (8, 4, 2, 1) code and the Gray code are not.

Table 2.12 Self-Complementing Property of Excess-3 Code

Decimal Digit	Excess-3 Code	Complement of Excess-3
0	0011	1100
1	0100	1011
2	0101	1010
3	0110	1001
4	0111	1000
5	1000	0111
6	1001	0110
7	1010	0101
8	1011	0100
9	1100	0011

Table 2.13 is a summary of the properties of the four codes we have just discussed. Note that all four possible combinations of the two properties are shown by the codes.

Table 2.13 Properties of the Four Codes Shown in Table 2.10

Code	Self-Complementing?	Weighted?
8, 4, 2, 1	No	Yes
Excess-3	Yes	Yes
Gray	No	No
Code X	Yes	No

Unit-Distance Codes

Consider an analog voltage being sampled by a digital voltmeter and the results transmitted to a digital computer. If the (8, 4, 2, 1) BCD code shown in Table 2.10 is employed by the voltmeter, as the analog voltage changes from 3 (0011) to 4 (0100) volts, bit 2 may change first, resulting in a temporary reading of 0111, which is 7. Thus, a considerable error will exist. (This type of error can occur whenever a mechanical or electrical analog quantity is converted to a digital representation.) The error, however, can be eliminated by the use of a code in which only one bit changes at a time. The class of codes with a one-bit change for adjacent integers is given the descriptive name: *unit-distance codes.*

Table 2.14 The Gray Code

Decimal Digit	Gray Code $y_3 y_2 y_1 y_0$
0	0000
1	0001
2	0011
3	0010
4	0110
5	1110
6	1010
7	1011
8	1001
9	1000

One code of this class is the Gray code shown in Table 2.14. Note that not only does one bit change at a time in the sequence 0 through 9, but the 9-to-0 transition also has only a one-bit change. The latter fact is important in BCD applications in that errors are eliminated when the count advances from 9 to 10 (i.e., 0000 1000 to 0001 0000). The Gray code of Table 2.14 has the property that it is *reflective*. This means that, except for bit 3, the second half of the table is a mirror image of the first half. (The reflective property is sufficient but not necessary for a unit-distance code, as the following discussion will show.)

In order to study unit-distance codes more fully, we now consider a graphic system for representing them. In Table 2.14, note that the first four lines form a separate two-bit Gray code. (In particular, the 3-to-0 transition is a one-bit change.) Using the two-bit Gray code for both axes, the discrete graph (or map) shown in Figure 2.1 is constructed for the four-bit Gray code. The reader should study the following features of the map:

1. Bits $y_3 y_2$ determine the vertical coordinate and bits $y_1 y_0$ determine the horizontal coordinate.

2. The binary digits associated with each row (column) are in correspondence with the $y_3 y_2 (y_1 y_0)$ labels, e.g., $y_3 y_2 = 10$ means $y_3 = 1$, $y_2 = 0$.

3. The map entry at the intersection of a row and column is the decimal equivalent of the row and column binary labels, e.g., $5 = y_3 y_2 y_1 y_0 = 1110$, which, of course, agrees with Table 2.14. (Observe that all the other entries also agree with the table.)

4. Since both coordinates change one bit at a time, adjoining squares in the vertical or horizontal direction are separated by only one bit in their binary descriptions. This is true even for the transition from the end of a row (column), e.g., the 9-to-0 transition is 1000 to 0000.

From the above discussion, it can be seen that the rules for forming a four-bit unit-distance code are:

1. Write the increasing decimal sequence for the desired code, within adjoining squares on the map.

2. Be sure that the last decimal entry in the sequence is effectively in an adjoining square to the first entry.

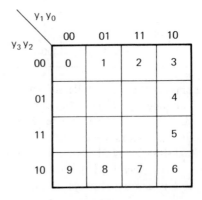

Figure 2.1 Map for the Gray Code.

To demonstrate how the map method can be employed in constructing codes, assume that we desire to construct a unit-distance BCD code which does not exhibit the reflective property. From Figure 2.1, note that the lower two rows form a mirror image of the upper rows. This is a requirement for a reflected code. The requirement for a unit code that is not reflected can now be solved if we satisfy the unit-distance rules but avoid a mirror image. Figure 2.2 is the map for the new code. As an exercise, construct the corresponding table for this code to further verify that it displays both the unit-distance and the nonreflected property. Another useful exercise is to place each of the codes from Table 2.10 on a map and show that, except for the Gray code, they do not exhibit the unit-distance property.

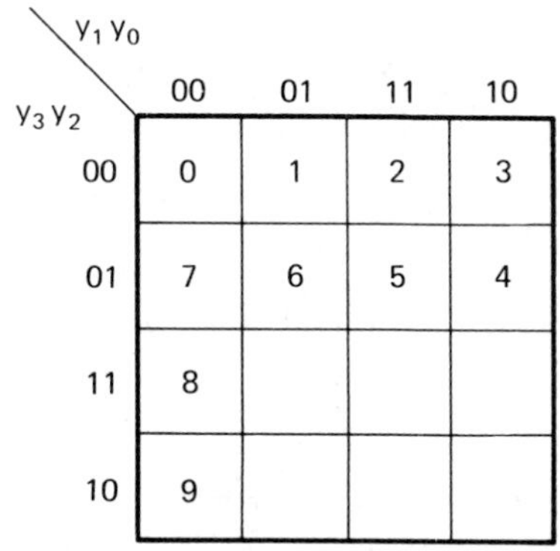

Figure 2.2 A Unit-Distance BCD Nonreflected Code.

The map method just discussed will be considerably amplified in Chapter 4 where it will be employed as a major tool in simplifying logical functions. At that point, it will be shown that the map is not limited to just four variables (bits), but that larger versions are easily constructed; thus, the method may be applied to unit-distance codes and other situations where more than four bits are required.

Error-Detecting Codes

For many computer applications, such as reading magnetic disk, paper tape, or telephone-transmitted data, it is important to detect any errors in the received information. When the probability of errors is not too large, a system called *parity checking* can be very effective. This method employs an extra bit, called the *parity bit*, whose value is selected to make the number of ones in the word even (odd) for an even (odd) parity code. For example, if even parity is employed and the parity bit, y_3, is to be added for the octal number $y_2 y_1 y_0 = 010$, the result will be $y_3 y_2 y_1 y_0 = 1010$. Here $y_3 = 1$ to yield the even number of ones required by even parity.

It is readily seen that a single-bit error in a word, including the parity bit, will always change the parity, which is easily detected. In the above even parity word, $y_3y_2y_1y_0 = 1010$, if the noise changes bit y_2, the result will be $y_3y_2y_1y_0 = 1110$, which is detected as a change to odd parity. On the other hand, if noise changes bit y_3, the result will be $y_3y_2y_1y_0 = 0010$, which again is detected as odd parity.

Even and odd parity codes for octal numbers are shown in Table 2.15. In each case, bit y_3 carries the parity information. The reader should check the accuracy of the parity bits in each code and note that bits y_2 through y_0 are the standard octal numbers.

Table 2.15 Octal Parity Codes

Decimal Digit	Even Parity $y_3y_2y_1y_0$	Odd Parity $y_3y_2y_1y_0$
0	0 0 0 0	1 0 0 0
1	1 0 0 1	0 0 0 1
2	1 0 1 0	0 0 1 0
3	0 0 1 1	1 0 1 1
4	1 1 0 0	0 1 0 0
5	0 1 0 1	1 1 0 1
6	0 1 1 0	1 1 1 0
7	1 1 1 1	0 1 1 1

Here again the map can give us additional insight into the nature of a code. Figure 2.3 is a map representation of the even parity code of Table 2.15. The fact that each map entry is separated from its neighbors by two squares means that noise could change *any single bit* and the new word would not be a valid even parity character. Thus all single errors can be detected.

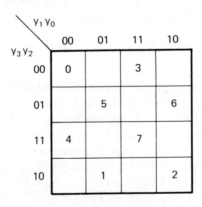

Figure 2.3 Even Parity Code.

Study of Figure 2.3 reveals that, in general, if all single errors are to be detected, each character must be separated from its nearest neighbor by at least two squares. Thus, a code such as that in Figure 2.3 is called a *minimum-distance-of-two code*. Figure 2.4 demonstrates that the parity property is not a necessary condition for an error-detecting code with a minimum distance of two. (The code for 1 has odd parity, and all the others have even.)

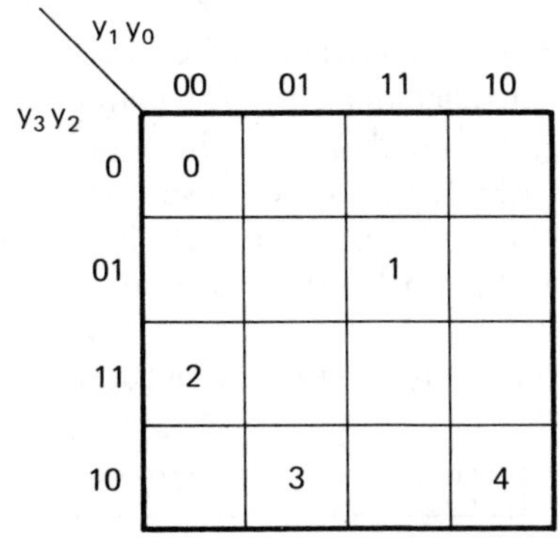

$y_3 y_2$ \ $y_1 y_0$	00	01	11	10
0	0			
01			1	
11	2			
10		3		4

Figure 2.4   A Nonparity Code with a Minimum Distance of Two.

The main advantage of the parity code of Figure 2.3 is that it is easy to construct a logic circuit to check the parity of a character. If it is not even parity, the circuit produces a signal which causes the data to be reread, or it can generate a warning signal. The nonparity code of Figure 2.4 does not possess this advantage.

Not only is it possible to detect errors, but with the proper code it is even possible to correct them. Error-correcting codes are beyond the scope of this text, but the interested reader is encouraged to consult References 4 and 5 in the Bibliography.

Special Codes

In addition to the basic codes just discussed, there is a special code class with which the computer engineer should become familiar. One major division of the class is the *alphameric codes*. These emphasize alphabetic and numeric information, but punctuation and control characters are also often included. Another major division of the special codes is the *display codes;* these control the formation of characters from segment or dot primitives for use in data output displays for devices such as digital voltmeters and pocket calculators.

Alphameric Codes. Teletypewriters, keyboards, paper tape readers, magnetic disk units, and punched-card readers are typical devices which employ alphameric codes. The American Standard Code for Information Interchange, ASCII, shown in Table 2.16, is a popular code of this type. Since the version shown contains seven bits, it encompasses $2^7 = 128$ characters. Note that the characters represented include uppercase letters, lowercase letters, numbers, punctuation marks, special symbols, and control characters. The word standard associated with ASCII is somewhat misleading since there is not just one standard alphameric code. On the contrary, EBCDIC, Extended Binary-Coded Decimal Interchange Code, an eight-bit code containing most of the same characters as ASCII, is widely employed, particularly in the IBM Systems 360 and 370. (The coding of ASCII and EBCDIC are much different, e.g., the number 2 has the hexadecimal representation F2 in EBCDIC and 32 in ASCII.) Moreover, even ASCII has six- and eight-bit versions which have different information-carrying capacities than the seven-bit version. (Of course, the coding is similar in the three versions.)

Table 2.16   Outline of the ASCII Seven-Bit Code

Seven-Bit Code (Hexadecimal)	Characters		Seven-Bit (Hexadecimal)	Characters	
00	Null	⎫ Special	41	A	⎫
. . .	. . .	⎬ control	42	B	⎬ Uppercase
20	Space	⎭ characters	. . .	. . .	⎱ letters
21	!	⎫ Punctuation	5A	Z	⎭
. . .	. . .	⎬ and special	5B	[	⎫
2F	/	⎭ symbols	. . .	. . .	⎬ Special
30	0	⎫	60	`	⎭ symbols
31	1	⎪	61	a	⎫
32	2	⎬ Numbers	62	b	⎬ Lowercase
. . .	. . .	⎪	. . .	. . .	⎱ letters
39	9	⎭	7A	z	⎭
3A	:	⎫ Punctuation	7B	{	⎫
. . .	. . .	⎬ and special	. . .	. . .	⎬ Special
3F	?	⎭ symbols	7E	~	⎭ symbols
40	@		7F	Delete	

MSD LSD

Bit arrangement for ASCII-7 code $\equiv \overbrace{b_5 b_5 b_4}\ \overbrace{b_3 b_2 b_1 b_0}$

Display Codes.   Again, there is not just one code employed to display information on cathode ray tubes, digital voltmeters, and other types of readouts. On the contrary, many schemes are in use, and no doubt many more will soon be invented. The basic principles of these are brought out by Figure 2.5, which shows a seven-segment display for decimal numbers. This type of display is extensively used in pocket calculators. Generally, numbers in a calculator or similar device are processed in another number system, such as BCD; but before they are displayed, there is a logic network which

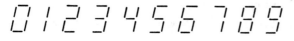

(a) Sample Numbers

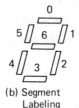

(b) Segment
Labeling

Decimal	BCD	Segments $S_6 S_5 S_4 S_3 S_2 S_1 S_0$
0	0000	0 1 1 1 1 1 1
1	0001	0 0 0 0 1 1 0
2	0010	1 0 1 1 0 1 1
3	0011	1 0 0 1 1 1 1
4	0100	1 1 0 0 1 1 0
5	0101	1 1 0 1 1 0 1
6	0110	1 1 1 1 1 0 1
7	0111	0 0 0 0 1 1 1
8	1000	1 1 1 1 1 1 1
9	1001	1 1 0 1 1 1 1

(c) Segment Coding

Figure 2.5   Numeric Display Code.

converts from the main code to the segment code in order tó produce the display. As an exercise, the reader should verify that the codes in the table of Figure 2.5 do indeed generate the sample numbers shown.

To control errors, a parity bit can be added to any of the special codes. Because they are usually transmitted over longer distances and through noisier environments, however, the alphameric codes are more likely to require parity protection than are display codes.

In later chapters, we will consider the implementation aspects of coding. For example, in Chapters 4 and 9, methods for designing logic to convert one code to another will be presented; in Chapter 7, a character generator will be described which employs a 4 × 6 array of dots to produce an alphameric display.

2.7 BIBLIOGRAPHY

1 BARTEE, THOMAS C., *Digital Computer Fundamentals* (3rd ed.). New York: McGraw-Hill Book Company, 1972.

2 DUDLEY, UNDERWOOD, *Elementary Number Theory*. San Francisco: W. H. Freeman and Company, 1969.

3 NEWMAN, JAMES R., *The World of Mathematics*, Vol. 1. New York: Simon & Schuster, Inc., 1956. (Especially pp. 430–520.)

4 PETERSON, W. W., *Error Correcting Codes*. New York: John Wiley & Sons, Inc., 1961.

5 PETERSON, W. W., "Error Correcting Codes." *Scientific American*, Vol. 215, No. 3, 1962, pp. 96–110.

6 RHYNE, V. THOMAS, *Fundamentals of Digital Systems Design*. Englewood Cliffs, N.J.: Prentice-Hall, Inc., 1973.

7 RICHARDS, R. K., *Arithmetic Operations in Digital Computers*. New York: Van Nostrand Reinhold Company, 1956.

8 WALTER, TERRY M., and WILLIAM W. COTTERMAN, *An Introduction to Computer Science and Algorithmic Processes*. Boston: Allyn & Bacon, Inc., 1970.

9 WARE, WILLIS H., *Digital Computer Technology and Design*, Vol. 1. New York: John Wiley & Sons, Inc., 1963.

2.8 EXERCISES

2.1 Write the octal numbers from 0 to 20_8.

2.2 Construct a table to show how counting is accomplished, between 0 and 40_{16}, in the hexadecimal system. Every number need not be expressed, but is important to show entries in the table for the hexadecimal number and its decimal equivalent at all possible points of confusion, e.g., at A_{16}, F_{16}, 10_{16}, $1A_{16}$, etc.

2.3 Convert the following binary numbers to their decimal equivalents:
   (a) 11010
   (b) 111011
   (c) 0.1101
   (d) 1101.00111

2.4 Convert the following binary numbers to their octal equivalents:
   (a) 110011010101
   (b) 0.01110101
   (c) 101111011100.0011110111

2.5 Convert the following octal numbers to their binary equivalents:
   (a) 5426
   (b) 0.7162
   (c) 4715.372

2.6 Convert the following octal numbers to their decimal equivalents:
   (a) 2654
   (b) 0.175

  (c) 7514

  (d) 5137.724

2.7 Convert the following decimal numbers to their binary equivalents:

  (a) 42

  (b) 93

  (c) 0.846

  (d) 39.725

2.8 Convert the following hexadecimal numbers to their binary equivalents:

  (a) C7F

  (b) 0.AF9

  (c) 8EA.DA2

2.9 Convert the following hexadecimal numbers to their decimal equivalents:

  (a) 3A8

  (b) 0.F2A

  (c) D9.7E

2.10 Convert the following decimal numbers to their hexadecimal equivalents:

  (a) 42

  (b) 93

  (c) 39.725

  (d) 46791

2.11 Prove the algorithm for decimal-to-binary conversion given in Section 2.2.

2.12 (a) Construct a complete addition table for octal analogous to Table 2.7.

  (b) Construct a complete multiplication table for octal analogous to Table 2.8.

2.13 Determine the following sums:

  (a) $11001.11 + 1011.01$

  (b) $10111.101 + 1010.01$

  (c) $65.27_8 + 54.67_8$

  (d) $AE.6F_{16} + C3.94_{16}$

2.14 Perform the following operations:

  (a) $1011.11 - 110.01$

  (b) $11011.10 - 1101.01$

  (c) $(1110.01)(101.1)$

  (d) $(1101.1)(110.01)$

  (e) $(1101) \div (101)$

  (f) $(1110.1) \div (1101)$

  (g) $(6473_8) \div (462_8)$

2.15 (a) Using the 1's complement method, with six-bit numbers, calculate $(6 - 10)_{10}$. Check the result by converting it back to a base-10 magnitude plus a sign.

    (b) Starting with the definition of 2's complement, prove a conversion method, based on complementing individual bits, similar to the 1's complement method.

    (c) Repeat (a) but use the 2's complement method.

    (d) Using first the 1's complement method and then the 2's complement method, calculate $-5_{10} + 5_{10}$. Comment on the results.

2.16 Construct a table similar to Table 2.10 for the (5, 4, 2, 1) and the (6, 4, 2, 1) codes.

2.17 How many punch positions (possible holes) exist per column on an IBM card? What is the maximum number of characters that could be represented by these punch positions in a single column? Using up to the maximum number of holes commonly punched per column, how many characters can be coded?

Table 2.17   Codes to be Analyzed

Decimal Digit	Code A	Code B	Code C	Code D	Code E	Code F
0	0000	0001	0000	0000	0000	0101
1	0001	0010	1101	0001	0011	0110
2	0010	0011	1000	0101	0101	0111
3	0011	0100	0100	?	0110	1000
4	0100	0101	0011	1111	1111	1001
5	1011	0110	1100	?	1000	1010
6	1100	1000	1011	?	*	1011
7	1101	1001	0111	?	*	1100
8	1110	1010	0010	1001	*	1101
9	1111	1100	1111	?	*	1110

*Not Required

Analyze the codes shown in Table 2.17 according to the following directions:

2.18 Is code A a weighted code? If so, what are the weights?

2.19 Repeat Exercise 2.18 for code B.

2.20 If code D is to have the unit-distance property, fill in the missing members.

2.21 Does code E possess an error-detection capability? Explain.

2.22 Add an even parity check bit to code A.

2.23 Repeat Exercise 2.18 for code C.

2.24 Which of the codes possess the self-complementing property?

2.25 Repeat Exercise 2.18 for code F.

2.26 Which of the codes in Table 2.10 show the self-complementing property?

2.27 Is the excess-3 code of Table 2.10 a weighted code? If so, what are the weights?

2.28 In the ASCII seven-bit code, what is the hexadecimal representation for the following symbols: D, e, 5, and 7? If an even parity bit were added as a new MSB to the code, what would the binary representation of the following symbols be: @, [, 2, and z?

2.29 Using the segment labeling of Figure 2.5, what would the codes be for the following alphabetic characters: A, C, F, H, and S?

3 SOFTWARE

Hardware refers to the electronic circuits, as well as the other electrical and mechanical components, which constitute the physical portions of a digital computer. By contrast, software refers to the programs and *algorithms* which are used by the hardware to solve problems and process data. (An algorithm is a mathematical procedure, a detailed set of directions, for the solution of a particular class of problems. The manual method for forming a quotient by long division is a type of elementary algorithm.) The main subject of this book is hardware design, but it is necessary to discuss various aspects of software, both in this chapter and in Chapter 10, so that the reader can more fully appreciate how hardware and software work together during a data-processing task. Until recently, these two facets of computer technology were developed relatively independently of each other during the design of a new machine. Engineers are beginning to realize, however, that a much more powerful system can be produced if hardware and software are developed in close relationship with each other. With a knowledge of software, the engineer can make better design decisions in all phases of hardware development. Moreover, an understanding of programming is a necessity in testing computers during development and in the field.

We are introducing programming concepts early in the text in order to give a clear and complete picture of the nature of a digital computer, which should accelerate understanding of the material in all later chapters. Initially, a rather elementary approach is taken since only the very basic concepts are required at this point and since these may be understood more easily if additional details are delayed until Chapter 10.

What is a program, anyway? It is the directions, or "recipe," that the hardware follows in processing the data it receives. You may already be familiar with a language like FORTRAN or PL/I. Experience with a higher-level language gives one an appreciation of the practical value of the computer and an understanding of its external operation. It does not, however, give the programmer much insight into machine organization, logic, or circuitry.

We begin this chapter with a discussion of organization, then present machine language programming, and end with a brief discussion of programming languages.

3.1 BASIC MACHINE ORGANIZATION

Later, a whole chapter will be devoted to machine organization, and several alternative configurations will be indicated; but at this point it is necessary to present enough fundamentals concerning one approach so that the reader will understand what portions of the computer are being controlled by particular machine language instructions. Also, it is important to keep the computer's structure in mind as the logic design techniques are developed in the next several chapters.

Figure 3.1(a) shows the basic block diagram for a digital computer, and Figure 3.1(b) shows an exploded view of the various sections in a modern minicomputer—the Hewlett-Packard 21MX. Note the correspondence between the actual hardware and the block diagram items in the two parts of the figure. The small size and the modularity of the 21MX system is typical of modern, well-designed, minicomputers. This type of construction facilitates tailoring the computer to a wide variety of applications.

The input device in Figure 3.1 typically receives information from punched cards or magnetic tape, converts it to electrical signals, and sends it to the *Control Unit* within the *Central Processing Unit*, CPU. From there the information is usually placed in magnetic cores or integrated circuits within the *Memory*. The Memory of a computer is divided into many individual storage locations, each containing one word (number). Every location is given a numerical address in a string configuration much like the address employed to locate a particular house on a long street. After the program and data have been transferred to the Memory, the Control Unit begins allowing one program instruction at a time, in sequence, to be executed. This is accomplished by initializing the *Program Counter*, within the control logic, to the starting address of the program and incrementing the counter by one each time an instruction is completed. Frequently an instruction will call for data to be removed from storage, processed in the Arithmetic Unit, and returned to storage. The address of the word being loaded or stored is determined by the *Memory Address Register*, MAR, which in turn receives its information either from the Program Counter or from the operand portion of the instruction being executed. The *Accumulator*, the main register in the arithmetic unit, always contains one of the numbers during execution of one or two operand arithmetic instructions. After the data have been processed, the results are read from Memory, under the supervision of the Control Unit, and presented to the user in a convenient form by the output device. This peripheral can be a teletypewriter, a cathode ray display, or one of a large number of other units currently available.

As is evident from the block diagram, the Control Unit is at the heart of the computer. It provides pulses (not indicated on the diagram) to synchro-

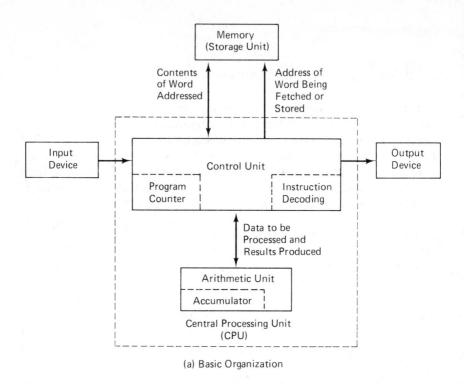

(a) Basic Organization

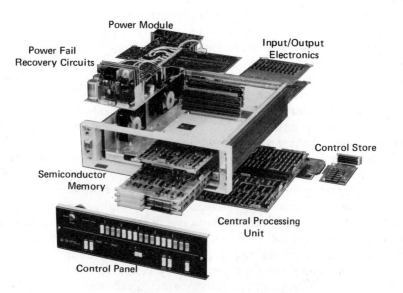

(b) The Hewlett-Packard 21MX Minicomputer
(Photo courtesy of Hewlett-Packard)

Figure 3.1 A Digital Computer System-Block Diagram and Hardware.

nize the operations of the other units; it decodes the program instructions; and it generally supervises the operation of all the other units. (A detailed discussion of machine organization is given in Chapter 7.)

3.2 MACHINE LANGUAGE PROGRAMMING

Machine language instructions directly control the various functions of a digital computer, and these may be divided into the following four classes.

1. Transfer—including instructions both for storing and fetching information from memory.

2. Arithmetic and Logic—including add, multiply, AND, and OR.

3. Branch and Control—including instruction sequence control functions such as jumps from one section of a program to another, halt the CPU, etc.

4. Input/Output—including operation of a teletypewriter, line printer, card reader, and cathode ray tube display.

Although it is theoretically possible to code any solvable problem through the use of a very small number of instructions, a desire to make programming easy and convenient may cause a designer to include more than 200 machine language instructions in the repertoire of a modern computer. This chapter is intended to provide an elementary understanding of machine language and to form an appreciation of the programmer's task, but not to actually develop expert programmers. Thus, only a limited number of instructions will be studied. In order to make our work more realistic, the instruction set presented here is taken from a well-known minicomputer family,* but the machine language operation of many other computers is quite similar. Incidentally, the LINC is a 12-bit machine, which means that all the information it processes and stores in its memory will be represented by 12-bit numbers. (The leftmost bit is for the sign, and the other 11 bits form the magnitude in a 1's complement system.)

The *add* instruction is very frequently needed in computer programs, and it can appear in one of several forms, e.g., ADD 235. The meaning of this instruction is: "add the number found in memory location 235 to the number currently in the main arithmetic register, called the accumulator, and leave the result in the accumulator." This is a so-called single-address instruction in that only one memory location is explicitly mentioned. Since addition requires at least two operands, the second operand must be implied, and this

---

*The classic LINC, Laboratory INstrument Computer, developed by Wesley Clark and his associates, and its descendent, the Digital Equipment Corporation PDP-12, are well-known members of this group.

is where the accumulator becomes necessary. (Computers have been built using two, three, and even four addresses. These variations will be discussed in more detail in Chapter 10.)

Let us consider a simple program for summing the numbers in locations 26, 27, and 30 and leaving the result in location 31. Table 3.1 shows the sequence of instructions starting at memory address 20.

The first instruction, CLR, clears the accumulator (sets it to zero). Then the numbers in locations 26, 27, and 30 are added in sequence, and the result appears in the accumulator after the instruction in 23 has been executed. Next the STC instruction S̲Tores* the sum in location 31 and C̲lears the accumulator. Finally, the computer is halted by means of the HLT instruction.

Table 3.1   A Program to ADD Three Numbers

Memory Address	Memory Contents
(Start) 20	CLR
21	ADD 26
22	ADD 27
23	ADD 30
24	STC 31
25	HLT
26 through 31	(Data and result)

Several points should be noted concerning Table 3.1:

1. Certain instructions, such as ADD and STC, have both an operation part and an address part.

2. Other instructions, such as CLR and HLT, have only an operation part.

3. The three-letter abbreviations for the instructions, called *mnemonics*, form a convenient shorthand.

Since all symbols and numbers in a digital computer are presented internally by means of binary numbers, the memory contents of Table 3.1 must be coded before it can be executed by the computer. As a convenience, these codes will be represented in octal, e.g., the code for STC 31 is 4031. Again, the number 4031 must code both the operator and the address. It is probably apparent that the first digit, 4, represents the operation, and the next three digits, 031, represent the address.

A brief list of LINC machine language instructions and their octal codes are given in Table 3.2. (For tutorial reasons, only a small fraction of the actual LINC instructions are listed here.)

Note: The underlined uppercase letters give the key to the name of the instruction.

Table 3.2 A Brief List of Pseudo-LINC* Instructions

Mnemonic	Machine Code Number (Octal)	Meaning
ADD XXX	2XXX	A D D the contents of the accumulator to the contents of location XXX and leave the result in the accumulator, using 1's complement addition.†
APO	0451	Skip the next instruction if the Accumulator is P Ositive, i.e., the sign bit equals zero.
CLR	0011	C LeaR accumulator.
COM	0017	C O Mplement the contents of the accumulator.
HLT	0000	The machine is ordered to HaL T.
JMP XXX	6XXX	JuM P to location XXX to obtain the next instruction.
MUL XXX	5XXX	M U Ltiply the contents of the accumulator by the contents of location XXX and leave the least significant 11 bits in the accumulator.†
NOP	0016 or YYYY	No O Peration. This instruction merely provides a delay before the next instruction is executed. The alternate code YYYY is *any* undefined code and has the same meaning as the standard code.
ROR n	0300 + n	R Otate Right. The contents of the accumulator are moved around to the right n places, with the least significant bit connecting into the most significant bit as shown:

12-bit accumulator

(move right n bits)

| STC XXX | 4XXX | S Tore the contents of the accumulator into location XXX and then Clear the accumulator. |

*These are called pseudo-LINC instructions because there are some minor variations between these instructions and those actually available on the LINC.
†The contents of location XXX are not changed.

Observe the following facts concerning the table:

1. The underlined uppercase letters in the last column give the key to the mnemonics.

2. The JMP and APO are branch instructions which furnish considerable programming power by allowing the normal sequence of operations to be modified. The APO instruction is particularly valuable in that it permits different actions to be taken, based on the sign of the accumulator. (Note that APO does not skip on −0.)

3. Because of the emphasis on fundamentals in this chapter, instructions such as *print*, which are rather specialized to a particular computer, are not presented. (Input/output functions are considered in Chapter 11.)

4. When arithmetic operations, such as ADD and MUL, employ one or more negative operands, the algebraic sign for the result is automatically placed in the leftmost bit. (This includes operations with ± 0.)

Example: Evaluating Expressions

Evaluate the expression $R = Y - X^2$. Assume that the number representing X is in location 100, and Y is in 101. We desire to have the result, $R = Y - X^2$, in location 102. The required program is shown in Table 3.3, with the results for $X = 3$, $Y = 4$, in the last column. In the program, note how subtraction is accomplished through the use of the complement operation, and that all numbers are given in octal.

Table 3.3 Program to Calculate $R = Y - X^2$

Memory* Address	Memory Contents Mnemonic	Machine Code	Accumulator Contents (After Execution of Instruction), and Remarks
(Start) 20	CLR	0011	0
21	ADD 100	2100	$X = 0003$
22	MUL 100	5100	$X^2 = 0011$
23	COM	0017	$-X^2 = 7766$
24	ADD 101	2101	$Y - X^2 = 7772 = -5$
25	STC 102	4102	0
26	HLT	0000	0
...	...	...	...
100	X	0003	
101	Y	0004	
102	R	7772	(Result after program runs)

*Note that each memory location contains a 12-bit word (see machine code column) and that the addresses are numbered in octal.

Example: Modification of Instructions

Sometimes a very concise program can be written by modifying an instruction during the actual running of the program. Assume that it is desired to test which of the numbers, 1 through 7, is present in register 100. If a 1 is present, a 1 is to be placed in location 201; if a 2 is present, a 1 is to be placed in location 202; etc. Finally, if a 7 is present, a 1 is to be placed in location 207. (The locations 201 through 207 are assumed to initially contain zeros.) The solution to this problem is given in Table 3.4, and you should study the program by verifying the accumulator contents (column 4) after each instruction is executed. In particular, observe how the instruction in address 25 is modified by the data in location 100 so that a 1 is stored in the proper location.

Table 3.4 Instruction Modification Program

Memory Address	Memory Contents Mnemonic	Machine Code	Accumulator Contents (After Execution of Instruction), and Remarks
(Start) 20	CLR	0011	0
21	ADD 100	2100	X = 0003
22	ADD 25	2025	4200 + X = 4203
23	STC 25	4025	0000
24	ADD 101	2101	0001
25	STC 200	4200	0000 (After instruction in 23 is executed, location 25 contains 4203)
26	HLT	0000	0000
...	...	...	...
100	X	0003	(Number being tested)
101	1	0001	(A constant)
...	...	...	
201	0	0000	
202	0	0000	
203	0	0000	(After instruction in 25 is executed, location 203 contains 0001)
204	0	0000	
205	0	0000	
206	0	0000	
207	0	0000	

In the program, the contents of location 25 act both as data and as an instruction, but to the computer hardware it merely looks like a number in both cases. This dual ability need not concern us, and as a matter of fact, it is very convenient. However, the inability of the computer to distinguish between instructions and data means that we must be careful to avoid accidental mixing of the two; otherwise, the computer might try to execute a piece of data in place of an instruction, resulting in a *program bug*.

Example: Summing the Numbers in a Group of Registers

Frequently it is necessary to perform a repetitive task or to iterate a particular mathematical operation. A very short program can usually be written in such cases by having the program loop through a particular group of instructions a specified number of times. (A DO LOOP in compiler language accomplishes the same thing.)

Assume that it is desired to sum the numbers in memory locations 200 through 225 and to place the result in location 300. Figure 3.2 shows a flow diagram for the desired iterative algorithm; note how concisely the desired processor action may be described.

The resulting machine language program is given in Table 3.5, with the Remarks column included as an aid to understanding the process. The trickiest part of a program like this is determining the correct value of the

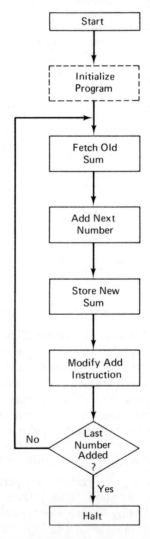

Figure 3.2 An Algorithm for Summing Numbers.

Table 3.5 Program to Sum the Numbers in Locations 200 through 225

Memory Address	Memory Contents Mnemonic	Machine Code	Remarks	1st time	2nd time	25th time	26th time
Start							
20	CLR	0011	⎫	0	0	0	0
21	ADD 200	2200	Sum and	$C(200)^*$	$C(201)$	$C(224)$	$C(225)$
22	ADD 300	2300	store	$C(200)$	$C(200)$ $+C(201)$	$\sum\limits_{200}^{224} C(i)$	$\sum\limits_{200}^{225} C(i)$
23	STC 300	4300	⎭	0	0	0	0
24	ADD 21	2021	⎫ Instruction	2200	2201	2224	2225
25	ADD 301	2301	modifi-	2201	2202	2225	2226
26	STC 21	4021	⎭ cation	0	0	0	0
27	ADD 21	2021	⎫	2201	2202	2225	2226
30	ADD 302	2302	Loop	7753	7754	7777	0001
31	APO	0451	counter	7753	7754	7777	0001
32	JMP 20	6020		7753	7754	7777	—
33	HLT	0000	⎭	—	—	—	0001
. . .							
200–225			Data (Numbers to be summed)				
. . .							
300	0	0000	Result (Initial value shown)				
301	+1	0001	⎫ Constants				
302	−(2225)	5552	⎭				

Accumulator After Execution of Instruction (spanning 1st, 2nd, 25th, 26th time columns)

*The notation $C(x)$ is frequently employed in computer technology and means the *contents of location x*.

constant in location 302 so that the program comes out of the loop after the proper number of iterations. In the current problem, after the 25th loop has finished, the number in location 225 has not yet been added to the sum; thus, a 26th loop is required. During that loop the accumulator first becomes positive at the APO instruction (location 31), causing the JMP 20 instruction to be skipped. This should be compared with loop 25, where the accumulator contains 7777. Since 7777 is negative zero, the skip does not occur.

Approximately twice as many instructions would have been required if the brute-force method of writing a sequence of ADD instructions for each of the locations 200 through 225 had been used. Moreover, the brute-force program grows linearly as the numbers to be summed increase, while the size of the Table 3.4 program is constant. In Chapter 10, we will find that by introducing an index instruction it is possible to make the program in Table 3.5 even more concise.

Often a program is loaded once into memory, and then several sets of data are processed, with the results being removed and new data inserted after each run. One difficulty, however, will be observed in the case of Table 3.5: the contents of memory addresses 21 and 300 are changed by the program. Consequently, these locations must be initialized before each run. The dashed section in Figure 3.2 shows how initialization fits into the flow diagram, and the added instructions are shown in Table 3.6. With the de-

Table 3.6   Initializing Routine for Use with Program in Table 3.5

Memory Address	Memory Contents Mnemonic	Machine Code	Remarks
13	ADD 200	2200	
(Start) 14	CLR	0011	
15	STC 300	4300	Initializing
16	ADD 13	2013	steps
17	STC 21	4021	
20	. . .	. . .	Program from Table 3.5

scribed improvement, each time the program is started at location 14, the two memory addresses in question are first initialized. Thus, the new program may be executed as many times as desired, and the correct instructions will always be present after the operation in location 17 is completed.

3.3   PROGRAMMING LANGUAGES[5,9,10]

There is a wide variety of programming languages in general use (actually hundreds), each of which is adapted to solving a particular group of problems. These programming languages can be arranged into a few classes, and this section presents a brief sketch of the classes and their basic properties. One scheme for partitioning languages into five main divisions is presented below:

Microprogram

Machine

Assembler

Macrodefinition

Compiler
     General-purpose or procedure-oriented
     Special-purpose or problem-oriented

Microprogram. This is the lowest-level programming language. It is used to compose conventional machine language instructions from more primitive operations, e.g., a multiply instruction can be constructed from a number of shift and add operations. Some contemporary machines, e.g., the PDP-8, are very flexible because they can be microprogrammed by the user, but in most instances this level of programming is either not available or is reserved for the expert programmer. (More details will be given in Chapter 10).

Machine Language. Programming in machine code was discussed at length in Section 3.2. It is the lowest-level language available to the programmer on most machines.

Assembler. Assembly language is sometimes called *symbolic machine language*, which provides a key to its nature. Machine language denotes the numerical codes that the computer executes in accomplishing a data-processing task. They are listed in the third column of most of the programs presented in Section 3.2 (see Table 3.3). Assembly language allows the programmer to enter on a card or other input medium the mnemonic ADD 100 instead of the machine code 2100. The symbols ADD 100, however, are not actually entered directly into the program to be run. Rather, the assembler interprets ADD 100 as 2100 and this is the number which is inserted in the proper memory location. The replacement of machine code by a mnemonic may not seem like much of an advance, but when we recall that the complete machine language may include 100 instructions and these may each have several variations, there actually is quite a benefit to the user. Moreover, assembly language provides a large number of other conveniences such as:

1. Symbolic references.
2. Relative addressing.
3. Editing of program manuscripts.*
4. Maintaining an index of programs which can be loaded by name.*
5. Error detection for illegal mnemonics, undefined symbols, etc.
6. Several other useful features such as printing a program listing which shows each memory address and the memory contents in mnemonics as well as in machine code.

Except for the first two, items in this list should be self-explanatory. *Symbolic references* means the use of a group of letters or a name to represent the contents of a particular memory location, instead of giving the actual

---

*These features will often be part of other sections of the software system.

address. The program in Table 3.7 shows how the calculation $R = X + Y$ is accomplished by referring to locations 27, 30, and 31 not by their actual addresses, but by their preassigned names, R, X, and Y. When the program is submitted to the assembler, the R in location 27 is recognized by the system as the convention for a symbolic reference.

Table 3.7   An Example of Symbolic Referencing

Memory Address	Memory Contents
(Start) 20	CLR
21	ADD X
22	ADD Y
23	STC R
24	HLT
25	. . .
26	. . .
27	R 0000
30	X 1053
31	Y 0261

Relative addressing is another way to avoid giving an absolute address in a program. Here, each instruction is considered to exist at P, the present address, and all other addresses are relative to p, e.g., $p - 2$ refers to the location two addresses before the current instruction. Table 3.8 shows the program of Table 3.7, but using relative addressing.

The advantage of both symbolic and relative addressing is that the same program may be located anywhere in memory without the need for changing

Table 3.8   An Example of Relative Addressing

Memory Address	Memory Contents
(Start) 20	CLR
21	ADD $p + 7$
22	ADD $p + 7$
23	STC $p + 4$
24	HLT
25	. . .
26	. . .
27	0000
30	1053
31	0261

the address references in the instructions. Also, they save time and trouble when instructions must be added or removed from an existing program.

Macrodefinition. A *macro* furnishes additional power to an assembly language by allowing a whole group of instructions to be automatically added to the program. It does this by merely inserting a name in the symbolic program. For example, consider the following macrodefinition:

$$*<\text{ADD } 2, R, X, Y$$
$$\text{CLR}$$
$$\text{ADD } X$$
$$\text{ADD } Y$$
$$\text{STC } R;$$

This is the definition of a procedure, called ADD 2, which adds two numbers and stores the result in location R. Once the procedure has been defined, it may be called (macrocall) simply by writing $<$ADD 2, A, B, C; the above four assembly language instructions will then be inserted in the final program. In this call, note the relationship between the symbolic references $A = R$, $B = X$, and $C = Y$.

Observe how the symbols $*<$ and $<$ were used to distinguish between a conventional instruction, a macrodefinition, and a macrocall. In addition, note the presence of a semicolon to indicate the end of the macrodefinition.

Compiler. A *compiler* is a program which translates <u>concise statements</u> in the source language to machine code. "Concise" implies that usually a <u>single statement</u> in the source language can replace many machine instructions. The source language should be very close to the notation of the problem and it should be machine-independent. For example, a FORTRAN IV program written for one machine should run, with few changes, on another machine; but the compiler, which provides the interface between the source language and the hardware, will be significantly different for each machine. A *procedure-oriented compiler* is one that allows the user to write programs in a language applicable to a rather broad class of problems, e.g., the language of mathematics. Typical compilers of this class are PL/I, FORTRAN IV, ALGOL, and BASIC.

With a well-designed *problem-oriented compiler*, it is easy to write a program for a fairly narrow class of problems. Thus, the trade-off between simplicity and generality is strongly weighed in favor of simple statements for a specialized class of problems, with a loss of ability to solve other problem classes. ECAP[8] (Electronic Circuit Analysis Program) is a typical example of a problem-oriented language. This system will analyze very complicated dc, ac, and transient networks and print out the results. The topology of the network and the values of the circuit elements are input to the computer by means of a simple notation. The person using ECAP needs to know almost nothing about computer programming, and the mathematical knowledge required is no higher than a first course in algebra. Engineers and scientists in many fields are now making considerable practical use of problem-orientated languages.[1,2] In contrast with its network capabilities, ECAP is not designed to handle programs requiring direct arithmetic calculations. On the other hand, FORTRAN can be used to make network calculations and a wide variety of other types of calculations. The solution of complicated electrical networks with FORTRAN, however, requires considerable mathematical as well as programming knowledge.

In order to expeditiously reach Chapter 4, the actual start of computer design, and yet have sufficient background for the next several chapters, we have presented a rather modest introduction to software. Chapter 10, however provides a major supplement to the fundamentals established here. It contains a large section on microprogramming and detailed consideration of several types of machine and assembly language instructions along with program examples. Moreover, both the Bibliography of this chapter and that of Chapter 10 contain ample modern references for supplemental reading.

3.4 BIBLIOGRAPHY

1 BOWERS, JAMES C., and S. R. SEDORE, *SCEPTRE: A Computer System for Circuit and Systems Analysis.* Englewood Cliffs, N.J.: Prentice-Hall, Inc., 1971.

2 CALAHAN, D. A., et al. (ed.), "Special Issue on Computers in Design." *Proc. IEEE*, Vol. 60, 1972, pp. 1–124.

3 DONOVAN, JOHN J., *Systems Programming.* New York: McGraw-Hill Book Company, 1972.

4 EADIE, DONALD, *Modern Data Processors and Systems.* Englewood Cliffs, N.J.: Prentice-Hall, Inc., 1971.

5 GALLER, B. A., and A. J. PERLIS, *A View of Programming Languages.* Reading, Mass.: Addison-Wesley Publishing Company, Inc., 1970.

6 HAUPT, FLOYD, *Elementary Assembler Language Programming.* Columbus, Ohio: Charles E. Merrill Company, 1972.

7 KNUTH, DONALD E., *Volume 1/Fundamental Algorithms, The Art of Computer Programming.* Reading, Mass.: Addison-Wesley Publishing Company, Inc., 1968.

8 LEVIN, HERMAN, *Introduction to Computer Analysis: ECAP for Electronic Technicians and Engineers.* Englewood Cliffs, N.J.: Prentice-Hall, Inc., 1970.

9 ROSEN, SAUL, *Programming Systems and Languages.* New York: McGraw-Hill Book Company, 1967.

10 SUMMET, J. E., *Programming Languages: History and Fundamentals.* Englewood Cliffs, N.J.: Prentice-Hall, Inc., 1969.

3.5 EXERCISES

3.1 Classify the LINC instructions in Table 3.2 according to the categories listed at the beginning of Section 3.2.

3.2 (a) List the sequence of instructions required to produce skip on negative accumulator.
(b) Repeat (a) for skip on zero accumulator. In parts (a) and (b), use only the versions of the LINC instructions given in Table 3.2.

3.3 After the machine language program shown has run, determine the following quantities using *octal* numbers:
(a) The mnemonics corresponding to the LINC codes and the magnitude and sign of any data contained in this program.
(b) The memory address where the program halts.
(c) The number in the accumulator when it halts.
The program starts in location 24, and all locations except those listed initially contain zero.

Memory Address	LINC Code	Mnemonics
24	0011	
25	2051	
26	0017	
27	2050	
30	4031	
31	6052	
32	0451	
33	0000	
34	6040	
. . .		
50	2327	
51	0277	

3.4 Write a program which searches memory locations 100 through 500 to try to find a number greater than the one stored in location 70. If a suitable number is found, halt the computer at the address immediately following that number. If no suitable number is located, halt the computer at address 501.

3.5 After the LINC program shown has run, determine the following quantities using *octal* numbers.

(a) The memory address where the program halts.

(b) The number of times the instruction in location 22 is executed.

(c) The address of any locations changed by the program and the final contents of those locations.

The program starts at address 20 and all locations except those shown initially contain zero.

Memory Address	LINC Code
17	7776
20	0011
21	2024
22	2017
23	4024
24	4040
25	2024
26	0451
27	6020
30	6050

3.6 After the LINC program shown has run, determine the following quantities using *octal* numbers:

(a) The memory address where the program halts.

(b) The number in the accumulator when it halts.

The program starts in location 23 and all locations except those listed initially contain zero. The starting contents of the accumulator is somewhere within ±3777.

Memory Address	LINC Code
23	4070
24	4033
25	2070
26	0017
27	2070
30	0451
31	6033
32	6040
33	6041

3.7 Write a program which converts 8, 4, 2, 1 code to excess-3 code. The original number is in location 100 and the result is to be placed in 200. The program should check that the content of 100 is a positive binary number between 0000_2 and 1001_2. If it is, halt the program at 350; otherwise, halt at 300.

3.8 Write a program which accepts 8, 4, 2, 1 code and converts it to code X, shown in Table 2.10. The number to be converted is found in location 100 and the result is to be placed in 200. (The content of 100 is a proper single BCD number.)

3.9 With the fewest possible instructions, modify the program in Table 3.4 so that it is self-initializing with respect to the data output locations as well as the program itself.

3.10 Using the fewest possible pseudo-LINC instructions, write a program to add the contents of locations 101 through 104. Normally, halt at 40 and leave the results in the accumulator; but if overflow occurs (the capacity of the accumulator is exceeded), halt at 50 and clear the accumulator. Assume that all numbers are positive.

3.11 Using the fewest possible pseudo-LINC instructions, write a program to determine whether the number initially in the accumulator is odd. If it is odd, halt at 100; otherwise, halt at 101.

3.12 After the program (employing pseudo-LINC instructions *only*) runs, determine the following quantities using octal numbers:
(a) Does the program halt? If so, at what memory address?
(b) The number of times the instruction in 20 is executed.

Memory Address	LINC Code
20	0011
21	0405
22	0451
23	0000
24	2021
25	2020
26	4021
27	6020
30	6100

The program starts at 20, and all locations, except those shown, initially contain zero.

3.13 After the machine language program shown has run on the LINC, determine the following quantities:
(a) The LINC mnemonics corresponding to the machine codes and the magnitude and sign of any data. (Fill in table below.)
(b) The memory address where the machine halts.
(c) The number of times the instruction in location 20 is executed. (Give results in octal.)

Memory Address	LINC Code	Mnemonics
20	0011	
21	2200	
22	2201	
23	4200	
24	2033	
25	2201	
26	4033	
27	2200	
30	2202	
31	0451	
32	6020	
33	6040	
. . .	. . .	
200	2300	
201	0001	
202	5437	
. . .	. . .	
300	0001	
301	0005	
. . .	. . .	

The program starts in location 20, and all locations except those listed initially contain zero.

4

LOGICAL DESIGN

To insure reliability and to satisfy other practical requirements, the signals in a digital computer consist only of two values e.g., -1.5 volts and $+4.0$ volts. These voltages have logical significance, with one value representing the existence of a particular condition and the other representing the absence of that condition. There is a branch of mathematics called *Boolean algebra* which is especially qualified to represent such situations. It was first developed by George Boole,[1] around 1850, in his analytical studies of the thought process. Although other philosophers and mathematicians employed Boole's work in the intervening years, it was not until 1937 that Claude Shannon[8] made a major application of it in engineering to describe digital circuits.

In this chapter we will develop the theorems of Boolean algebra and will show how they can be employed in the design of digital systems. Also of major interest will be the use of map methods for the simplification of logic circuits. We will then present the techniques for realizing logical expressions with standard electronic circuits, called *gates*. Finally, we will extend these design methods to higher-level logic blocks which allow complex design tasks to be easily accomplished with only a few electronic packages.

It is important to obtain a thorough understanding of this chapter because the fundamentals developed here will be applied in *all* later chapters.

4.1 BASIC LOGICAL OPERATIONS

Digital computer logic can be expressed through combinations of three operations: AND, OR, and NOT.

AND. The AND operation requires that all input conditions are simultaneously true before the output is true. Thus, the circuit in Figure 4.1 requires that both switch A AND switch B be operated before the light comes on.

In Boolean algebra, the following notation is employed for any variable, X:

Input X operated is represented by $X =$ true or $X = 1$.

Input X unoperated is represented by $X =$ false or $X = 0$.

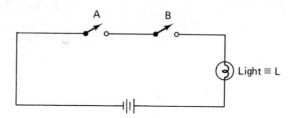

Figure 4.1 The AND Operation.

Thus, the binary values 0 and 1 represent a convenient shorthand notation for the logical state of a variable.

Another way to express the AND circuit of Figure 4.1 is by means of a truth table such as that shown below. Note that this table presents the outputs for all possible values of the inputs, *A* and *B*.

Table 4.1   Truth Table for the AND Operation

A	*B*	*L*
0	0	0
0	1	0
1	0	0
1	1	1

Boolean algebra allows us to represent the AND operation for Table 4.1 in a very concise way by the equation:

$$L = A \cdot B = AB$$

Note that the symbol "·" is read as AND and that it is the same as the multiply operation of ordinary algebra. Moreover, the electronic circuit, or gate, which implements the AND operation is symbolized below:

OR. The OR operation is developed in a parallel way to that used for AND. The output of an OR circuit, however, is on if at least one of the switches is operated. This is represented by switches in parallel in Figure 4.2.

The truth table corresponding to Figure 4.2 is shown in Table 4.2. Again Boolean algebra allows the operation to be represented by a concise equation:

$$L = A + B$$

This time the symbol "+" is read as OR and it corresponds to the addition

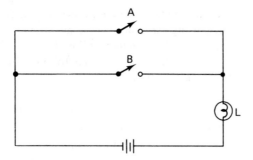

Figure 4.2 The OR Operation.

Table 4.2 Truth Table for the OR Operation

A	B	L
0	0	0
0	1	1
1	0	1
1	1	1

operation of ordinary algebra.* In Boolean algebra, however, it should be observed that $1 + 1 = 1$ (not $1 + 1 = 2$), i.e., a true variable ORed with a true variable results in the output being true.

The electronic gate symbol for OR is shown below:

NOT. The last of the three fundamental operations, which is variously called NOT, negation, complementing, or inversion, is really simpler than the other two. Its switching circuit representation is shown in Figure 4.3. Note

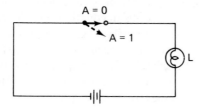

Figure 4.3 The NOT Operation.

*We are using engineering notation for these operations, but the corresponding mathematical symbols are:

$$OR \equiv \vee, \quad AND \equiv \wedge, \quad and \quad NOT \equiv \sim$$

that the switch A is normally closed. When the switch is unoperated ($A = 0$), its contacts are closed, causing the light to come on; when it is operated ($A = 1$), the contacts are open, causing the light to be off.

The very simple truth table corresponding to this circuit is shown in Table 4.3, and the resulting equation is:

$$L = \bar{A}$$

Table 4.3   Truth Table for the NOT Operation

A	L
0	1
1	0

Note the use of the bar over the A to represent the NOT operation; this is the general engineering notation for taking the complement whether a single variable or a large expression is being considered. The electronic gate that performs the NOT operation is called an *inverter* and is represented by the symbol:

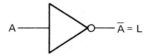

The triangular portion is the symbol for an amplifier, but the small circle is the general representation for inversion. It will be used later in conjunction with AND and OR gates to form NAND and NOR gates, which are very popular in integrated-circuit technology.

4.2   THEOREMS[2,4,5]

In the design of digital systems, Boolean algebra provides a way of describing various complex operations in equation form by means of the AND, OR, and NOT parameters. Moreover, there is a group of theorems in the algebra which allow the equations to be manipulated into forms that are more advantageous for circuit mechanization. Frequently, the theorems also allow an original expression to be simplified by eliminating the occurrence of a variable or a group of variables. A convenient feature about this new algebra is that most of the laws of ordinary algebra can be shown to be valid; e.g., the commutative, associative, and distributive laws hold.

Table 4.4 contains a group of postulates from which all the theorems of Boolean algebra can be derived. (Using the previous definitions of the primitive operations, the reader should critically examine Table 4.4.)

Table 4.4   The Postulates of Boolean Algebra

AND Operations:	$0 \cdot 0 = 0$ $0 \cdot 1 = 1 \cdot 0 = 0$ $1 \cdot 1 = 1$
OR Operations:	$0 + 0 = 0$ $0 + 1 = 1 + 0 = 1$ $1 + 1 = 1$
NOT Operations:	$\overline{1} = 0$ $\overline{0} = 1$

Applying the above postulates and the fact that any variable, e.g., A, can only take on the two values, 0 and 1, it is easy to prove the elementary theorems shown in Table 4.5.

Table 4.5   The Elementary Theorems of Boolean Algebra

AND Operations:	$A \cdot 1 = A$ $A \cdot 0 = 0$ $A \cdot A = A$ $A \cdot \overline{A} = 0$
OR Operations:	$A + 0 = A$ $A + 1 = 1$ $A + A = A$ $A + \overline{A} = 1$
NOT Operations:	$\overline{\overline{A}} = A$

In the above theorems, as well as in the others we shall develop, the replacement can go in either direction, e.g., occasionally it is convenient to replace A by $A + A$.

Perfect Induction.   We will now discuss a general though frequently brute-force way to prove theorems. First recall that, by definition, a theorem is true only if both sides always assume the same value for a particular combination of variables. Usually in mathematics, any variable can take on an infinite number of values, but in Boolean algebra it can take on only two values, 0 and 1. Thus, it is feasible here to prove theorems by comparing both sides for all combinations of the variables, and this method is known as *perfect induction.* Suppose we desire to prove what will be called theorem 1, which is $A + AB = A$. We can do this conveniently with perfect induction by placing the information in the configuration shown in Table 4.6.

Table 4.6   Proof by Perfect Induction

A	B	A + AB
0	0	0 + (0)(0) = 0
0	1	0 + (0)(1) = 0
1	0	1 + (1)(0) = 1
1	1	1 + (1)(1) = 1

We complete the proof by simply checking for agreement between entries in the individual lines of the first column and the corresponding lines in the last column.

It is easy to show by perfect induction that the commutative, associative, and distributive laws are valid, and these are explicitly stated in Table 4.7. Note that the distributive law provides a means for transforming between

Table 4.7   Three Familiar Laws of Algebra

Law	AND Form	OR Form
Commutative	$AB = BA$	$A + B = B + A$
Associative	$A(BC) = (AB)C$	$A + (B + C) = (A + B) + C$
Distributive	$(A + B)(A + C) = A + BC$	$AB + AC = A(B + C)$

product-of-sums, POS, and *sum-of-products*, SOP, expressions. These two basic forms lead to different but equivalent circuits, which will become obvious in Section 4.5.

Once the distributive law is available, theorem 1 can be proved algebraically as follows:

$$A + AB = A \quad \text{(Statement of theorem 1)}$$
$$A(1 + B) = A \quad \text{(The distributive law)}$$
$$A(1) = A \quad \text{(Elementary theorem)}$$

Therefore $\quad A = A$

Since any variable can represent a compound expression, a good way to remember theorem 1 is:

A quantity ORed with the product of that same quantity and another expression can be simply reduced to the original quantity.

Frequently electronic components can be saved, and generally a better design produced, if a given function can be simplified. *Simplification* is defined to be a process for the elimination of variables or terms from the original expression without changing the form of the expression, i.e., by staying either in SOP or POS form. (The reduction of functions by changing the form of the expression is discussed in Section 4.5.)

Example

Simplify $f = x + \bar{y} + (x + \bar{y})(\bar{w} + z)$ using theorem 1 with $x + \bar{y} = A$ and $(\bar{w} + z) = B$:

$$f = x + \bar{y}$$

By comparing the original expression with the final one, we see that considerable simplification has taken place. The most complex term, $(x + \bar{y})(\bar{w} + z)$, was completely elimintated and SOP form was maintained.

In addition to the elementary theorems and the three familiar laws of algebra, there is a set of four very useful theorems, occurring both in SOP and POS forms, which are not at all obvious. These are presented in Table 4.8. (Note that the theorem numbers for the POS form carry an asterisk to

Table 4.8   Main Theorems of Boolean Algebra

Sum-of-Products Form	Product-of-Sums Form
1. $A + AB = A$	1.* $A(A + B) = A$
2. $A + \bar{A}B = A + B$	2.* $A(\bar{A} + B) = AB$
3. $AB + A\bar{B} = A$	3.* $(A + B)(A + \bar{B}) = A$
4. $AB + \bar{A}C + BC = AB + \bar{A}C$	4.* $(A + B)(\bar{A} + C)(B + C) = (A + B)(\bar{A} + C)$
5. $\overline{A + B} = \bar{A} \cdot \bar{B}$	5.* $\overline{A \cdot B} = \bar{A} + \bar{B}$

distinguish them from corresponding theorems in SOP form.) It is certainly not self-evident that the two forms of theorems with the same number are closely related, but this will be proved later in this section.

Any of the theorems in Table 4.8 could be proved by perfect induction, but in order to develop facility with the mathematics, let us perform some algebraic proofs. Since theorem 1 has already been proven, consider theorem 2.

$A + \bar{A}B = A + B$	(Statement of theorem)
$A + \bar{A}B = (1)(A + B)$	(Elementary theorem)
$A + \bar{A}B = (A + \bar{A})(A + B)$	(Elementary theorem)
Therefore $A + \bar{A}B = A + \bar{A}B$	(Distributive)

The above work is typical in that there are few set ways to approach a proof, but imagination and experience often lead to very concise arguments.

De Morgan's Theorem. The two forms of theorem 5 are known as *De Morgan's theorem*, and together they allow any complemented expression to be reduced to a form with, at most, individual variables complemented. (Theorem 5 can easily be proved by perfect induction.) As an example, consider $(\overline{\bar{X}Y + \bar{W}})Z$. Since this is basically a product-of-sums expression, we first apply theorem 5*, with the result $(\overline{\bar{X}Y + \bar{W}}) + \bar{Z}$. The left term is now in sum-of-products form, so we apply theorem 5 to it: $(\overline{\bar{X}\,Y})\bar{W} + \bar{Z} = (\overline{\bar{X}\,Y})W + \bar{Z}$. Now we apply theorem 5* to the bracketed term; $(\bar{X} + \bar{Y})W + \bar{Z} = (X + \bar{Y})W + \bar{Z}$. Thus $(\overline{\bar{X}\,Y + \bar{W}})Z = (X + \bar{Y})W + \bar{Z}$. This same method of alternately using the two forms of theorem 5 can be applied to any complemented expression in order to simplify it.

The Dual Operation. The *dual* is similar to the complement except that the variables are left unchanged. (The operations $(+, \cdot)$ and the constants $(0, 1)$, however, are complemented.) For example:

$$\text{dual}\,[X \cdot 1 + (\bar{Y} + Z)\bar{W}] = [\text{dual}\,(X \cdot 1)][\text{dual}\,(\bar{Y} + Z)\bar{W}]$$
$$= [X + 0][\text{dual}\,(\bar{Y} + Z) + \bar{W}] = X[\bar{Y}Z + \bar{W}]$$

Consider theorem 3 which is:

$$AB + A\bar{B} = A \tag{4.1}$$

Since this is a general theorem, it also holds for the complemented variables, i.e., with the substitutions $\bar{A}$ for A, $\bar{B}$ for B, and B for $\bar{B}$.

$$\bar{A}\bar{B} + \bar{A}B = \bar{A}$$

Complementing both sides of the equation,

$$\overline{\bar{A}\bar{B} + \bar{A}B} = \bar{\bar{A}}$$

Therefore

$$(A + B)(A + \bar{B}) = A \tag{4.2}$$

Now we take the dual of Equation (4.1)

$$(A + B)(A + \bar{B}) = A \tag{4.3}$$

Note that Equations (4.2) and (4.3) are exactly the same. It can be shown that, in general, substituting variables for their complements and then taking the complement of both sides of a theorem is equivalent to taking the dual of both sides of the theorem. Usually the dual is a much simpler operation.

Thus, once a theorem is proved, a second form of that theorem is immediately available through the dual operation. We now recognize that the SOP and POS farms of the theorems in Table 4.8 are duals of each other, and only one of them need be proven.

Proof and Examples for Theorem 4. Theorem 4 is very useful for simplifying Boolean expressions, as will soon be seen; but first let us prove the theorem.

$$AB + \bar{A}C + BC = AB + \bar{A}C \qquad \text{(Statement)}$$
$$= (1)AB + AB + \bar{A}C \qquad \text{(Elementary theorem)}$$
$$= (C + \bar{C})AB + AB + \bar{A}C \qquad \text{(Elementary theorem)}$$
$$= ABC + \bar{A}C + AB\bar{C} + AB \qquad \text{(Distributive)}$$
$$= C(AB + \bar{A}) + AB\bar{C} + AB \qquad \text{(Theorems 1 and 2)}$$

Therefore $AB + \bar{A}C + BC = AB + \bar{A}C + BC$, which proves the theorems.

1. Suppose it is desired to simplify the expression:

$$f = \bar{X}Y + XZ\bar{Y} + YZ$$
$$= \bar{X}Y + (X\bar{Y} + Y)Z \quad \text{(Distributive followed by theorem 2)}$$
$$= \bar{X}Y + XZ + \cancel{YZ} \quad \text{(Theorem 4)}$$

Therefore the simplified expression is:

$$f = \bar{X}Y + XZ$$

2. Simplify

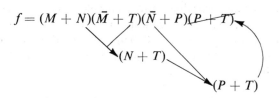

$$f = (M + N)(\bar{M} + T)(\bar{N} + P)(P + T)$$

Therefore

$$f = (M + N)(\bar{M} + T)(\bar{N} + P)$$

In this example, we achieved simplification by applying theorem 4*
backwards two times until the redundant term $(P + T)$ was discovered.
Problems such as this demonstrate the importance of the right-to-left version
of the theorems.

4.3 THE CANONICAL EXPANSION OF FUNCTIONS

In the field of digital computers, the term *canonical expansion* relates to a
standard form of expansion, which we will soon see has both practical and
theoretical significance. Consider the truth table for the OR function shown
in Table 4.9. Note that the extra column labeled Minterms contains a product
term which is the only combination of complemented and uncomplemented
variables that equals 1 for the values of the variables in that row. Thus,
reading the $f = 1$ minterms from the table, we obtain the following expression:

$$f = \bar{X}Y + X\bar{Y} + XY \tag{4.4}$$

This is the conventional method for reading a function from a truth table in
minterm expansion form. Equation (4.4) can be simplified to the usual expression for OR by the following procedure:

$$f = \bar{X}Y + X\bar{Y} + XY \quad \text{(Original expression)}$$
$$f = \bar{X}Y + XY + X\bar{Y} + XY \quad \text{(Elementary theorem)}$$
$$f = Y(\bar{X} + X) + X(\bar{Y} + Y) \quad \text{(Distributive)}$$
$$f = Y + X \quad \text{(Elementary theorem)}$$

Table 4.9 The OR Functions and Minterms

X	Y	f	Minterms
0	0	0	$\bar{X}\bar{Y}$
0	1	1	$\bar{X}Y$
1	0	1	$X\bar{Y}$
1	1	1	XY

By way of a formal definition, a *minterm* is a term in a sum-of-products form which contains all variables, either complemented or uncomplemented.

An alternate method of reading a function from a truth table is to select the minterms that go with $f = 0$. This, of course, results in the $\bar{f}$ expression. Let us use the above approach on the EXCLUSIVE-OR function shown in Table 4.10.

Table 4.10 The EXCLUSIVE-OR Function

X	Y	f
0	0	0
0	1	1
1	0	1
1	1	0

Here $\bar{f} = XY + \bar{X}\bar{Y}$. We complement both sides in order to obtain f:

$$f = \bar{\bar{f}} = \overline{XY + \bar{X}\bar{Y}} = (\bar{X} + \bar{Y})(X + Y) \tag{4.5}$$

A term such as one of those in Equation (4.5), which contains all variables, either complemented or uncomplemented, in a product-of-sums expression is called a *maxterm*. (This definition is analogous to the minterm definition.) Of course, a sum-of-products expression in minterm form could have been obtained from Table 4.10 by reading the minterms with $f = 1$, but another method is to apply the distributive law to Equation (4.5), which results in $f = \bar{X}X + \bar{X}Y + \bar{Y}X + \bar{Y}Y$. Therefore, $f = \bar{X}Y + X\bar{Y}$ (elementary theorem). This sum-of-products form for the EXCLUSIVE OR function is probably more common than the product-of-sums form in Equation (4.5), but the minterm expansion and the maxterm expansion are equally valid. Incidentally, neither of these forms can be simplified.

Function of n Variables. Considerable time has already been devoted to discussing the common functions—AND and OR. Also some other two-variable functions such as the EXCLUSIVE-OR have been mentioned. We now investigate all the possible functions of two variables, as well as their nature. To show this, Table 4.11 has been constructed much like a usual

Table 4.11 Listing of All Functions of Two Variables

X	Y	F_0	F_1	F_2	F_3	F_4	F_5	F_6	F_7	F_8	F_9	F_{10}	F_{11}	F_{12}	F_{13}	F_{14}	F_{15}
0	0	0	1	0	1	0	1	0	1	0	1	0	1	0	1	0	1
0	1	0	0	1	1	0	0	1	1	0	0	1	1	0	0	1	1
1	0	0	0	0	0	1	1	1	1	0	0	0	0	1	1	1	1
1	1	0	0	0	0	0	0	0	0	1	1	1	1	1	1	1	1

Common names for functions:

- F_0 — NULL
- F_1 — NOR
- F_2 — INHIBIT ($\underline{X}$)
- F_3 — INHIBIT
- F_4 — $\underline{Y}$
- F_5 — EXOR (Exclusive OR)
- F_6 — NAND
- F_7 — AND
- F_8 — EXNOR or EQUALITY (Exclusive NOR)
- F_9 — Y
- F_{10} — IMPLICATION
- F_{11} — X
- F_{12} — IMPLICATION
- F_{13} — OR
- F_{14} — IDENTITY
- F_{15} —

73

truth table, except that all possible two-variable functions are represented. Note that the functions are arranged in order of increasing binary numbers, with the least significant bit being at the top. The subscript at the head of each column represents the decimal equivalent of the binary number. The most common operations are labeled with their names at the bottom of the table. For example, NOR is $F_1 = \bar{X}\bar{Y} = \overline{X + Y}$. This is OR followed by NOT, but the "N" is placed first for easy pronunciation. As would be expected, the gate symbol is:

$$\overline{A + B} \equiv NOR$$

Similarly, the NAND is:

$$\bar{f} = xy$$

Therefore $f = \bar{\bar{f}} = \overline{xy}$. Here, the minterm associated with $\bar{f}$ rather than f was selected from the table because there is only one $f = 0$ minterm, while there are three $f = 1$ terms. However, if the three minterms had been selected, the resulting function could have been reduced to $f = \overline{xy}$. The corresponding NAND gate symbol is:

$$\overline{AB} \equiv NAND$$

Now let us determine the number of minterms there are for n variables. Since each variable in a minterm can be either complemented or not, there are two possibilities, but each variable is independent of the others. Thus, the total minterms are equal to

$$\underbrace{2 \cdot 2 \cdot 2 \ldots 2 \cdot 2}_{n \text{ factors}} = 2^n$$

Each of the 2^n minterms can either be in f or not; thus the total possible functions are equal to

$$\underbrace{2 \cdot 2 \cdot 2 \cdot \ldots \cdot 2 \cdot 2}_{2^n \text{ factors}} = 2^{2^n}$$

Although the equation for the number of functions looks innocent enough, it yields very large numbers even for fairly small values of n, as Table 4.12 shows.

Table 4.12 Minterms and Functions of n Variables

n	Number of Minterms	Functions of n Variables
1	2	4
2	4	16
3	8	256
4	16	65,536
5	32	4,294,967,296

4.4 THE KARNAUGH MAP[3,4,5]

The *Karnaugh* map* is one of the most important tools in digital system design. It is very useful in simplifying Boolean functions, and it forms an important part of many logic circuit synthesis techniqies, such as those to be introduced in Chapter 5. We have already employed a rudimentary version in Chapter 2 for the analysis of codes.

The essence of the map method is that the 2^n possible minterms of n variables are represented by means of separate squares, or cells, on the map. Thus, it contains the same information found in a truth table. If a minterm is present in a particular function, the corresponding square contains a 1; otherwise, it contains a 0. Once all the minterms of the function have been placed on the map, there are some fairly simple rules for reading the map in terms of a simplified representation of the function.

Consider a four-variable situation ($n = 4$). Here $2^n = 2^4 = 16$; thus the map will contain 16 squares. Such a map is shown in Figure 4.4. Note how the bracket is drawn adjacent to the last two rows to indicate the region where $w = 1$. (In turn, the top two rows represent the region of $w = 0$.)

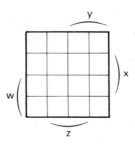

Figure 4.4   Four-Variable Karnaugh Map.

*The name Karnaugh is used in honor of one of the map's inventors; however, the name *Veitch diagram* is also frequently used in recognition of its other inventor.

This, of course, is repeated in a similar way for the other variables. Another way to represent the regions on the map is to label each column and row separately. Both methods are shown in Figure 4.5, along with a representation of the function,

$$f_1 = \bar{w}\bar{x}\bar{y}\bar{z} + wx + yz$$

One of the requirements on Karnaugh map labeling is evident in Figure 4.5: the fact that adjacent columns (rows) must differ only in the value of one variable. This is done so that minterms which may be combined are contiguous on the map, e.g., note how the minterms $\bar{w}\bar{x}yz + \bar{w}xyz + wxyz + w\bar{x}yz$ combine to form the reduced term yz.

Plotting the 1-terms of a function on the map is a fairly simple task if we remember that the result of ANDing a group of variables on the map results in a term which is the intersection of the map regions represented by the individual variables. An example of this is the way the term wx is formed in Figure 4.5 from the intersection of the w and x regions. (If you are thinking that operations on a Karnaugh map are very similar to those with Venn diagrams, you are entirely correct.) Now return to Figure 4.5 and verify the labeling of regions $\bar{w}\bar{x}\bar{y}\bar{z}$ and yz.

Similarly, when several terms are combined through the OR operation, the resulting map coverage is the union of the various terms. The function, f_1, above is formed by such an operation.

Now that we can see how to plot a sum-of-products function on the map through the AND and OR operations, let us ask ourselves how a reduced function can be read from the map. First observe that when 2^m squares combine to form a single group, m variables are eliminated. An example of

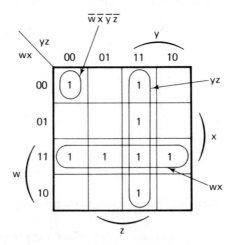

Figure 4.5 Two Methods for Labeling a Karnaugh Map. (Zeros on the map are frequently represented by blank squares.)

this is shown in Figure 4.5 where the four squares forming yz have a description with $m = 2$ variables missing. Thus, each coverage on a map must involve a number of squares which is equal to a power of 2.

The following sequence of rules for grouping 1-terms on the map should be observed:

1. First cover, circle, all isolated 1's. These will be described by all n variables and require no further attention.

2. Next, consider each remaining term separately. If it can be grouped in more than one way, go to another term; otherwise, include it in the *largest possible rectangular coverage that is a power of 2.*

3. Following exhaustive application of the above rule, if at least one term remains which can be grouped in more than one way, arbitrarily select one of the ways and return to rule 2.

4. The simplified function is complete as soon as all terms are covered; but in the process of making a group as large as possible, it is permissible to use previously covered terms.

Figure 4.6(a) shows an example of a reduced expression for the function, $f_2 = x\bar{y}z + \bar{w}yz + \bar{w}\bar{x}y\bar{z} + wx\bar{y}\bar{z}$, formed by using the above rules. Note that, according to rule 2, the squares containing the * symbol are the only valid places to begin the coverage—the other terms can be grouped in more than one way.

Examination of Figure 4.6(a) reveals that not only is there a single change of variables between adjacent columns (rows) but that the first and last columns (rows) also have only a single variable change. This leads us to suspect that the edges are effectively joined, and that minterms in the same row in both end columns are not isolated but join to form a common term. (A similar situation exists between the top and bottom rows.) Figure 4.6(b) is a representative example of the joining of edge terms.

One may be tempted to choose the largest possible coverage as one of the terms in a simplified function. This may be incorrect, however, if alternate coverages are possible, as Figure 4.7(a) shows. Here, according to rule 2, the squares containing the * symbol should be selected first, since they are the only minterms which may be grouped in only one way. Once these solid coverages are made, the largest (dashed) group has already been covered in an indirect way, and Rule 4 indicates that the function is complete. (To add another term would clearly be introducing redundancy.)

In contrast, Figure 4.7(b) shows a function with only one minterm different from that in part (a) of the figure; but here the resulting function contains the group of four squares because the minterm in the lower right-hand section can be covered in only one way. Again the * symbol is employed to indicate which minterms may be considered first. Another interesting thing

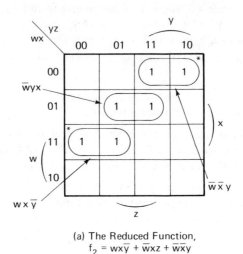

(a) The Reduced Function,
$f_2 = wx\bar{y} + \bar{w}xz + \bar{w}\bar{x}y$

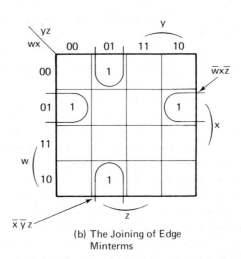

(b) The Joining of Edge
Minterms

Figure 4.6 Karnaugh Map Examples.

about Figure 4.7(b) is that it demonstrates rule 4 through the multiple use of minterms.

An application of rule 3 is demonstrated through Figure 4.7(c). Here, each of the minterms can be grouped in two ways; thus, one arbitrary selection must be made. If the selection of the solid coverage at the * minterm is made, all the other groups are fixed, and the function $f_3 = \bar{A}B\bar{C} + \bar{A}CD + \bar{A}\bar{B}\bar{D}$ results. On the other hand, if the selection of the dashed coverage at the * minterm is made, a different fixed grouping is produced, resulting in

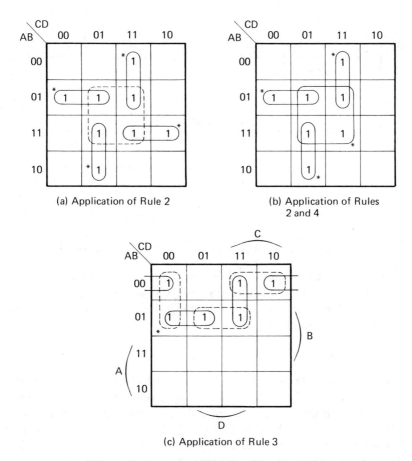

(a) Application of Rule 2

(b) Application of Rules
2 and 4

(c) Application of Rule 3

Figure 4.7   Examples of the Map Coverage Rules.

the function $f_4 = \bar{A}\bar{C}\bar{D} + \bar{A}BD + \bar{A}\bar{B}C$. Now, functions f_3 and f_4 are obviously equal. They contain the same minterms, and they are both simplified representations of the original functions; yet not one term between the two functions is identical.

Of course, the map method is not limited to only four-variable functions. Problems with fewer than four-variables are a trivial variation of the four-variable case. Problems with more than four variables are handled by using sufficient four-variable maps to achieve the 2^n squares required by the n-variable function. In the groups of four-variable maps representing an n-variable function, corresponding squares on adjacent maps may be combined to form a simpler term, just as the adjacent squares on a single map may be combined. Figure 4.8 shows how the five-variable function

$$f_5 = B\bar{C}\bar{D}\bar{E} + \bar{C}D\bar{E} + \bar{B}CD\bar{E} + \bar{A}BC\bar{D} + \bar{B}DE$$

may be simplified by using a pair of four-variable maps.

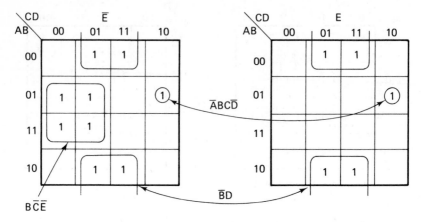

Figure 4.8   The Simplified Function, $f_5 = B\bar{C}\bar{E} + \bar{B}D + \bar{A}BC\bar{D}$.

In the above five-variable problem, observe how the fifth variable is used to label the two maps. Also, the variable E is eliminated to create the term $\bar{A}BC\bar{D}$ formed between corresponding squares on the two maps. (The term $\bar{B}D$ is even a more interesting example of the same thing.)

Don't-Care Conditions

There are physical constraints in practice which either prevent a particular group of minterms (e.g., $\bar{w}xy$) from occurring or which make their occurrence in the function optional. These are called *don't-care* conditions. Constraints of this type can be very helpful in that the don't-care minterms can then be included with either the 0 or 1 Karnaugh map coverage, depending on which leads to the simplest function. To represent the above situation, a dash is entered into each don't-care Karnaugh map square. The map is then covered following the same rules as previously employed.

As an example, consider $f_6 = \bar{w}\bar{x}yz + wyz$, with the don't-care term $\bar{w}xy$. The resulting Karnaugh map is shown in Figure 4.9. Note that the simplified function $f_6 = yz$ can be obtained by grouping one of the don't-care squares with the 1's and allowing the other to be grouped with the 0's. We will make extensive use of don't-care terms when designing sequential circuits in Chapter 5.

The Complementary Approach

Sometimes it is more convenient to read the zero coverage from a Karnaugh rather than reading the 1's. This is known as the *complementary approach*. It involves the same map rules as employed previously, but the resulting function is $\bar{f}$, and f is then obtained from De Morgan's theorem.

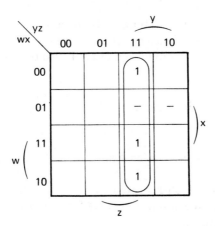

Figure 4.9 Simplification Through the Use of a Don't-Care Term.

Consider the three-variable map shown in Figure 4.10(a). The dashed coverage yields the function $\bar{f}_7 = \bar{x}\bar{y}z$; therefore, $f_7 = \bar{\bar{f}}_7 = \overline{\bar{x}\bar{y}z} = x + y + \bar{z}$. The solid coverage on the map leads to the same function, but it requires more effort. For larger Karnaugh maps, the complementary coverage of a single zero minterm, or a small group, leads to an even greater savings in effort.

Another facet of the complementary approach is the ability to place product-of-sums functions on the Karnaugh map. There are many ways of

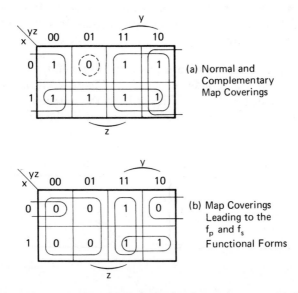

Figure 4.10 Comparison of the Normal and Complementary Methods.

working with product-of-sums functions. The most straightforward method, however, it to find the complement of the function, $\bar{f}$, and place this as a zero coverage on the map. The following example demonstrates this approach.

Example

Simplify the following function:

$$f = (x + z)(x + y + \bar{z})(\bar{x} + y)$$

Now $\bar{f} = \bar{x}\bar{z} + \bar{x}\bar{y}z + x\bar{y}$, which leads to the Karnaugh map shown in Figure 4.10(b).

Once the map is established, we obtain the sum-of-products function by covering the 1's, or the product-of-sums function by covering the 0's and then taking the complement.

The resulting sum-of-products function is:

$$f_s = xy + yz$$

and the resulting product-of-sums function is:

$$\bar{f}_p = \bar{y} + \bar{x}\bar{z}, \qquad f_p = (\overline{\bar{y} + \bar{x}\bar{z}}) = y(x + z)$$

In this case, the product-of-sums form, f_p, can easily be expanded to the sum-of-products form, f_s.

$$f_p = y(x + z) = xy + yz = f_s$$

In some other cases, the expansion is not so obvious. It should be clear, however, that one can always obtain a product-of-sums function by reading $\bar{f}$ from the map and then applying De Morgan's theorem.

The presence of don't-care terms may lead to f_s and f_p functions from the same map which are actually unequal. In Figure 4.11(a), the don't-care minterm is included with the 1's, resulting in the function $f_s = \bar{B} + \bar{A}C + A\bar{C}$. On the other hand, part (b) of the figure shows the don't-care minterm

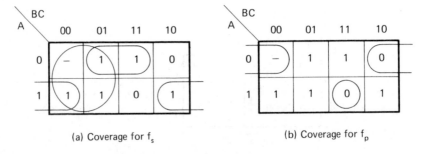

(a) Coverage for f_s (b) Coverage for f_p

Figure 4.11   An Example Where $f_s \neq f_p$.

being included with the 0's, resulting in the function

$$f_p = \overline{\overline{AC} + ABC} = (A + C)(\bar{A} + \bar{B} + \bar{C}) = A\bar{B} + A\bar{C} + \bar{A}C + \bar{B}C$$
$$= A\bar{B} + \bar{A}C + A\bar{C}$$

where the $\bar{B}C$ term was eliminated by use of theroem 4. Both maps were properly covered following the rules, yet it is obvious that $f_s \neq f_p$. The reason for this is the fact that the don't-care minterm was included in both f_s and f_p. Thus, anytime the minterms grouped with the 1's include some terms also grouped with the 0's, there will be a difference between f_s and f_p.

We have already completed many important facets of the Karnaugh map method. More examples, however, and further developments of it as a design tool are covered in Sections 4.6, 4.8, 5.2, and 5.3.

Other Simplification Methods

Although the map method is quite popular, there are other good manual and computer techniques for simplifying Boolean functions. The most notable of these is the Quine–McCluskey method, explained in References 4 through 7, which involves a tabular reduction of the original function into a simplified one somewhat similar to the Karnaugh map process. The tabular method has the advantage that it may be easily programmed, and the reader is encouraged to consult Rhyne's work in Reference 7 for a very interesting FORTRAN version.

4.5 LOGIC DIAGRAMS

Since in most cases pictures are a more efficient way to convey design information, the logic diagram is developed in this section as an alternate way to describe Boolean functions. We have already indicated that any function can be constructed with only AND, OR, and NOT primitives.

Consider the equation $f = \bar{A}B + C\bar{D}$. Since it is in sum-of-products form, the output gate will be an OR with AND gates feeding it, as Figure 4.12 shows.

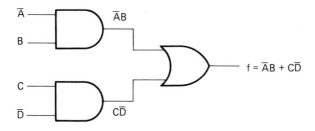

Figure 4.12 Logic Diagram for a Sum of Products.

One measure of circuit complexity, and thus cost, is the number of gate inputs in the circuit. It is easy to count them directly from the logic diagram (e.g., there are six inputs in Figure 4.12). However, the same number can be obtained by inspection of the equation, and the reader should take note of this. Another fact that should be observed concerning Figure 4.12 is that each input passes through two *levels* of gating before the output is reached. The number of levels of gating in a digital circuit is an important parameter because it is a measure of electronic delay. Again observe that the number of levels required by the circuit can be obtained from the equation by inspection. For the above function, it was assumed that the complemented quantities, $\bar{A}$ and $\bar{D}$, were available. In the following example, however, inverters are required to generate the complements.

Example

Draw the logic diagram for the function

$$f = \bar{A}C\bar{D} + \bar{A}B\bar{D}\bar{G} + \bar{A}EF + \bar{A}B\bar{D}G \qquad (4.6)$$

By inspection of Equation (4.6), we see that 3 inputs are required to invert A, D, and G; 14 inputs are required to generate the product terms; and 4 inputs are needed to combine them into the output f. Thus, as it stands, 21 inputs are required. Now before drawing the diagram, we check to see if f can be simplified. If we use theorem 3, the second and fourth terms in Equation (4.6) may be combined. Therefore,

$$f = \bar{A}C\bar{D} + \bar{A}B\bar{D} + \bar{A}EF \qquad (4.7)$$

The result requires 14 inputs. By factoring Equation (4.7), however, we obtain the following new equation requiring only 12 inputs:

$$f = \bar{A}[\bar{D}(C + B) + EF] \qquad (4.8)$$

The corresponding circuit representation is shown in Figure 4.13.

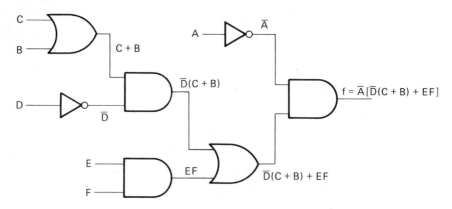

Figure 4.13   Mechanization of Equation (4.8).

In comparing Equations (4.7) and (4.8), we observe that the number of literals has been reduced from nine to six. (Every occurrence of a variable, whether complemented or uncomplemented, is defined as a *literal*.) However, the process of reducing the number of literals has changed the form of the expression in that Equation (4.8) is no longer a sum of products. This process of reducing the number of literals by changing the form of an equation should be contrasted with the process of simplification such as that used in reaching Equation (4.7). There, the form of the equation stayed the same, but the number of terms was reduced as was the number of variables. An elementary example of reducing the number of literals by changing the form of an equation is shown below:

$f = (A + B)(B + C)$ (Four literals in product-of-sums form.)

$f = AC + B$ (Three literals in sum-of-products
form after applying the distributive law.)

4.6 NAND-NOR LOGIC[4]

The basic *NAND* and *NOR* gates were mentioned in Section 4.3, but their current importance demands that design details be presented at this point.

First, study the two-level NOR circuit shown in Figure 4.14. Note that Figure 4.14 gives the same results as the two-level OR-AND circuit in Figure 4.15. Thus, two levels of NOR gating are equivalent to OR followed by AND, i.e., it realizes a product-of-sums function.

Figure 4.14   Two-Level NOR Circuit.

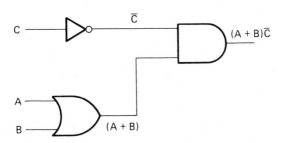

Figure 4.15   Two-Level OR-AND Circuit.

It should be observed that since the variable C in Figure 4.14 passes through only one level of gating, it is inverted in the output. Similarly, it is easy to show that a two-level NAND configuration is equivalent to AND followed by OR.

Example

Using either NANDs or NORs draw a logic diagram for the function $f = xy + \bar{x}\bar{z} + y\bar{z}$.

First we try to simplify the expression:

$$f = xy + \bar{x}\bar{z} + y\bar{z} \quad \text{(By theorem 4)}$$

$$y\bar{z}$$

Therefore $f = xy + \bar{x}\bar{z}$, and the resulting logic diagram is shown in Figure 4.16.

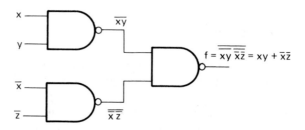

Figure 4.16   NAND Realization of $f = xy + \bar{x}\bar{z}$.

It is known that a sum-of-products equation can always be realized with a two-level NAND circuit; thus, the term written at the output of each gate is merely used for checking.

Since all functions can be reduced to sum-of-products and product-of-sums forms, the above method gives a straightforward, two-level circuit realization for a function with either NANDs or NORs. As was observed in Section 4.5, however, simpler circuits frequently result if neither a pure sum of products nor product of sums is used. Consider the function $f = AC + A\bar{D} + BE + BF$. As it stands, a circuit requiring 5 gates and 12 inputs is needed. Moreover, the function cannot be simplified, but the number of literals can be reduced if we change the form of the expression through factoring as follows

$$f = A(C + \bar{D}) + B(E + F) \quad (4.9)$$

This equation is not a pure sum of products; therefore, our previous design technique is not applicable. However, the expression may easily be mechanized by AND-OR gates as shown in Figure 4.17.

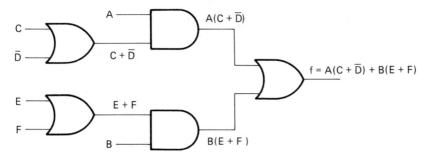

Figure 4.17 AND-OR Realization of Equation (4.9).

Consider the NAND gate, shown below, which realizes the function $f = \overline{AB}$.

Carrying out the complement operation in the expression, we get $f = \bar{A} + \bar{B}$, and this expression leads to the symbol:

This is a second, though less familiar, form of the NAND symbol. Now let us replace the AND gates in Figure 4.17 by NAND gates of the first form, and replace the OR gates by NAND gates of the second form, with the results shown in Figure 4.18. Note that all complements between gates occur in pairs and are effectively canceled, but the circuit input variables must be complemented from those appearing in Figure 4.17 to account for the inverter present on each gate input. If desired, all gates in Figure 4.18 can now be

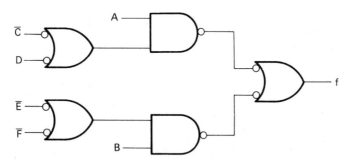

Figure 4.18 NAND Replacement of the Gates in Figure 4.17.

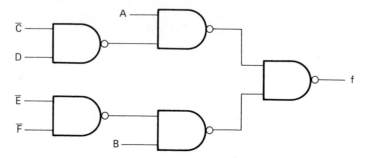

Figure 4.19   Realization of Equation (4.9) with Common NAND Gates.

replaced by the common NAND symbol (the first form), which results in Figure 4.19.

The reader should now show that the output of Figure 4.19 does indeed agree with Equation (4.9).

Instead of NAND gates, suppose NOR gates had been required. The common form of NOR gate is given by $f = \overline{A + B}$:

Again carrying out the complement operation, we get $f = \overline{A}\overline{B}$, which leads to the symbol:

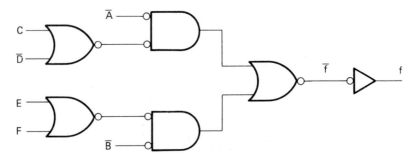

Now let us replace the OR gates in Figure 4.17 by the first NOR form above and replace the AND gates by the second NOR form. The resulting diagram is presented in Figure 4.20.

Again the complements between gates occur in pairs, so that they are

Figure 4.20   NOR Replacement of the Gates in Figure 4.17.

canceled. The input variables A and B are complemented because of the inversions where they enter, and an inverter is inserted at the output gate so that f instead of $\bar{f}$ is produced. If desired, the NOR gates in Figure 4.20 can now be replaced by common NORs.

The above NAND (NOR) procedure is a general one that we will call the *replacement method,* and its steps are summarized as follows:

1. Simplify the given function.

2. Place the function in a form with minimum literals.

3. Draw the AND-OR block diagram.

4. Use the two forms of NAND (NOR) gates to replace the ANDs and ORs in the original block diagram. Make sure that any inverters on the input or output of the circuit are properly canceled.

5. If desired, replace all gates by those of the common form.

Example

Employing either all NOR or all NAND gates, synthesize a two-level logic circuit with fewest possible gate inputs for the function:

$$f = (B + C + D)(\bar{A} + \bar{C})(\bar{B} + \bar{C} + \bar{D})(A + B + \bar{C} + D)$$
$$\times (A + \bar{B} + C + \bar{D})$$

(Assume complemented and uncomplemented input variables are available.)

Solution: Now

$$\bar{f} = \bar{B}\bar{C}\bar{D} + AC + BCD + \bar{A}\bar{B}C\bar{D} + \bar{A}B\bar{C}D$$

which results in the zero entries shown on the map of Figure 4.21.

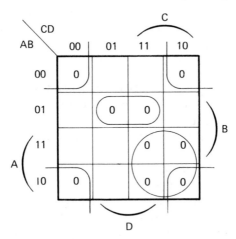

Figure 4.21 Optimum Zeros-Coverage for the Function f.

The same map with the optimum coverage of the 1's is shown in Figure 4.22, and the resultant two-level function is:

$$f = \bar{A}\bar{B}D + \bar{A}B\bar{D} + A\bar{C}D + AB\bar{C}$$

Since a sum-of-products expression can be realized with two levels of NAND gates, and since each product term requires three gate inputs and the sum requires four gate inputs, a total of $T = 4 \times 3 + 4 = 16$ gate inputs are required.

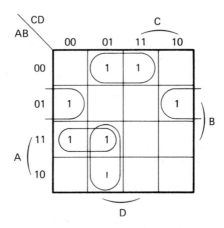

Figure 4.22 Optimum 1's Coverage for the Function f.

As an alternate approach, we return to Figure 4.21 and read the optimum coverage shown there for $\bar{f}$:

$$\bar{f} = \bar{A}BD + \bar{B}\bar{D} + AC$$

therefore, $f = (A + \bar{B} + \bar{D})(B + D)(\bar{A} + \bar{C})$. Here the total requirements for a two-level NOR network will be:

$$T = 3 + 2 \times 2 + 3 = 10 \text{ gate inputs}$$

In yet another approach, the above product-of-sums function is realized in Figure 4.23 with NAND gates. Observe, however, that the NAND design necessitates that one gate be employed as an inverter at the circuit output, which makes the circuit in Figure 4.23 a three-level circuit. Two additional inputs are also used, resulting in a total of 12 gate inputs required.

Thus, it appears that the best circuit design for the given function is the product-of-sums realization employing NOR gates. Incidentally, if the original function, without simplification, had been realized with a two-level NOR circuit, a total of 21 gate inputs would have been needed. Consequently, our design has realized more than a factor-of-2 savings in gate inputs.

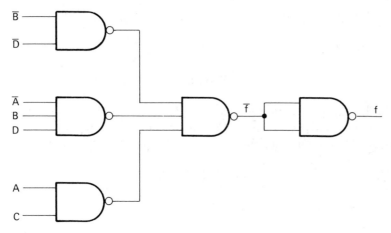

Figure 4.23 NAND Gate Realization of the Function f.

This example points out the importance of understanding all facets of NAND-NOR design. In also emphasizes the value of considering alternate approaches before deciding on the final realization of a particular function.

4.7 EXCLUSIVE-OR LOGIC[4]

Probably there will always be a need for lower-level *primitives** such as NANDs and NORs to implement certain logical functions. The birth of integrated circuits, however, has initiated a continuing trend toward a greater emphasis on higher-level primitives. This section and the next two sections will explore this trend, first with a medium-level primitive, the EXCLUSIVE-OR gate, and then with high-level primitives, the decoder and the multiplexer.

Among the two variable functions listed in Table 4.11 are the EXCLUSIVE-OR, EXOR, and its dual the EXCLUSIVE-NOR, EXNOR. (The EXNOR function is often called the EQUALITY function because it is true when the two inputs match.) Although these functions are not as widely used as NAND and NOR, they are very handy in concisely performing certain arithmetic and coding operations, as we shall soon see in Section 4.8. Initially, it is our intention to define some new notation and then to present some of the properties of the EXOR function. (Although it will not be emphasized, the EXNOR has similar properties to the EXOR.)

*The word primitive has a special connotation in computer technology. It specifies a basic building block which is employed in the development of more complicated structures.

The EXOR and the EXNOR have been assigned the following special symbols:

$$\text{EXOR} \equiv \oplus$$

$$\text{EXNOR} \equiv \odot$$

Thus, in place of writing $f = A\bar{B} + \bar{A}B$, it is more concise to write $f = A \oplus B$. (Similarly, $f = \bar{A}\bar{B} + AB$ can be more concisely written as $f = A \odot B$.) In addition, the following gate notation has been widely adopted:

EXOR: $A \oplus B$

EXNOR: $A \odot B$

The first thing to observe concerning the properties of EXOR is that it obeys the following laws of algebra:

Commutative: $A \oplus B = B \oplus A$

Associative: $(A \oplus B) \oplus C = A \oplus (B \oplus C) = A \oplus B \oplus C$

Distributive: $A(B \oplus C) = (AB) \oplus (AC)$

Now let us prove the distributive law for the EXOR function:

$A(B \oplus C) = (AB) \oplus (AC)$	(Original statement of the law)
$A(B \oplus C) = AB\overline{AC} + \overline{AB}AC$	(Definition of EXOR)
$A(B \oplus C) = AB(\bar{A} + \bar{C}) + (\bar{A} + \bar{B})(AC)$	(DeMorgan's law)
$A(B \oplus C) = AB\bar{A} + AB\bar{C} + \bar{A}AC + \bar{B}AC$	
$A(B \oplus C) = 0 + AB\bar{C} + 0 + A\bar{B}C$	(Elementary algebra)
$A(B \oplus C) = A(B\bar{C} + \bar{B}C)$	
$A(B \oplus C) = A(B \oplus C)$	(Definition of EXOR)

(The other laws can be proved in a similar way.)

Table 4.13 lists a number of other important properties of the EXOR function. The proofs for these are straightforward and will be left as exercises.

Table 4.13   Important Properties of the EXOR Function

Property Number	Statement
1	$1 \oplus 0 = 1$
2	$0 \oplus 0 = 1 \oplus 1 = 0$
3	$A \oplus 0 = A$
4	$A \oplus 1 = \bar{A}$
5	$A \oplus A = 0$
6	$A \oplus \bar{A} = 1$
7	$\bar{A} \oplus \bar{B} = A \oplus B$
8	$\overline{A \oplus B} = A \odot B$

(Each of the properties in Table 4.13 has a corresponding EXNOR form, but these will not be considered here.)

How does one quickly evaluate an EXOR function of more than two variables if the truth values are given? Consider the following function:

$$f = A \oplus B \oplus C \quad \text{(With } A = 1, B = 0, C = 1\text{)}$$
$$f = 1 \oplus 0 \oplus 1 \quad \text{(Values substituted)}$$
$$f = (1 \oplus 0) \oplus 1 \quad \text{(Associative law)}$$
$$f = 1 \oplus 1 \quad \text{(Property 1)}$$
$$f = 0 \quad \text{(Property 2)}$$

Thus, an even number of 1's in this special case of a three-variable function has lead to $f = 0$. Actually, it can easily be shown that an EXOR function, with any number of variables, will always equal 0 if there are an even number of variables which are true; the function will equal 1 if there are an odd number of variables which are true. For these reasons, EXOR is described as a modulo-2 function, meaning its output is determined by counting the true variables in base 2 without carry. This is a formal way of saying the function is true if an odd number of the input variables are true.

The EXOR function has an interesting pattern on a Karnaugh map. For example, consider the function $f = A \oplus B \oplus C \oplus D$. By the modulo-2 property, this function will be true if one or three of the variables are true. Thus, it is easy to plot the function on a Karnaugh map by noting the number of 1's in the description of each square. For example, the square $AB = 01$, $CD = 11$ has three true variables; therefore $f = 1$ for that square. The complete K-map for the function f is shown in Figure 4.24. Observe that the result looks like a checkerboard. This is characteristic of all EXOR functions. Previously, when such a map appeared, the only way to describe it was by laboriously listing the individual minterms. It can now be described very concisely by means of an EXOR function.

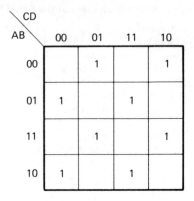

Figure 4.24 Map of the Function $f = A \oplus B \oplus C \oplus D$.

How may the gating circuit be drawn for a map like that in Figure 4.24? After recognizing it as an EXOR function, we may express it in the following ways:

$$f = (A \oplus B) \oplus (C \oplus D) \quad \text{or} \quad f = [(A \oplus B) \oplus C] \oplus D$$

where the brackets show how the function is assembled from two-variable gates. The resulting two circuits are shown in Figure 4.25. The tree-type circuit gets its name from its symmetrical branch-like appearance, which of

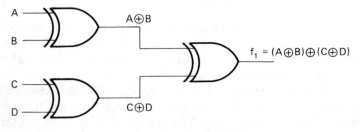

(a) Tree-Type Circuit

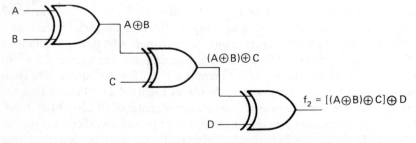

(b) Unbalanced Circuit

Figure 4.25 Two Circuit Realization for $f = A \oplus B \oplus C \oplus D$.

course is even more obvious for a six- or eight-variable function. One advantage of the tree-type circuit is that it requires fewer levels of gating than the unbalanced circuit. In Chapter 6 we will see that an increase in the number of gate levels has the detrimental effect of slowing the circuit's response.

4.8 IMPLEMENTATION OF COMPUTER CIRCUITS[7]

In Section 2.6 we developed a number of codes and considered several algorithms useful in processing numbers. Now, as a practical application of EXOR gates, we will present the circuits for implementing some of these processes.

Conditional Complementor

When BCD numbers are processed, the 9's complement technique is frequently employed for subtraction, which means that a method of complementing the bits of the subtrahend is required. Since each bit is independent of the others, a conditional complementing circuit can be developed for a single bit and then duplicated as many times as required. Table 4.14

Table 4.14 Bit Logic Required by the Conditional Complementor

B_i (Input)	C (Control)	R_i Result
0	0	0
0	1	1
1	0	1
1	1	0

shows the logic requirements where B_i is one bit of the number being complemented, C is the control signal ($C = 1$ for complement desired; $C = 0$ otherwise), and R_i is the resulting output signal. It is immediately evident that $R_i = \bar{B}_i C + B_i \bar{C}$, which, of course, is the EXOR function $R_i = B_i \oplus C$. The resulting circuit for an n-bit conditional complementor is shown in Figure 4.26. Note that when $C = 0$, the modulo-2 property requires that each input bit simply be transmitted through to the output without change; but when $C = 1$, each bit must be inverted.

Parity Circuits

When parity is employed, circuits are required to generate the parity bit from the source register and to check the parity at the receiving register. Figure 4.27 shows a system of this type employing EXOR gates. (The numbers

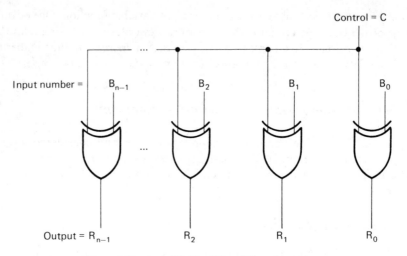

Figure 4.26   An n-Bit Conditional Complementor.

next to the bit lines represent the logic values for a particular example. Note how the parity information is created in the parity generator and how it is processed at the receiving end by the parity checker circuit.) Assuming even parity, the sending circuit should detect any word whose data bits do not have an even number of ones and compensate by making the parity bit one. Otherwise, the output of the parity generation should be zero, causing the parity bit to be zero. Due to the modulo-2 property of the EXOR function, the sending circuit requires the parity (source) function to be

$$P_s \equiv B_0 \oplus B_1 \oplus B_2 \oplus B_3$$

where B_i is the ith bit of the source register. It is clear that an odd number of ones in the data word shown in Figure 4.27 results in $P_s = 1$. Now, if P_s is employed as a new bit, bit 4, added to the data word, the augmented word will have even parity. The opposite case of an even number of ones in the original data word results in $P_s = 0$, which maintains even parity in the augmented data word. When the signal arrives at the receiver, all five bits of the augmented word are checked with the parity (receiving) function, and

$$P_r \equiv B_0 \oplus B_1 \oplus B_2 \oplus B_3 \oplus B_4$$

If no bits are in error, $P_r = 0$ (as it is in Figure 4.27). Should one (or any odd number) error occur, however, then $P_r = 1$, which can be used to sound an alarm so that corrective action can be taken.

As an exercise, the reader should now change one of the bits in the received signal, e.g., make $B_3 = 0$. This will change the outputs of some of the EXOR gates in the parity checker and will eventually result in $P_r = 1$, which indicates that the error has been properly detected.

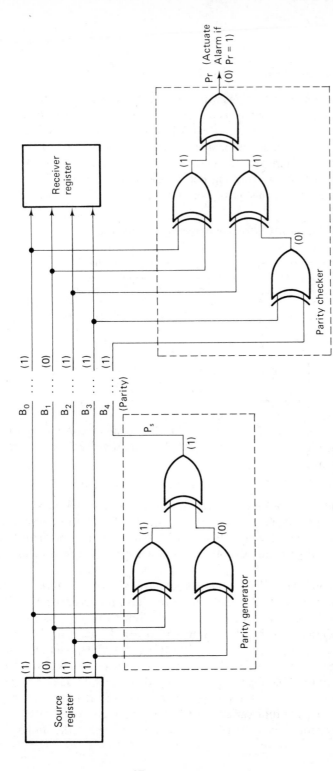

Figure 4.27 A Small Even-Parity Error Detection System.

Code-Conversion Logic

There are many situations where it is desired to convert from one code to another within a computer system. For example, the information originating in a digital-to-analog converter is often in Gray code; before it can be processed in the arithmetic unit, conversion to another code is required.

To better understand the code-conversion problem, consider a simple example—that of transforming octal code to Gray code. The relationship between these two codes is shown in Table 4.15, and it is evident that the

Table 4.15   Relation Between the Octal and Gray Codes

Decimal	Octal B_2	B_1	B_0	Gray G_2	G_1	G_0
0	0	0	0	0	0	0
1	0	0	1	0	0	1
2	0	1	0	0	1	1
3	0	1	1	0	1	0
4	1	0	0	1	1	0
5	1	0	1	1	1	1
6	1	1	0	1	0	1
7	1	1	1	1	0	0

conversion requires that each bit in the Gray code be considered a function of all three bits of an octal digit. Thus, we require three functions in the following form:

$$G_0(B_0, B_1, B_2), \qquad G_1(B_0, B_1, B_2), \quad \text{and} \quad G_2(B_0, B_1, B_2)$$

These functions can be easily determined if we place the Gray code truth values for each bit on a separate Karnaugh map. The three resulting independent maps are shown in Figure 4.28. Note that two of these functions require EXOR gates and the other is obtained directly from bit 2 of the octal code.

The circuit demanded by these functions is shown in Figure 4.29. (In this case, it just happens that a very concise result is produced with EXOR gates; usually other gate types will be employed.) Since a very simple example was just considered, a word of caution is in order. In general, the conversion from one code to another, usually four bits to four bits, requires a multiple-output logic circuit in which the optimum solution demands that some of the terms generating one function be employed to generate one or more of the other functions. Thus, the process of selecting the coverage for, say, four Karnaugh maps of four variables each is often a more complicated task than we have previously considered. The reader is referred to the Bibliography—

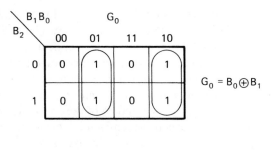

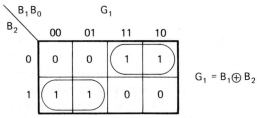

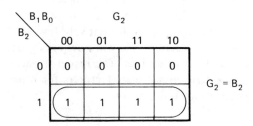

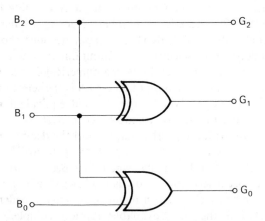

Figure 4.28 Octal to Gray Code Conversion Functions.

Figure 4.29 Octal to Gray Code Conversion Circuit.

particularly Hill's and Peterson's text[2]—for a formal method which guarantees an optimum solution. On the other hand, by employing our usual method of covering each Karnaugh map independently, we will obtain a reasonable suboptimum function. Moreover, once some tentative coverages have been placed on each map, it is often possible by inspection to make the final selections in such a way that an optimum or near-optimum set of functions result. In Chapter 9, we will return to the code-conversion problem and find that the advent of inexpensive semiconductor memories has in many cases greatly simplified the problem of code conversion.

4.9 SPECIAL LOGIC BLOCKS

The logical operations discussed in Sections 4.5 and 4.6, AND, OR, NOT, NAND, and NOR, are low-level primitives in that many are required to generate a complex function. With the development of integrated circuits, however, a pool of higher-level primitives is now available, making it very easy to accomplish many complex logical design tasks. The EXOR and EX-NOR gates, discussed in Sections 4.7 and 4.8, represent a beginning of the higher-level primitives; but they are still relatively low level compared to those to be presented in this section. We will now briefly describe some of the more popular higher-level primitives. The reader is encouraged to consult catalogs from integrated circuit manufacturers for other primitives as well as additional applications information.

Decoders

A *decoder* is a circuit which permits one of N possible outputs to be selected, where $N = 2^n$, and n is the number of bits in a binary word specifying the selected output. The BCD-to-decimal decoder shown in Figure 4.30 is a typical decoder for selecting one output line from ten available lines. It is frequently employed to select digits in a numerical display such as found in digital voltmeters. The dashed input labeled S is the select line. When it is high, the output developed will be the decimal equivalent of the binary number $DCBA$ at the input. For example, when $SDCBA = 10010$, gate 2 will be the only gate with a logical 1 output, and the decimal number 2 will be decoded. When $SDCBA = 00010$, none of the gates will be high because the decoder select bit is off. Finally, when $SDCBA = 11010$, none of the gates will be high because $1010_2 = 10_{10}$ is not a legal BCD number.

Decoders are extensively used in memory systems in order to select the word of interest from the many available. A more complete discussion of several types of decoders is found in the memory section in Chapter 9.

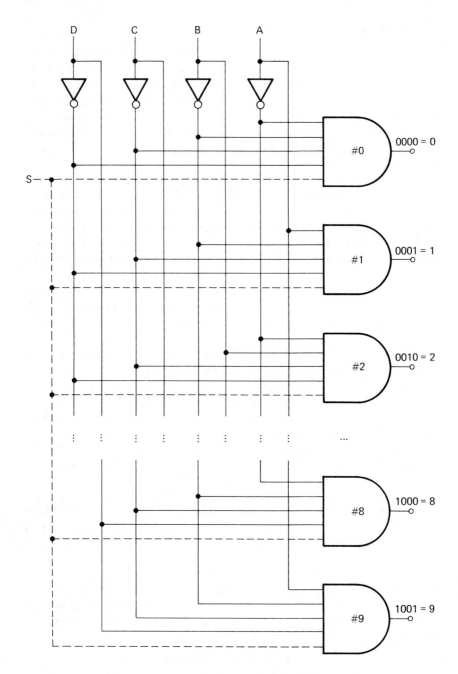

Figure 4.30 A Parallel BCD-to-Decimal Decoder.

Multiplexers and Demultiplexers

A decoder may be converted to a *demultiplexer* if the select terminal is used as a signal input. When the circuit in Figure 4.30 is employed in this way, the S input stands for signal and one of the ten gates (0 through 9) is selected by means of the input address terminals A, B, C, and D. The signal will pass through only the selected gate; all others will remain at zero. Thus the demultiplexer is a circuit for selecting one of N receivers for the input signal. For example, if line 1 is to become a logical 1, the demultiplexer should receive $SDBCA = 10001$.

The purpose of a *multiplexer* is just the opposite of a demultiplexer, i.e., the multiplexer is a circuit for selecting one of N input data sources to be transmitted on a single output channel. Figure 4.31(a) shows the detailed logic for a 4-to-1 multiplexer, and Figure 4.31(b) shows the corresponding circuit symbol. In Figure 4.31(a), note the function of the enable, the select, and the data input sections of the circuit. For example, to pass the information on input X_2 through to the output, it is necessary to select output gate 10, which means the control signals become $EBA = 110$. It is important to understand that when the enable signal is low ($E = 0$), the output will remain at zero, independent of the data or select inputs.

Expansion Circuits.   Since the multiplexer is a primitive, it is not surprising to find that it can be combined with itself and with other primitives to perform a large number of different operations. One of the simplest operations is *expansion*, which combines two 4-to-1 multiplexers to form the 8-to-1 multiplexer shown in Figure 4.32. Note how the enable input is employed to control which multiplexer is active. In this case, in order to pass the data on y_6 through to the output, the control signals must be $CBA = 110$. (Further attention will be given to expansion circuits in the Exercises at the end of the chapter.)

Realization of Boolean Functions.   Multiplexers provide a very efficient way to realize Boolean functions. Consider Figure 4.31(a). When $E = 1$, the function realized by that circuit is:

$$f(B, A) = X_0 \bar{B} \bar{A} + X_1 \bar{B} A + X_2 B \bar{A} + X_3 B A$$

With X_0 through X_3 set to the proper combinations of zeros and ones, $f(B, A)$ generates the minterm expansion of any two-variable function. For example, to generate the EXOR function, it is only necessary to make the following selections:

$$X_0 = X_3 = 0 \quad \text{and} \quad X_1 = X_2 = 1$$

The corresponding circuit is shown in Figure 4.33. Thus, we see that there is no difficulty in generating any of the 16 functions of two variables with a 4-

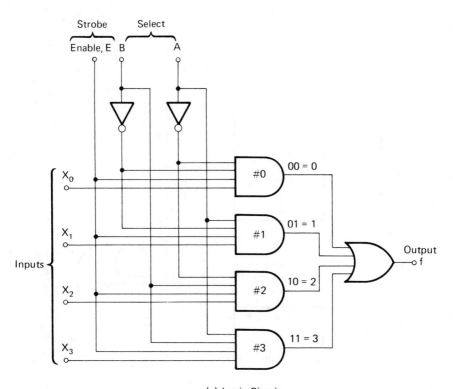

(a) Logic Circuit

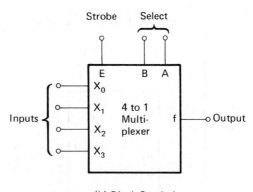

(b) Block Symbol

Figure 4.31 A 4-to-1 Multiplexer.

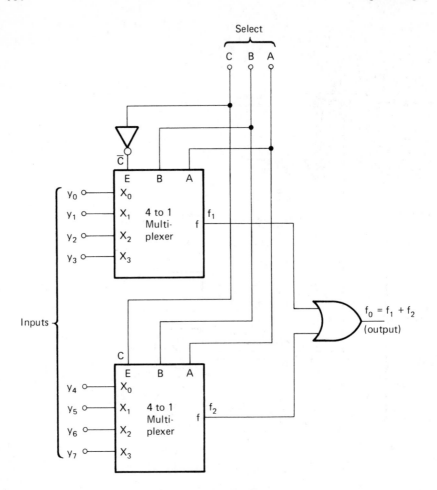

Figure 4.32   An Expansion Circuit.

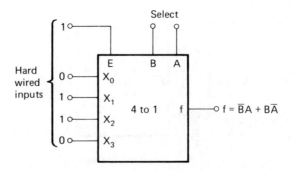

Figure 4.33   Multiplexer Realization of EXOR Function.

to-1 multiplexer. Of course, an 8-to-1 multiplexer would be required for the 256 functions of 3 variables, etc.; but in principle, any function can be generated with a multiplexer. Moreover, instead of hard-wiring the X inputs as shown in Figure 4.33, we could connect them to a set of switches or to another circuit so that the function generated could be changed quite rapidly.

Multiplexers Employed As Read-Only Memories. When a simple *read-only memory* is required, it is frequently very convenient to do the design by employing multiplexers as primitives. Read-only memories will be discussed in detail in Chapter 9. It is sufficient here to say that they are a type of memory which contains permanent (or semipermanent) information such as a table of trigonometric functions. As an example, assume that the function $R = t^2 + 5$ (where $0 \le t \le 3$) is to be used very frequently in a special-purpose digital system. Instead of performing the calculation in the arithmetic unit each time a new value of R is required, a read-only memory could be constructed as specified in Table 4.16. The possible input numbers are shown first in

Table 4.16 Specifications for a Multiplexer Read-Only Memory

Number		Results $R = t^2 + 5$						
t_{10}	t_2	R_{10}	R_2				X_i
	BA		d	c	b	a	(Ref.)
0	00	5	0	1	0	1	X_0
1	01	6	0	1	1	0	X_1
2	10	9	1	0	0	1	X_2
3	11	14	1	1	1	0	X_3

decimal, t_{10}, and then in binary, t_2, in columns one and two. A similar tabulation of the results is given in the next two columns. In the case of the binary results, R_2, note that four bits, a through d, are required; each of these bits represents a different function generated by a separate multiplexer. Thus, in response to $t_2 = BA = 00, 01, 10, 11$, the multiplexer for bit d generates the bit sequence 0, 0, 1, 1, which corresponds to the function $f(B, A) = B\bar{A} + BA$. The complete multiplexer circuit may be obtained by inspection of Table 4.16, and it is left as a problem at the end of the chapter.

In addition to the applications just discussed, there is a wide variety of other important uses of multiplexers. Whenever one needs to generate a series of Boolean functions or to perform a complex switching task, there is a good chance that a multiplexer circuit will be applicable. Again, the reader is referred to manufacturers' catalogs on integrated circuits for many additional practical applications.

Other special logic blocks and higher-level primitives will be considered later in Sections 6.5, 8.1, and 9.5.

4.10 BIBLIOGRAPHY

1 BOOLE, G., *An Investigation of the Laws of Thought*. New York: Dover Publications, 1954.

2 HILL, FREDRICK J., and GERALD R. PETERSON, *Introduction to Switching Theory and Logical Design* (2nd ed.). New York: John Wiley & Sons, Inc., 1974.

3 KARNAUGH, M., "The Map Method for Synthesis of Combinational Logic Circuits." *Trans. of AIEE, Communications and Electronics*, Vol. 72, Pt. 1, 1953, pp. 593–99.

4 KOHAVI, ZVI, *Switching and Finite Automata Theory*. New York: McGraw-Hill Book Company, 1970.

5 MARCUS, MITCHELL, *Switching Circuits for Engineers* (2nd ed.). Englewood Cliffs, N.J.: Prentice-Hall, Inc., 1967.

6 McCLUSKEY, E. J., *Introduction to the Theory of Switching Circuits*, New York: McGraw-Hill Book Company, 1965.

7 RHYNE, V. THOMAS, *Fundamentals of Digital Systems Design*, Englewood Cliffs, N.J.: Prentice-Hall, Inc., 1973.

8 SHANNON, CLAUDE, "A Symbolic Analysis of Relay and Switching Circuits." *Trans. AIEE*, Vol. 57, 1939, pp. 713–23.

4.11 EXERCISES

Simplify the functions in Exercises 4.1 through 4.6 using the theorems.

4.1 $A\bar{B}C + \bar{A}\bar{C}D + A\bar{C}$

4.2 $(A + B + CD)(\bar{A} + B)(\bar{A} + B + E)$

4.3 $DEH + \bar{E}G\bar{H} + \bar{H}E + HF\bar{E} + \bar{H}EJ$

4.4 $AB + \bar{A}C + \bar{B}CD$

4.5 $AB + \bar{A}C + \bar{B}D + CD$

4.6 $(A + B)(\bar{A} + C)(B + C + D)$

4.7 Draw a diagram similar to Figure 4.2 for the NOR operation. Employ the following switch configuration:

Figure E4.7

Simplify the functions in Exercises 4.8 and 4.9 using the Karnaugh map.

4.8 (a) $A\bar{B}C + \bar{A}\bar{C}D + A\bar{C}$

(b) $B\bar{C} + \bar{A}B + BC\bar{D} + \bar{A}\bar{B}D + A\bar{B}\bar{C}D$

(c) $\bar{A}B\bar{C}D + AC\bar{D}\bar{E} + ABC\bar{D}E + A\bar{B}C\bar{D}E + \bar{B}\bar{C}\bar{D}E$

(d) $\bar{A}\bar{C} + AB\bar{C}\bar{D} + \bar{A}BD + \bar{A}\bar{B}C\bar{D} + A\bar{B}\bar{C}D$

(e) $\bar{A}\bar{B} + AB + A\bar{B}C + \bar{A}B\bar{C}$

(f) $(A + B)(\bar{A} + C)(B + C + D)$

4.9 (a) $B\bar{C} + \bar{A}D + \bar{A}B + A\bar{B}\bar{C}D$, in the f_p form.

(b) $\bar{A}\bar{B}C + \bar{A}\bar{B}\bar{C}\bar{D} + \bar{A}BC$, in f_s form with the don't-care terms: $AB\bar{C}\bar{D}$, $\bar{A}B\bar{C}\bar{D}$, $A\bar{B}\bar{C}D$.

(c) Exercise 4.2 in f_s form.

4.10 Prove the theorem $(Z + X)(Z + \bar{X} + Y) = (Z + X)(Z + Y)$. Do *not* use perfect induction.

4.11 Complement: $S[\bar{W} + I(T + \bar{C})] + H$.

4.12 Using the fewest possible gates, realize the function $f = A(C + \bar{D}) + B(C + \bar{D})$ first with NORs, then with NANDs.

4.13 Find the function realized by the gate:

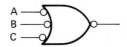

Figure E4.13

4.14 Using only NORs, synthesize a gate circuit with the fewest possible gate inputs for the functions listed below:

(a) $f = \bar{A}CD + \bar{A}BD\bar{G} + \bar{A}EF + \bar{A}BDG$

(b) $f = \bar{A}CD + BD + EF$

(c) $f = \bar{X}\bar{Y}\bar{Z} + XYZ, g = \bar{X}\bar{Z} + YZ$ (both functions with one circuit)

Both complemented and uncomplemented inputs are available.

4.15 Simplify the following functions using the map method:

(a) $f = X\bar{Y} + X\bar{Z} + \bar{X}Y + \bar{X}Z$

(b) $f = (\bar{A} + C + \bar{D})(A + \bar{B} + \bar{C} + D)(\bar{B} + \bar{C} + \bar{D} + E)$
$\times (\bar{A} + B + \bar{C} + \bar{D})(A + C + \bar{D} + \bar{E})(A + \bar{B} + \bar{C} + \bar{E})$

Present the resulting f both as a sum of products and as a product of sums. [Required for part (b) only.]

4.16 Simplify using the theorems:

(a) $f = \bar{A}C\bar{D} + B(DG + A\bar{E}) + E + D(\bar{C} + \bar{A}C)$

(b) The function in 15(a)

4.17 Prove the properties listed in Table 4.13.

4.18 Prove that any EXOR function is true if an odd number of its variables take on true values.

4.19 (a) As concisely as possible, express the function shown on the map.

(b) Using only EXOR gates, draw the circuit for the function in part (a).

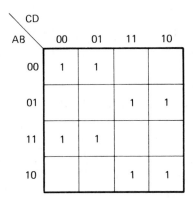

Figure E4.19

4.20 (a) Using EXOR gates, draw the tree-type logic diagram for the function

$$f = A \oplus B \oplus C \oplus D \oplus E \oplus F \oplus G \oplus H$$

(b) Redraw the circuit in part (a) using the fewest possible two-input EXNOR gates.

4.21 If $A \oplus B = C$, find $B \oplus C$ in simplified form.

4.22 Design a logic circuit to convert from BCD to the (2, 4, 2, 1) code.

4.23 How should Figure 4.27 be changed if odd parity is employed?

4.24 It can be shown that a Gray-to-binary conversion algorithm for n-bit numbers is given by $B_i = G_i \oplus B_{i+1}$, where G_i is the ith bit of the Gray code and B_{i+1} is the bit of the binary word that is to the left of the bit B_i. The conversion proceeds from the most significant bit, B_{n-1}, toward the least significant bit, B_0, and it is assumed that $B_n = 0$.

(a) Check the method for the Gray-code elements 0110, 1100, and 0100.

(b) Employing an algebraic method, find the formula for converting from binary to Gray code.

(c) Check the method in part (b) for the binary elements: 0011, 0110, and 1001.

(d) For $n = 4$, draw the circuits for both conversion systems.

4.25 Using the fewest possible multiplexers (no gates), draw a circuit to realize the expression $R = t^2 + 5$ shown in Table 4.16.

4.26 Using the fewest possible multiplexers (no gates), draw a circuit to realize the expression $E = x^2 + 3x + 1$, where x and E are the input and output binary numbers, and $0 \leq x \leq 3$.

4.27 Repeat Exercise 4.26, with $E = 2x + 1$ and $0 \leq x \leq 7$.

4.28 Repeat Exercise 4.26, with $E = -2x^2 + 15x + 32$ and $0 \leq x \leq 7$.

4.29 Using 4-to-1 multiplexers only, draw the circuit for a 16-to-1 multiplexer. Provide an enable input for the new circuit.

SEQUENTIAL CIRCUITS

We have been dealing exclusively with *combinational circuits*, where the output depends only on the present inputs. Digital computers, however, contain another broad class called *sequential circuits* to which counters, registers, accumulators, and many other subsystems belong. Here, the output depends on input history (memory) as well as on the present inputs themselves.

The term *synchronous* is applied to the set of sequential circuits considered at the beginning of this chapter because circuit action is synchronized by means of an input pulse, often a pulse from an electronic clock. The elements of our circuit model are shown in Figure 5.1. Note that network inputs* and outputs can exist in both pulse and level forms. The combinational network not only generates the output signals, but it also sends signals, levels, and/or pulses into the memory and changes its content under certain

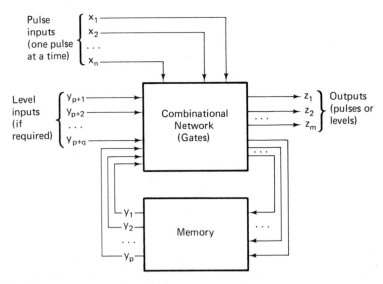

Figure 5.1 Basic Block Diagram of a Synchronous Sequential Circuit.

*Although *n* pulse input lines exist, simultaneous pulses on two or more lines are not allowed in the class of circuits considered here.

conditions. The memory output signals (levels) give information about the *state* of the memory and are labeled y_k, where $k = 1, 2, \ldots, p$. As implied in Figure 5.1, the state of the memory can be visualized simply as a binary number, consisting of p bits, which indicates the history of signals the memory has received. Note that the input levels are represented by the same symbols as the memory states. This is done to emphasize that, for design purposes, the input levels are treated as if they were another part of the memory. Of course, a synchronous sequential circuit can exist without any level inputs, but there must always be at least one pulse input and at least two memory states.

It is our objective to study some general methods for the design of both the combination network and the memory connections needed in sequential circuits. An introductory computer text, however, can not hope to cover all important aspects of every topic, and this is particularly true of a subject as extensive as sequential design. Thus, it is expected that serious students of computer hardware will later take an advanced course devoted so switching theory. Meanwhile, those desiring more details will want to consult the Bibliography.

This chapter begins with the development of formal synthesis techniques, follows with consideration of heuristic methods, and concludes with a discussion of cost and circuit speed as design criteria.

5.1 DESCRIPTION OF SEQUENTIAL CIRCUITS

Our sequential design method requires, as a starting point, a concise tabular description of the required action. Usually, this table is constructed in a multistep process starting with a *state diagram*, which is a geometric picture of the requirements.

State Diagram

Basically a state diagram consists of several numbered nodes which represent the individual memory states previously mentioned. A transition produced by an input pulse, x_i, is indicated by an arrow, and the presence of an output pulse is represented by the variable Z.

Figure 5.2 illustrates the state diagram for a circuit which accepts x_1 and x_2 input pulses and produces an output whenever the sequence $x_2 x_1 x_1$ occurs. Note the following facts concerning the diagram:

1. All inputs are possible at each memory state. Thus, the complete set of x_i signals are shown leaving each node.

2. When an input does not cause a state change, its arrow forms a loop at the original state, e.g., the x_2 input loops at state 1.

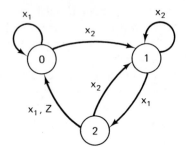

Figure 5.2   State Diagram for the Sequence $x_2x_1x_1$.

3. The diagram contains sufficient elements so that any aribtrary sequence will terminate at a unique node; but only the specified sequence will produce the output pulse.

4. This circuit is nonterminal, i.e., it will continue to produce an output pulse each time the proper sequence is received.

5. Under certain conditions, many input pulses (e.g., a string of x_1's) can be received with no output produced.

6. An output pulse is produced only when a Z appears adjacent to the transition arrow, and at all other times the circuit output is zero.

7. Physical meaning may be attached to each state:
   (a) State 0—a new sequence has not yet started.
   (b) State 1—an x_2 has occurred.
   (c) State 2—an x_2x_1 sequence has occurred.

8. The x_1 and x_2 pulses must not occur simultaneously. (Figure 5.1 will not allow it.)

Consider the state diagram of a modulo-3 counter which has the ability to be cleared, i.e., it can be reset to zero. (This counter produces the following sequence: 0, 1, 2, 0, 1, etc., and the output, Z, occurs when the counter makes the transition $2 \rightarrow 0$.) In Figure 5.3 note that the pulses to be counted

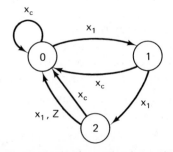

Figure 5.3   State Diagram for the Modulo-3 Counter.

are labeled x_1, and the clear pulse is labeled x_c. As we would expect, the counter may be cleared from any state; but in contrast to receiving an x_1 input when in state 2, clearing from state 2 produces no output.

The State Table and State Assignment

Once the requirements of a problem have been expressed in the state diagram, the information is usually translated into a *state table*. This is the next step in the systematic procedure developed for determining an economical circuit for exciting the system's memory. The state table contains a row for each state in the state diagram and a column for each system input. The entry at the intersection of a particular row and column represents the next state the system will enter upon the occurrence of the input at the top of the column. Table 5.1 is the state table for Figure 5.3.

Table 5.1   State Table for the Modulo-3 Counter

Present State	Next State	
	x_1	x_c
0	1	0
1	2	0
2	0, Z	0

Now let us determine how much memory is needed to satisfy the example given in Table 5.1. Since we are working in bits, an n-bit memory has 2^n possible states. But our example requires three states; thus, two bits are needed, and there will be one unused state which will lead to the presence of some don't-care map entries. (Unused states $= 2^2 - 3 = 1$.)

The state numbers of Table 5.1 must now be coded in binary. This process is known as the *state assignment problem*. The present assignment is very straightforward because it has previously been specified that the memory follows the binary code: 00, 01, 10, etc. Thus, the state coding is: $0 = 00$, $1 = 01$, $2 = 10$. Table 5.2 is now constructed using this assignment, where the table entry $y_2 y_1$ represents the memory state variables, and y_1, of course, is the least significant bit. (Since the counter should never enter the $y_2 y_1 = 11$ state, it is shown as a don't-care.)

Table 5.2   State Table—Binary Code Assignment

Present State $(y_2 y_1)$	Next State $(y_2 y_1, Z)$	
	x_1	x_c
00	01	00
01	10	00
10	00, Z	00
11	—	—

We are fortunate that, in the design of many sequential networks for digital computers, an natural state assignment is frequently obvious. An example occurs when a 3-bit counter following the conventional binary code is required. Here, the natural assignment is state $0 = 000$, State $1 = 001$, etc., and finally, state $7 = 111$. In other cases where a natural assignment is not obvious, an arbitrary assignment can be made and at least a suboptimum solution will be obtained. Some other assignment, however, may lead to a more economical circuit. A general solution for the optimum state assignment is very difficult, but additional information on the subject is found in Exercise 5.7, References 3 and 6, and Section 5.8.

Elimination of Redundant States[3,5]

Sometimes a state table for a particular sequential problem will contain redundant states, i.e., more states are employed than the minimum required to describe the circuit Redundant states should be avoided because they often result in a larger momory than necessary (more bits), and they usually lead to a more complicated design by reducing the number of available don't-care conditions. Consider the situation shown in Table 5.3(a); states 2 and 3 are clearly equivalent since the same inputs lead to the same states and the same outputs. With $2 \equiv 3$, the circuit description may be redrawn as shown in Table 5.3(b). Here, one state is saved and the memory requirement drops from two bits to one bit.

Table 5.3   Equivalent State Descriptions

(a) Original			(b) Reduced Version		
Present State	Next x_1	State x_2	Present State	Next x_1	State x_2
1	1	3	1	1	2
2	2	1, Z	2	2	1, Z
3	2	1, Z			

In general, how may equivalent states be recognized? The simplest approach is to assume that two states are equivalent, unless it depends on a proven nonequivalence. A proven nonequivalence can be either a direct difference in outputs or a dependence on two different states which can be proven nonequivalent.

Table 5.4 is an example of *chain dependence*, where finally the proven nonequivalence of states 3 and 4 results in none of the states being equivalent. In contrast, Table 5.5(a) demonstrates chain dependence where there are no proven nonequivalent states. Thus, all states are equivalent and the result [shown in Figure 5.5(b)] is a single state which can be realized entirely by a combinational circuit, without requiring any memory bits.

Table 5.4 Nonequivalent States

Present State	Next x_1	State x_2
1	2	1, Z
2	3	1, Z
3	4	1, Z
4	2, Z	1, Z

Table 5.5 Equivalent State Descriptions

(a) Original			(b) Reduced Version		
Present State	Next x_1	State x_2	Present State	Next x_1	State x_2
1	2	1, Z	1	1	1, Z
2	3	1, Z			
3	4	1, Z			
4	1	1, Z			

The above discussion shows that considerable simplification can result from the recognition and elimination of equivalent states. Also, the burden on the designer in constructing the original state table from a verbal specification is reduced since it is not necessary to insure that each added state is really unique. In fact, a good rule to follow when trying to decide whether a particular state should be added is: if in doubt add an extra state. One must be sure, however, to always check each state table for redundancies before making the state assignment.

5.2 CIRCUIT SYNTHESIS[3, 5, 6]

Special Karnaugh maps, called *memory excitation maps*, are next constructed so that the Boolean functions required to drive the memory may be determined. To make our work more specific, from this point on, the more restricted term, *flip-flop*, is used to represent each memory bit. A flip-flop is an electronic circuit whose output state, y, can assume one of two logic levels (0 or 1), and it has one or more inputs which can cause the output to change. For tutorial reasons, our discussion begins with the S-R flip-flop, whose standard symbol is shown in Figure 5.4. Later, however, several other types will be presented. (The arrows in Figure 5.4 merely indicate the direction of signal flow for this diagram; they will not be employed in later figures.)

In Figure 5.4, S is the *set* input, R is the *reset* input, and the state of the flip-flop, y_1, and its complement, $\bar{y}_1$, are available.* When a pulse occurs at

*Observe how the flip-flop is labeled with the standard engineering notation: 1 represents the y output and 0 represents the y output.

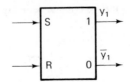

Figure 5.4 The *S-R* Flip-Flop Symbol.

the set input to the flip-flop, the state of the flip-flop, y_1, is *set* to 1, independently of its previous value. Similarly, the reset input permits changing the flip-flop's state to 0. These operations are summarized by the *timing diagram* given in Figure 5.5. Note how the occurrence of a pulse on the *S* Input at time t_1 causes the circuit to be set ($y_1 = 1$), and how the pulse into *R* at t_2 causes the circuit to be reset ($y_1 = 0$). The class of flip-flop considered here is negative-edge triggered. This means that the state change occurs at the negative-going edge of the pulse ($1 \rightarrow 0$ transition),* or more accurately, a short delay follows the negative edge (usually within a few nanoseconds). This type of flip-flop has the characteristic that the next state is determined only by the inputs *before the negative edge*, and it is not influenced by inputs from other flip-flops once the edge occurs. Thus, it is not influenced by the memory transient states. In other words, the order in which various flip-flops in the memory change state is unimportant, as long as all flip-flops have reached steady state before the next input occurs. Figure 5.5 (at t_4) demonstrates that, once the flip-flop is set, another set pulse has no effect on its state.

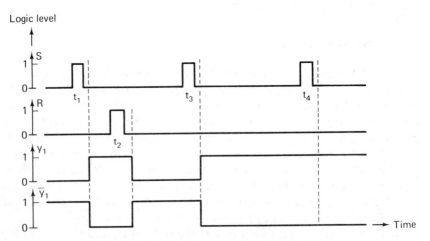

Figure 5.5 Timing Diagram for an *S-R* Flip-Flop.

*The process mentioned is the most common, but others will be considered later.

We may now ask: "If the flip-flops are in a particular state and a given input occurs, what signal should be received by each flip-flop?" This question can be answered by means of a group of memory excitation maps, Karnaugh maps, whose coordinates (y_i values) represent the present state and whose entries represent the next state for each flip-flop, with columns for each pulse input and rows for each flip-flop. Such a group of maps for the modulo-3 counter is shown in Figure 5.6.

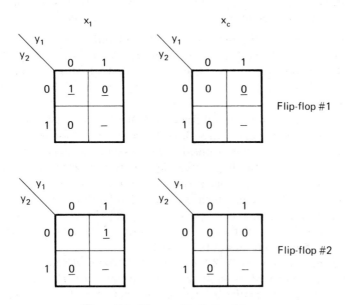

Figure 5.6 Memory Excitation Maps.

In addition to the don't-care symbol, there are other special symbols appearing on the excitation map to label the required transition for each flip-flop. These are summarized in Table 5.6.

Table 5.6 Meaning of Map Symbols

Required Flip-Flop Transition	Map Symbol
$0 \longrightarrow 1$	1
$1 \longrightarrow 0$	0
$0 \longrightarrow 0$	0
$1 \longrightarrow 1$	1
Optional	—

The following facts should be observed concerning Figure 5.6; they are representative of all such maps:

1. Each map possesses a square for every possible memory state, map coordinates $y_2 y_1$.

2. Each row of maps is for a particular flip-flop, and each column for a particular input.

3. Memory states $y_2 y_1 = 11$ possess don't-care entries throughout because the present state 11 does not exist in Table 5.2.

Thus, the map in the second row, first column of the figure describes what happens to flip-flop 2 when the input x_1 occurs. Moreover, if that flip-flop had been in state $y_2 y_1 = 10$, Figure 5.6 shows that it would have been reset by the x_1 input, i.e., a $1 \rightarrow 0$ transition, represented by $\underline{0}$, would have occurred.

We next inquire how the information from Table 5.2 can be translated into Figure 5.6 and, particularly, how the entry discussed in the previous paragraph can be obtained. Consider the present state $y_2 y_1 = 10$ in Table 5.2, the x_1 input, and flip-flop 2 (this flip-flop is represented by y_2, the first member of each table entry pair). Here, the y_2 transition is from the present state 1 to the next state 0, which is represented by the symbol 0. Thus, the entry $\underline{0}$ is placed on the Karnaugh map for flip-flop 2, x_1 input, and 10 memory state. In a similar way, the translation of the remaining data from Table 5.2 to Figure 5.6 should be verified.

Let us review how the *S-R* flip-flop behaves when it receives an input.* Table 5.7 summarizes all possible cases: S is the set input, R is the reset input, and y_n is the state of the flip-flop before the input. $(S = 1, R = 0)$ represents a pulse into S and no pulse into R.

Table 5.7 Operating Characteristics of the *S-R* Flip-Flop.

Input		Next State
S	R	(y_{n+1})
0	0	y_n
0	1	0
1	0	1
1	1	Not allowed

From the properties given in Table 5.7, a new table can be constructed which allows the Karnaugh map entries in Figure 5.6 to be translated in such a way that the S and R excitation equations can be obtained. These

---

*Discussion of the internal construction of flip-flops will be delayed until Section 6.3 where their realization from gate primitives will be presented.

flip-flop input requirements are listed in Table 5.8 and may be understood through the following reasoning. When a 1-term appears on the map, that flip-flop is to be set. But Table 5.7 tells us that $(S = 1, R = 0)$ will produce the required action in an S-R flip-flop; thus, these entries are shown on the second line of Table 5.8. When a 0-term appears on the map, that flip-flop is to stay at zero. But Table 5.7 tells us that either $(S = 0, R = 1)$ or $(S = 0, R = 0)$ will satisfy the requirement; thus, on the third line Table 5.8 shows $S = 0$ and $R = $ —, which is consistent with the don't-care condition of $R = 0$ or $R = 1$. The other entries in Table 5.8 follow from Table 5.7 in a similar way, and the reader should verify them as an exercise.

Table 5.8   Required Inputs for the *S-R* Flip-Flop

Karnaugh Map Entry	Required Inputs	
	S	R
0̲	0	1
1̲	1	0
0	0	—
1	—	0
—	—	—

Consider the 1̲-line in Table 5.8. The $(S = 1, R = 0)$ entries have their usual logical meaning, i.e., in order to satisfy the Karnaugh map, the 1̲ *must* be included in the map coverage for the set function but it *must not* be in the coverage for the reset function. On the other hand, the 1-line in Table 5.8 tells us that the 1-entry on the Karnaugh is a don't-care for the set function. In any case, it must not be included in the reset function. The other lines of the table follow in a similar way. (It is important to understand that *all* under-lined terms *must be covered*, with 1̲ in S and 0̲ in R.)

From the information given in Figure 5.6, let us now obtain the required logic equations to drive the flip-flops. First consider the set equations. Once the symbols in Figure 5.6 are properly interpreted as either a 1, 0, or—according to the S column in Table 5.8, the required map coverage can be made by simply following the usual Karnaugh map rules developed in Chapter 4 for combinational circuits. Remember, however, that the individual maps are actually isolated, e.g., the rows of maps are independent because they belong to different flip-flops, and the inputs occur separately. Thus, within a row of maps, the coverages must be ORed. Moreover, the coverage from each map must be ANDed with the X_i input at the head of its column. The resultant map coverages for the set inputs are shown labeled S_1 and S_2 in Figure 5.7. In the figure, consider the set equation for flip-flop 1, which comes

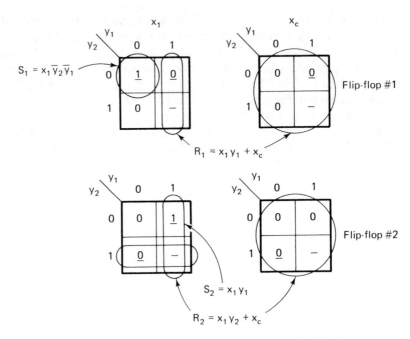

Figure 5.7 Excitation Maps of Figure 5.6 Showing the S-R Coverage.

from the first row of maps. There is only a single $\underline{1}$ on these two maps, and the squares contiguous to it are $\underline{0}$ and 0, which are both translated as 0 by Table 5.8. Thus, the symbol $\underline{1}$ is circled as an isolated quantity and the resulting coverage is read as $\bar{y}_2\bar{y}_1$ which, when combined with the input, yields $S_2 = x_1\bar{y}_2\bar{y}_1$. In the case of flip-flop 2, the don't-care term can be used; this results in $S_2 = x_1 y_1$. The process for the reset equations is similar except that the R column of Table 5.8 is employed to interpret the map symbols, and the resultant map coverage is labeled R_1 and R_2 in Figure 5.7. When considering the R_1 equation, note that the upper right-hand map contains two 0 squares, which are translated as don't-cares by Table 5.8. Thus, the optimum coverage is the entire map, which is read as 1, and it yields x_c when ANDed with the input. Finally, the result is ORed with the coverage from the left-hand map, yielding $R_1 = x_1 y_1 + x_c$. The set and reset equations, along with the equation for the output, are summarized below. Figure 5.8 shows the corresponding logic diagram.

$$S_1 = x_1\bar{y}_2\bar{y}_1, \qquad R_1 = x_1 y_1 + x_c$$
$$S_2 = x_1 y_1, \qquad R_2 = x_1 y_2 + x_c$$
$$Z = y_2\bar{y}_1 x_1 + y_2 y_1 x_1 = y_2 x_1$$

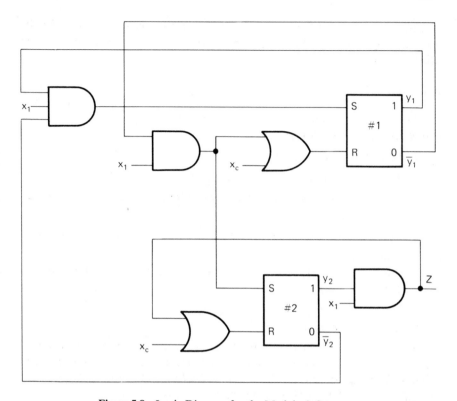

Figure 5.8 Logic Diagram for the Modulo-3 Counter.

Several facts should be observed concerning these equations:

1. Subscripts 1 and 2 on S and R refer to flip-flops 1 and 2.

2. The equation for Z is obtained by inspection of Table 5.2, i.e., Z occurs only when the flip-flops are in the state $y_2 y_1 = 10$ or in the $y_2 y_1 = 11$ don't-care state and an x_1 occurs. More complicated cases could be evaluated simply by use of another Karnaugh map.

3. One gate may be saved in implementing these equations if a common AND gate output is used to provide the $x_1 y_1$ term in both the circuit for S_2 and that for R_1. A similar saving is realized for Z and the first term of the R_2 equation.

4. While one may be tempted to write the simpler equation $R_2 = y_2$ in order to reset flip-flop 2, this is incorrect because circuit action must only occur in synchronization with either an x_1 or x_c input.

One of the general and very useful characteristics of the above technique is that one set of maps may be used to determine the excitation equations for a large variety of flip-flop types. All that is needed is a new Table 5.8 for each different type of flip-flop. These tables and further examples will be discussed later.

Summary

The circuit synthesis method, which has just been presented, may be summarized by the following six steps:

1. Draw a state diagram to meet the requirements of the circuit.

2. Translate the diagram into a state table.

3. Make a state assignment. (An arbitrary one is acceptable, but it may not be optimum.)

4. Translate the data from step 3 into the required memory excitation maps.

5. Employing a table of required inputs for the desired flip-flop (such as Table 5.8), select the minimal map covering and read the flip-flop excitation equations.

6. Draw a logic diagram for the circuit.

As long as the sequential circuit requirements fit the broad model of Figure 5.1 (e.g., only one pulse input occurs at a time), the synthesis method is quite general and, in principle, is not limited by such things as the number of flip-flops or the type of state diagram specified.

Table 5.9   State Table for a Three-Bit Up-Counter

Present State	x Input, Next State
0	1
1	2
2	3
3	4
4	5
5	6
6	7
7	0

5.3 SYNTHESIS OF COUNTERS

Counters belong to one of the most important groups of circuits found in digital computers. They are particularly common in the control and arithmetic sections, where they are used to keep track of the sequence of instructions in a program (an instruction counter), to totalize the number of shifts in multiply and divide operations (a step counter), to distribute the sequence of timing signals (often a ring counter), and to accomplish a host of similar applications. Counters are of the proper degree of difficulty (not trivial, but also not overly complex) for presenting synthesis techniques (and later heuristic design methods). Moreover, an in-depth look at the design of one class of circuits is an excellent way to learn the basic engineering principles which apply to all digital logic.

Example: Synthesis of a Three-Bit Up-Counter

The toggle, T, flip-flop is introduced in this example, together with one of the most frequently used counters. After we construct the state diagram (Figure 5.9) and the state table (Table 5.9) and make a conventional binary assignment for a three-bit up-counter, we obtain Table 5.10. The following facts should be observed from the table:

1. The counter is of the nonterminal type, i.e., with an X input, the next state following $y_3y_2y_1 = 111$ is 000.

2. Although it is not explicitly shown, an output would be developed for at least one state; otherwise, the counter would not have any practical use. Because of the chain dependence found in the table and because of the implicit output(s), there are no redundant states. (This is also true of *all* the following counter circuits.)

3. For simplicity, the circuit is not externally cleared.

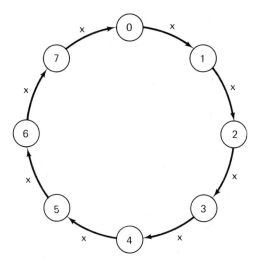

Figure 5.9 State Diagram for a Three-Bit Up-Counter.

Table 5.10   State Table for a Three-Bit Up-Counter

Present State $(y_3y_2y_1)$	x Input, Next State $(y_3y_2y_1)$
000	001
001	010
010	011
011	100
100	101
101	110
110	111
111	000

The excitation maps corresponding to Table 5.10 are shown in Figure 5.10.

In order to interpret the maps, we must understand the operating characteristics of the T flip-flop. These are presented in Table 5.11, and the resulting map requirements are given in Table 5.12. (As an exercise, the reader should derive the requirements of Table 5.12 from the facts given in Table 5.11.)

Table 5.11   Operating Characteristics of the T Flip-Flop

Input	Next State (y_{n+1})
0	y_n
1	$\bar{y}_n$

Table 5.12   Map Covering Demanded by the T Flip-Flop

Karnaugh Map Entry	Required Inputs T
$\underline{0}$	1
$\underline{1}$	1
0	0
1	0
—	—

When reasoning similar to that previously used for the S-R flip-flop is applied to the T flip-flop, and Table 5.12 is employed to read the maps in Figure 5.10, the resulting equations are:

$$T_1 = x, \qquad T_2 = xy_1, \qquad T_3 = xy_1y_2$$

These lead to the circuit diagram in Figure 5.11.

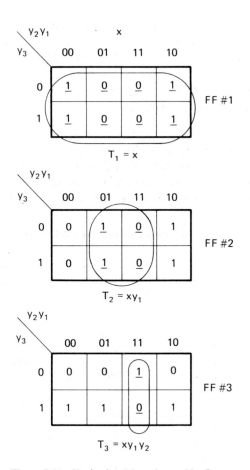

Figure 5.10 Excitation Maps for an Up-Counter.

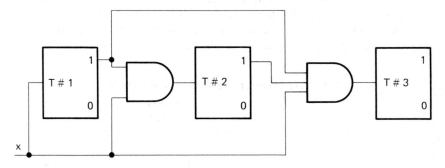

Figure 5.11 Three-Bit Up-Counter.

Inspection of the excitation equations and the figure leads us to the conclusion that this circuit could be extended to as large a counter as desired, with the excitation for the nth bit being given by $T_n = xy_1y_2 \ldots y_{n-1}$.

Example: Synthesis of One of a BCD Counter With Clocked J-K Flip-Flops

Having gained a basic understanding of sequential circuits through design with the elementary *S-R* and *T* flip-flops, we now move to the more practical clocked *J-K* type, which combines some characteristics of both previous types. One stage of a binary-coded decimal, BCD, counter is to be designed. (This circuit counts from 0 to 9 in binary, and then returns to 0.)

Table 5.13   State Table for a BCD Counter

Decimal Number	Present State $(y_4y_3y_2y_1)$	x Input, Next State $(y_4y_3y_2y_1)$
0	0000	0001
1	0001	0010
2	0010	0011
3	0011	0100
4	0100	0101
5	0101	0110
6	0110	0111
7	0111	1000
8	1000	1001
9	1001	0000
10–15	1010 to 1111	Don't-care

The required state conditions are shown in Table 5.13, and the excitation maps corresponding to Table 5.13 are shown in Figure 5.12.

Table 5.14 presents the characteristics of the *J-K* flip-flop, and the resulting map requirements are given in Table 5.15.

Table 5.14   Characteristics of the J-K Flip-Flop

Inputs J	K	Next State (y_{n+1})
0	0	y_n
0	1	0
1	0	1
1	1	$\bar{y}_n$

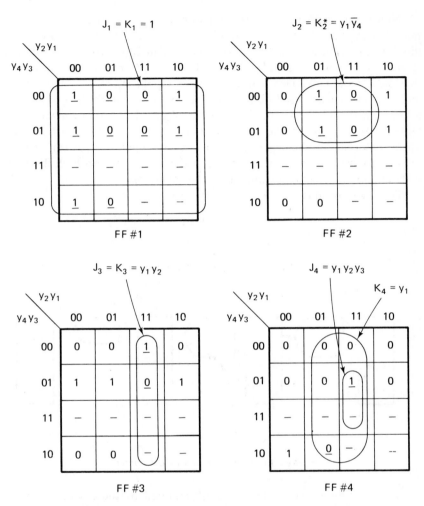

Figure 5.12   Flip-Flop Excitation Maps—BCD Counter.*

*The coverage $K_2 = y_1$ is also suitable. It does not save any gate inputs, however, because the $y_1 \bar{y}_4$ term must still be generated for J_2.

Table 5.15   Map Covering Demanded by the *J-K* Flip-Flop

Karnaugh Map Entry	Required Inputs	
	J	K
$\underline{0}$	—	1
$\underline{1}$	1	—
0	0	—
1	—	0
—	—	—

Now we are ready to obtain the equations to satisfy Figure 5.12; however, note that clocked flip-flops are required in this example. The use of clocked flip-flops simply means that the actions indicated in Table 5.14 are not allowed to occur until a pulse is supplied to the clock input, C. Effectively, this means that the C input is ANDed with the J and K inputs. (In most systems, the J and K inputs must be stabilized at the values shown in Table 5.14 at least a few nanoseconds *before* the clock pulse occurs. The action shown in the table takes place when the *clock pulse* makes the $1 \rightarrow 0$ transition, i.e., operation occurs on the negative edge.) Since the input pulse, x, can be used to synchronize the circuit action through the clock input on each flip-flop, it should no longer appear explicitly in the equations.

The resulting equations for this example are:

$$J_1 = K_1 = 1, \qquad J_2 = K_2 = y_1\bar{y}_4, \qquad J_3 = K_3 = y_1y_2,$$
$$J_4 = y_1y_2y_3, \qquad K_4 = y_1$$

These lead to the circuit diagram shown in Figure 5.13.

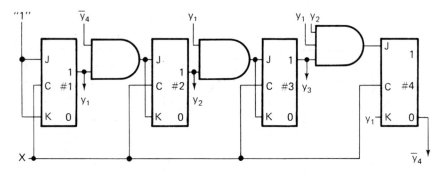

Figure 5.13   One Stage of a BCD Counter.

Note that only seven gate inputs are required by the circuit, which is much fewer than would have been required had the clock input not been available. The equations $J = K = 1$ mean that flip-flop 1 will change state each time the trailing edge of the clock pulse occurs; thus, a clocked J-K flip-flop wired in this way behaves like a T flip-flop. As a matter of fact, the T flip-flops appearing in Figure 5.11, and all other occurrences of them, can be easily realized with clocked J-K flip-flops wired as shown in Figure 5.13. In the future, however, we will continue to employ T flip-flops where appropriate, since they uniquely (and concisely) represent a particular mode of flip-flop operation. On the other hand, in actual circuit construction, the more powerful clocked J-K flip-flops would usually be used. Incidentally, clocked flip-flops usually possess a direct reset input which is independent of the clock circuit; therefore, counters and other systems constructed from these devices may be cleared without additional gating. (This, and the possibility of a direct set input will be considered more fully in Chapter 6.)

‡The Ripple Method

All previous designs in this chapter employed synchronous circuits—all flip-flop action was individually synchronized by means of a clock pulse. We now introduce an asynchronous method, which will be further developed in Chapter 10. Here an attempt will be made to eliminate all gates by connecting the T input of one flip-flop directly to the output of the previous flip-flop in a chain-like arrangement so that each flip-flop will cause the next one in line to operate.

Example: Synthesis of a Down-Counter

A three-bit down-counter is to be designed, with T flip-flops, which will decrement by one each time an input pulse occurs. After we construct the state diagram with a decreasing binary assignment, the following state table results:

Table 5.16   State Table for a Three-Bit Ripple Counter

Present State ($y_3y_2y_1$)	x Input, Next State ($y_3y_2y_1$)
111	110
110	101
101	100
100	011
011	010
010	001
001	000
000	111

The excitation maps corresponding to Table 5.16 are shown in Figure 5.14. The information needed to design a ripple counter is contained in these excitation maps; however, we must take a somewhat different approach in order to interpret the maps. With our knowledge of the characteristics for the T flip-flop and particularly the fact that it changes state on the trailing edge of the input pulse, Table 5.17 may be prepared. Consider the lower

‡This double-dagger symbol will be employed exclusively throughout the remainder of the text to indicate a section which may be omitted, or not studied in detail, without seriously affecting the understanding of later material. (The symbol refers to an *entire numbered section*; but if the section is unnumbered, the symbol refers only to the subsection itself and any lower-level subsections which it contains. For example, the current symbol applies up to the beginning of Section 5.4.)

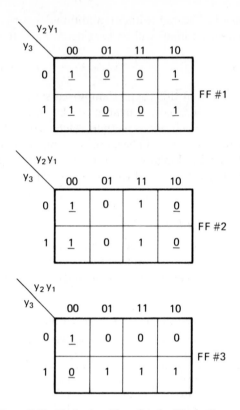

Figure 5.14 Excitation Maps for the Ripple Counter.

Table 5.17   Map Requirements for the *T* Flip: Flop-Ripple Method

Corresponding Map Entries for Flip-Flop Numbers		Required Circuit Connection Between FF i and input T_j
i	j	
$\underline{1}$	0, 1, or −	$T_j = \bar{y}_i$:
Anything but $\underline{1}$	0, 1, or −	
$\underline{0}$	0, 1, or −	$T_j = y_i$:
Anything but $\underline{0}$	0, 1, or −	

half of the table; here $T_j = y_i$. This means that when FF*i* undergoes the transition $1 \rightarrow 0 = \underline{0}$, FF*j* will change state; but this is exactly the desired action when the *j* map contains $\underline{0}$ or $\underline{1}$. Moreover, when FF*i* undergoes any other transition (i.e., $\underline{1}$, 1, or 0), FFj will remain unchanged; but this is exactly the desired action when the j map contains 0 or 1. Thus, we have verified the lower half of the table. The upper half can be verified in the same way, but the inversion caused by $T_j = \bar{y}_i$ must be taken into account.

The equations for T_2 and T_3 are easily obtained from Table 5.17 and Figure 5.14 when we compare the corresponding squares of the i and $j = i + 1$ maps. We find that, for T_2 and T_3, the upper half of Table 5.17 is satisfied, resulting in the equations $T_2 = \bar{y}_1$ and $T_3 = \bar{y}_2$. Note that we can obtain the equation for T_1 from the conventional map covering demanded by the T flip-flop or by observing that the FF 1 excitation map contains nothing but $\underline{1}$ and $\underline{0}$, which are satisfied by the trailing edge of each x pulse. By either method, the equation is $T_1 = x$. Therefore, the resulting circuit for the above equations is as shown in Figure 5.15.

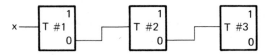

Figure 5.15   Ripple Down-Counter.

Observe how much simpler the three-bit ripple counter is than the ones designed by our previous method. The ripple counter is said to be *asynchronous* in that the individual flip-flops do not operate in synchronization with the x input, but rather (especially when an up-counter changes from 111 to 000) each flip-flop triggers the next in the chain like a row of dominoes falling. The corresponding term for the previously discussed counters is *synchronous* since each flip-flop operates in synchronization with the input signal.

To further explain the ripple method, consider the flip-flop excitation maps shown in Figure 5.16. The requirements for flip-flops 1 and 2 can be satisfied in a straightforward way by the equations $T_1 = x$ and $T_2 = x$; but it is not so easy to satisfy the requirements for flip-flop 3. If an attempt is made to connect the output of FF 2 into the T input of FF3, contradictions occur: The second column of the FF3 excitation map requires $T_3 = \bar{y}_2$, but the first column will not allow it. The third column requires $T_3 = y_2$, but the the fourth column will not allow it. On the other hand, if the FF1 output is connected into FF3, the resulting equation is $T_3 = y_1$. This equation is consistent, and the resulting ripple circuit is shown in Figure 5.17. Thus, the ripple method is actually more general than it at first appeared, since not only may successive flip-flops be connected together, but also the input of any flip-flop may be connected to the output of any other.

It should be mentioned that the ripple conditions (Table 5.17) are rather specialized, and only if they are satisfied for an entire map can a connection

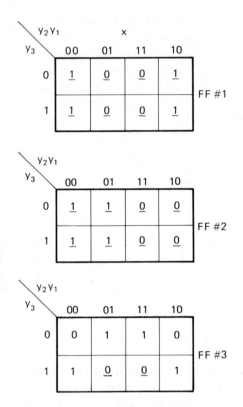

Figure 5.16   An Example Demonstrating the Ripple Method.

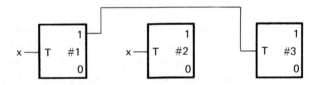

Figure 5.17   Ripple Circuit for Figure 5.16.

without gates be used for that flip-flop. On the other hand, the fact that important circuits found in computers frequently allow the ripple method makes it an important design technique.

An example of a set of flip-flop excitation maps, which can be only partially satisfied by the ripple method, is shown in Figure 5.18. There, $T_3 = \bar{y}_1$ or $T_3 = \bar{y}_2$, but gates are required to satisfy the needs of the other two flip-flops. These equations are $T_1 = \bar{y}_2 + xy_1$, or $T_1 = x(y_1 + \bar{y}_2)$ and $T_2 = x\bar{y}_1$. (The first term in the first equation for T_1 is produced by the ripple method, but in this case neither gates nor gate inputs are saved.)

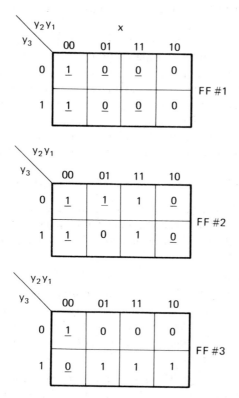

Figure 5.18 An Example Allowing Only a Partial Solution by the Ripple Method.

5.4 HEURISTIC DESIGN METHODS

In principle, one could use the synthesis method to obtain any required sequential circuit; however, the method becomes unwieldy when many flip-flops are involved. On the other hand, the design effort required by the method to be introduced here is not nearly so strongly influenced by the number of flip-flops. In practice, solutions are often obtained that can be easily extended to as many flip-flops as desired. Moreover, it is desirable to have quicker techniques for common circuits as well as for circuits that are fairly simple in structure. Rather than presenting another formal technique such as the circuit synthesis method developed in Section 5.2, the object of this section is to give the reader insight into some of the subjective techniques employed by the computer engineer. This should enable the novice to develop his own design skills more rapidly.

The circuits in Section 5.3 suggest a number of intuitive design techniques which, though fairly obvious, are nevertheless important. Such intuitive techniques, which do not guarantee an optimum solution (or for that matter any kind of solution), but which are frequently successful and fast, are called *heuristics*. (An example of a common heuristic in mathematics is the concept of converting all functions to sines and cosines when proving trigonometric identities.) Although the number of flip-flops may be large (10 or 20 or more), the inherent structure of most individual computer circuits is surprisingly simple. For this reason, heuristic methods have been very successful and are widely used in computer design (although they are not always called by that name).

The following list summarizes some of the heuristic techniques:

1. *Iteration.* Once a good circuit has been developed, it can frequently be iterated to fulfill the requirements of a much larger circuit. (An example is the iteration of one stage of a ripple counter to produce a 12-bit unit.)

2. *Module Synthesis.* The synthesis method itself may be one of the most important heuristics because it can be used to optimize a portion of a circuit, a module. This module may then be iterated to produce a very large system. (An example is the synthesis of one stage of a BCD counter, which is then iterated to produce a counter with 10 decimal places.)

3. *Information Analysis.* Through analysis of the information needed to excite a particular flip-flop, the necessary gating to obtain this information from the other flip-flops is often apparent. (The way a single AND gate couples the information into the nth T flip-flop from all the others in a synchronous up-counter is representative of this concept.)

4. *Evolution (ANDing or ORing).* Often we have had experience with a circuit which almost satisfies the current requirements. The idea here is to complete the design by modifying the original circuit in a fairly straightforward way or by combining two or more circuits. Of course, the obvious way to combine circuits is to use the AND or OR function. A simple example of ORing is shown in the design of the modulo-3 counter in Section 5.2. There, a "clear" capability was given to the counter simply by ORing the clear pulse, x_c, with the reset equations for all of the flip-flops. Depending on the emphasis which appears to be most appropriate, the name *evolution*, *ANDing*, or *ORing* will be applied to this heuristic.

5. *Requirements Segmentation.* By segmenting the circuit requirements into small manageable sections, we can often determine the gating

and flip-flops necessary to realize the individual sections by inspection or by using one of the other heuristics. Two major segments which are frequently employed are flip-flop excitation and output generation. In the first of these, the designer concentrates on the inputs and current flip-flop states to generate all the next flip-flop states. In the other, emphasis is placed on inputs and the current flip-flop states to generate the outputs.

Taking a more formal approach, we can divide the segmentation heuristic into the following components:

(a) Break the requirements into the smallest identifiable entities for individual realization and later recomposition.

(b) Try to use circuit entities already mechanized as components in those not yet realized.

(c) Work on the most primitive requirements first, with the hope that these may later be used in satisfying more complex requirements.

(d) Look for regularity and symmetry in the overall requirements, and take as much advantage of these as possible.

(e) Where possible, use circuits or parts of circuits which have already proved their worth as a base for designing new circuits.

The above heuristics will now be employed in the design of several practical computer circuits. The emphasis will be on heuristics 1 through 4, because these apply to circuits which essentially have only a single requirement and may be illustrated in a page or two. Heuristic 5 has its major power in the design of large computer blocks where there is a multiplicity of requirements, and it will be mainly applied in later chapters, particularly in our study of the control system (Chapter 7). However, certain aspects of heuristic 5, particularly (b), (d), and (e), also apply at lower levels.

Example: Heuristic Design of a Ring Counter

Let us now attempt to use the heuristic methods to design a ring counter, which passes a single binary 1 from flip-flop to flip-flop until the last stage is reached, at which point the 1 is passed back to the first stage. Consider clocked *J-K* flip-flops and a four-bit system.

From the statement of the problem, and using the information heuristic, it is apparent that, whatever state the nth flip-flop is in, the next one should be in that same state after the input pulse has passed. This and the properties of the clocked *J-K* flip-flop lead us to the circuit for one stage in Figure 5.19.

Using the iteration heuristic, we obtain the complete circuit for the ring counter, as shown in Figure 5.20.

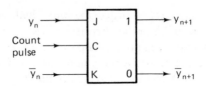

Figure 5.19 One Stage of a Ring Counter.

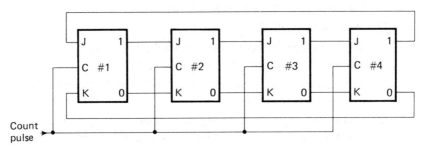

Figure 5.20 A Four-Stage Ring Counter.

Example: Heuristic Design of a Modulo-5 Counter

Suppose we desire to design a modulo-5 counter with T flip-flops. From experience or from simple heuristic considerations, we can construct a three-bit up-counter (see Figure 5.11). This will be identical to the modulo-5 counter until the state $y_3 y_2 y_1 = 100$ is reached. However, the next state must be 000. The AND heuristic is used to detect the 100 state and the occurrence of the input, x, through the function $T_3 = y_3 y_3 \bar{y}_2 \bar{y}_1 x$. (On further consideration, this may be reduced to $T_3 = y_3 x$, since the only time $y_3 = 1$ occurs is when the counter is in the state 100.) If the system is in state 100 and an input occurs, two things must happen: FF3 must be reset and FF1 must not be complemented. The shaded gates at the input to FF3 in Figure 5.21 show how the former requirement is satisfied through the AND and OR heuristics. Similarly, the shaded gate feeding FF1 uses the AND heuristic, function $T_1 = x\bar{y}_3$, to allow the input pulse to complement FF1 in all states except 100.

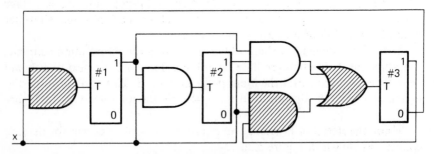

Figure 5.21 Modulo-5 Counter.

Actually, one gate input may be saved in this design if we use the function $T_3 = x(y_3 + y_1 y_2)$ instead of the one shown. In principle, all such functions which are constructed heuristically should be checked for possible simplifications.

Example: Heuristic Design of a Gray-Code Counter

We will now attempt a fairly challenging problem—the design of a Gray-code counter using T flip-flops. This code follows the sequence shown in Table 5.18.

Table 5.18   The Gray Code

Decimal	Gray Code $(y_4 y_3 y_2 y_1)$
0	0000
1	0001
2	0011
3	0010
4	0110
5	0111
6	0101
7	0100
8	1100
9	1101
10	1111
11	1110
12	1010
13	1011
14	1001
15	1000

The equation for T_4 is fairly simple. We see by inspection of Table 5.18 that an input causes FF4 to change state when $y_4 y_3 y_2 y_1 = 0100$. This results in $T_4 = x \bar{y}_4 y_3 \bar{y}_2 \bar{y}_1$. (*Note*: It is assumed here that the counter is not designed to turn over, i.e., no transition from 1 to 0 is required for FF4.) On the other hand, how can the equation for T_4 be generated? Must the states $y_4 y_3 y_2 y_1 = 0000, 0011, 0110, \ldots, 1001$ be detected, resulting in a function with eight terms? (Incidentally, these are eight isolated terms on a Karnaugh map—no simplification is possible.) Before allowing the circuit to become this complicated, let us use the information heuristic and ask ourselves what is regular about the requirements on y_1. Inspection of Table 5.18 reveals that y_1 changes state on the odd-numbered inputs. Certainly, a very easy way to detect this condition is to run the input through an auxiliary flip-flop. Thus, there is a tradeoff between the cost of an extra flip-flop and the cost of many gates. It appears that, under most conditions, the auxiliary flip-flop method would be favored.

After the auxiliary bit is included, the code is as shown in Table 5.19. Is there any additional regularity about these data which may be used to further simplify our design? Careful examination of the table reveals that the y_n bit changes state only if $y_{n-1} = 1$ and the other less significant bits equal zero.

Table 5.19   The Gray Code Including the Auxiliary Bit, A

Decimal	Code ($y_4 y_3 y_2 y_1 A$)
0	00001
1	00010
2	00111
3	00100
4	01101
5	01110
6	01011
7	01000
8	11001
9	11010
10	11111
11	11100
12	10101
13	10110
14	10011
15	10000

Thus, the equation for the ith flip-flop is $T_i = y_{i-1} \bar{y}_{i-2} \ldots \bar{y}_1 \bar{A} x$, where $1 < i \leq 4$. (For $i = 1$, $T_1 = Ax$.) The resulting circuit is shown in Figure 5.22. (Temporarily omit the dashed line bypassing the shaded gate.)

If we desire to have the counter return to zero on the 16th input pulse, we can use the OR heuristic to find the term which must be added to the T_4 input. From Table 5.19, this term is $x \bar{y}_3 \bar{y}_2 \bar{y}_1 \bar{A}$. From Figure 5.22, the original input to T_4 was $x y_3 \bar{y}_2 \bar{y}_1 \bar{A}$; therefore, $T_4 = x \bar{y}_3 \bar{y}_2 \bar{y}_1 \bar{A} + x y_3 \bar{y}_2 \bar{y}_1 \bar{A} = x \bar{y}_2 \bar{y}_1 \bar{A}$, which means that the dashed line may be used to replace the shaded gate as the input to FF4. Thus, requiring the counter to reset on the 16th pulse actually allows a gate to be eliminated from the circuit.

Example: Heuristic Design of a Multifunction Register
*   with* D-*Type Flip Flops*

The register to be designed must allow:

1. Parallel-input loading from another register.

2. Serial output by shifting the data toward the least significant bit, where it can be extracted.

3. A register clear capability.

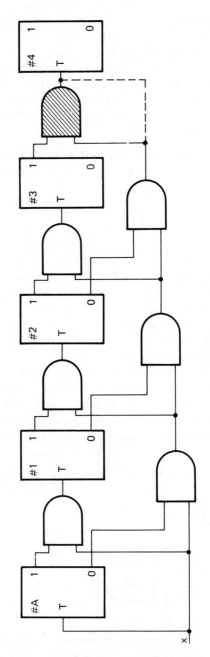

Figure 5.22 Heuristic Design for a Gray Code Counter.

The functions are selected individually by having a logical-1 level appear on one of three lines, and the actual operation is synchronized by means of a clock pulse. Circuits of this type are extensively used in practice, and a typical realization is in the medium scale integration (MSI) package 8274 manufactured by Signetics Corporation.

The type-D flip-flop is introduced here since it is very important in practice and quite convenient in the design of this class of circuit. The type-D flip-flop has only two inputs, the data or D terminal and the clock terminal. The logic level, 1 or 0, existing at the D input is transferred to the flip-flop immediately following the trailing edge of the clock pulse. This characteristic is represented in Table 5.20.

Table 5.20   Characteristics of the Type-D Flip-Flop

Input D	Next State (y_{n+1})
0	0
1	1

Having defined the problem and the nature of the D flip-flop, we now seek a heuristic approach to the design. Consider heuristic 5, requirements segmentation. It is apparent that there are three identifiable tasks (load, clear, and shift) that can be realized separately and then combined with the OR heuristic. Moreover, the symmetry in each of these three tasks indicates that module synthesis can be employed for the ith bit, and then iteration can be employed for as many bits as desired.

From information analysis, construction of Figure 5.23 showing the shift and load circuit for the ith bit is straightforward. Assume that it is desired to load data from a source register, S. A logical-1 level will appear at the load input to AND gate 2 a short time before the clock pulse occurs. If the input from bit i of the source register is a 1, it will be passed through the

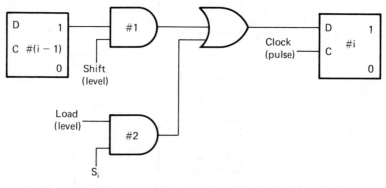

(Level from ith bit of source register.)

Figure 5.23   Multifunction Shift Register—Partially Complete Design.

AND gate and the OR gate. Then, when the clock pulse occurs, a 1 will be deposited in the *D* flip-flop. If the input from bit *i*, S_i, of the source register is a 0, AND gate 2 will have a 0 output, as will the OR gate. Thus, a 0 will be correctly placed in the flip-flop by the following clock pulse. (The shift operation is accomplished in a similar way.)

Now consider the clear function—a 0 is to be placed in the flip-flop when the clear level is 1. Figure 5.23 shows that when the shift and load levels are 0, the output of the OR gate will be a 0. Thus, this condition alone is sufficient to cause the clock pulse to deposit a 0 in the flip-flop, and the design seems to be complete.

When the circuit is tested, however, a flaw will be discovered—if the shift, load, and clear levels are all 0 and a clock pulse occurs, the flip-flop will be cleared. Therefore, the circuit fails to hold the data until one of the three functions is selected. Information analysis of the hold requirement reveals that when neither shift, load, nor clear is true and bit *i* is in the 1 state, the *D* input to the flip-flop must receive a 1 so that a 1 will be retained in the flip-flop following the clock pulse. The equation for this task is:

$$D_n = y_n\overline{(\text{shift} + \text{load} + \text{clear})}$$
$$= y_n\overline{(\text{shift})}\,\overline{(\text{load})}\,\overline{(\text{clear})}$$

Figure 5.24 shows this addition to the original circuit. Testing under all of

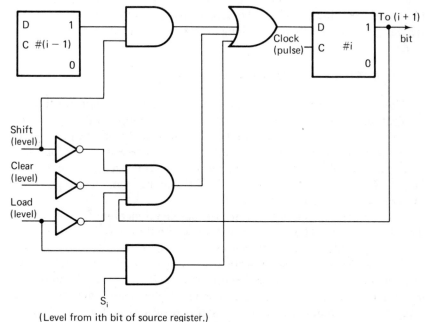

(Level from ith bit of source register.)

Figure 5.24 Multifunction Shift Register—Complete Design for Module *i*.

the specified conditions will verify correct operation of the circuit. Moreover, iteration may be employed to construct a register with as many bits as desired; the only addition that might be required is some special circuitry for shifting in and out at the two ends of the register.

Summary:

The following four steps, which should be executed in sequence, serve as a broad summary of the heuristic approach:

(a) Concisely describe the desired circuit action. A state diagram or other graphical representation is frequently helpful here.

(b) Design the described circuit using the heuristics.

(c) Check the combinatorial portions for possible simplifications using the theorems or a Karnaugh map.

(d) Test the logic by trying critical input–memory state combinations. (Even here, heuristics can be applied to take advantage of symmetry and the like in avoiding brute-force exhaustive testing of all combinations.)

The above method is open-ended in that professional designers are constantly adding new heuristics to their repertoire and striving to improve old ones. Moreover, once mastered, techniques of this class often are employed almost subconsciously. Thus, this section has provided an introduction to the design art rather than furnishing an algorithmic procedure.

5.5 COST AND SPEED AS DESIGN CRITERIA

How can one choose from several competing circuit designs for the same application? The obvious answer is to select the optimum design, but this depends on your definition of optimum. The optimum could be based on the fastest system, the one with lowest cost, the most reliable, the one most easily maintained, a weighted combination of several criteria, or many other factors. To give a brief idea of how designs may be compared, we will limit this consideration to cost and speed.

Before integrated circuits became the dominant medium for constructing digital computers, the number of gates and flip-flops required by a system was a reasonable universal measure of cost. With the advent of complex logic blocks (such as those in Sections 4.9 and 10.4), it's not nearly so easy to formulate a concise universal cost criteria. If the system requirements match the characteristics of an available module or small group of modules, a very complex requirement may be satisfied for only a few dollars. Here, cost is closely related to module, package, count rather than gate and flip-flop count. On the other hand, special requirements sometimes dictate that circuits be constructed from fairly low-level primitives. In this case, the gate and flip-flop count may again be a valid criteria.

Since a more complex analysis is beyond the scope of this text, let us assume that system requirements force low-level primitives to be employed in the following design. The basic methods employed here, however, are still valid when higher-level primitives are permitted. (For a detailed discussion of cost minimization through the use of higher-level primitives, see the text by Blakeslee.[1])

Assume the following parameters concerning cost and circuit speed:

1. Each flip-flop costs 0.2 unit.

2. Gate costs $= 0.05 + (0.01)(n)$ units per gate, where n is the number of gate inputs.

3. Flip-flops require a delay of 30 ns to settle to their final value.

4. Each gate requires a delay of 15 ns.

These are arbitrary cost units; but they are of the correct order of magnitude when treated as dollars for current computer technology. The same numbers are used to compare a number of designs; thus, the relative ordering is not sensitive to the cost assumptions. Moreover, our ultimate purpose is not so much to present data on various counters but to give the reader insight into engineering design methods.

Using the above information, we see that a five-bit synchronous (parallel) up-counter constructed from clocked J-K flip-flops (an extension of the three-bit counter shown in Figure 5.11)* requires five flip-flops and four gates. Since each C input in this counter receives an input from each lower bit, the number of gate inputs follows the arithmetic progression:

$$\text{Inputs} = 2 + 3 + 4 + 5 = 14$$

In addition, if we want to decode each of the $2^5 = 32$ counter combinations so that only 1 out of 32 lines contains a logical 1, 32 AND gates are needed. Each of these has 5 inputs, and $(32)(5) = 160$ total inputs. A diagram of this circuit is shown in Figure 5.25, and the resulting cost equation becomes:

$$\text{Cost} = 5(0.2) + (4 + 32)(0.05) + (14 + 160)(0.01)$$
$$= 4.54 \text{ units}$$

What is the maximum frequency at which this counter can operate? Assuming that the output from the decoder is to be sampled in coincidence with the next pulse being counted, the delay of the gate driving each flip-flop,

---

*Because we are considering practical circuit realization, use is made of the previously discussed equivalence between properly wired, clocked J-K flip-flops and T flip-flops in extending the circuit of Figure 5.11.

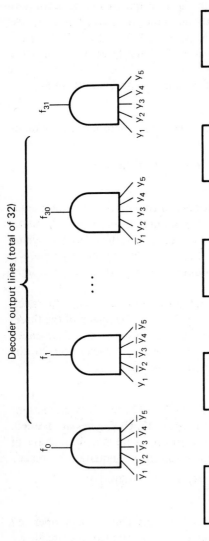

Decoder output lines (total of 32)

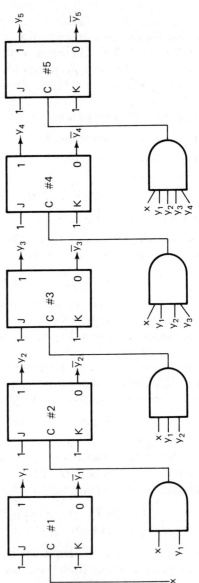

Figure 5.25 A Five-Bit Parallel Counter with Its Decoder.

the delay of the flip-flop itself, and the delay of the decoder must all be combined to give the total delay.

$$\text{Delay} = 15 + 30 + 15 = 60 \text{ ns}$$

This leads to the following maximum counting frequency:*

$$\text{Maximum frequency} = \frac{1}{60 \times 10^{-9}} = 16.7 \text{ MHz}$$

Before the synchronous counter design is compared with several others, the ripple-carry counter shown in Figure 5.26 will be introduced. This circuit is actually a combination of the parallel and ripple counters. Note that the only way the C_5 input can receive a pulse is for all the AND gates to have an output of 1. But, since the gates are in series, the final output of any gate waits for the propagation delay of all the previous gates, which produces a ripple when all the previous flip-flops are in the 1 state. Therefore, the maxi- delay in the four AND gates is (4)(15) = 60 ns.

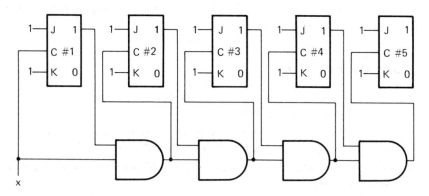

Figure 5.26 Ripple-Carry Counter (decoder not shown).

Table 5.21 contains the results of cost and maximum frequency calculations similar to those completed for the synchronous counter. Note that two different applications were considered and that, in each, there is a trade-off between operating frequency and cost, with the highest-frequency counter costing the most. In the 10-bit counter, comparing the ripple method with the parallel method, it is interesting that a 50% increase in counter cost (2.00 to 2.99 units) leads to almost a 600% increase in maximum operating frequency. While such calculations as shown in Table 5.21 are instructive and useful, they certainly are not absolute since other design parameters such as reliability and maintainability may be the deciding factors. On the other hand, some basic ideas have been presented for evaluating competing

---

*The quoted frequency is optimistically high since the width of the input pulse (flip-flops operate on the trailing edge) was neglected and no safety factor was allowed.

**Table 5.21   Performance Comparison for Several Counter Types
in Two Different Applications**

Type of Counter	Flip-Flops Required	Gates (Inputs) Counter Only	Decoder Gates (Inputs)	Maximum Operating Frequency MHz	Total Cost
		Five-Bit Up-Counter with Decoder			
Ring	32	0	0	33.3	6.40
Parallel	5	4 (14)	32 (160)	16.7	4.54
Ripple-Carry	5	4 (8)	32 (160)	9.5	4.48
Ripple	5	0	32 (160)	6.1	4.20
		Ten-Bit Up-Counter (no Decoder)			
Parallel	10	9 (54)	0	22.2	2.99
Ripple-Carry	10	9 (18)	0	6.6	2.63
Ripple	10	0	0	3.3	2.00

logic designs, and it is now more obvious why the computer engineer needs to be aware of a variety of logic circuits.

5.6 BIBLIOGRAPHY

1 BLAKESLEE, THOMAS R., *Digital Design with Standard MSI and LSI*. New York: John Wiley & Sons, Inc., 1975.

2 HUFFMAN, D. A., "The Synthesis of Sequential Circuits." *Journal of the Franklin Institute*, Vol. 257, Nos. 3 and 4, 1954, pp. 161–90 and 275–303.

3 KOHAVI, ZVI, *Switching and Finite Automata Theory*. New York: McGraw-Hill Book Company, 1970.

4 MALEY, G. A., *The Logic Design of Transistor Digital Computers*. Englewood Cliffs, N.J.: Prentice-Hall, Inc., 1963.

5 MARCUS, MITCHELL, *Switching Circuits for Engineers* (2nd ed.). Englewood Cliffs, N.J.: Prentice-Hall, Inc., 1967.

6 MCCLUSKEY, E. J., *Introduction to the Theory of Switching Circuits*. New York: McGraw-Hill Book Company, 1965.

7 MEALY, G. H., "A Method for Synthesizing Sequential Circuits." *Bell System Tech. Jour.*, Vol. 34, 1955, pp. 1045–79.

8 MINSKY, MARVIN, *Computation Finite and Infinite Machines*. Englewood Cliffs, N.J.: Prentice-Hall, Inc., 1967.

9 MOORE, E. F., "Gerdanken-Experiments on Sequential Machines." *Automata Studies, Annals of Math. Studies*, No. 34, Princeton, N.J.: Princeton University Press, 1956, pp. 129–53.

10 RHYNE, V. THOMAS, *Fundamentals of Digital Systems Design*. Englewood Cliffs, N.J.: Prentice-Hall, Inc., 1973.

11 SOBEL, HERBERT S., *Introduction to Digital Computer Design*. Reading, Mass.: Addison-Wesley Publishing Company, Inc., 1970.

12 WICKES, WILLIAM, *Logic Design with Integrated Circuits*. New York: John Wiley & Sons, Inc., 1968.

5.7 EXERCISES

5.1 (a) Construct the state diagram and state table, using the binary code assignment, for a modulo-7 counter. Allow the counter to be cleared from any state.

(b) Repeat part (a) for the same counter, but with the maximum count resulting in a terminal state.

5.2 Construct a state diagram and state table, using the binary code assignment, for a modulo-4 up-down counter. This device becomes an up-counter whenever the pulse x_2 is received, and it continues in that mode until an x_3 pulse is received, at which time it becomes a down-counter, etc. Each x_1 pulse increments or decrements the counter's total, depending on whether it is in the up or down mode.

5.3 Synthesize a three-bit up-counter using S-R flip-flops.

5.4 Using the synthesis method, T flip-flops, and the binary code assignment, determine the logic circuit for the flow table shown.

Present State	Next x_1	State x_2
0	1	0
1	2	0, Z
2	2	2

5.5 Using the synthesis method and J-K flip-flops (not clocked), determine the logic circuit for a modulo-6 counter.

5.6 Using the synthesis method and clocked J-K flip-flops, determine the logic circuit for a modulo-7 counter.

5.7 Determine which assignment leads to the simplest sequential circuit for the flow table shown. The three unique assignments appear in the table, and it can be shown that, for two secondary variables, all others are equivalent to these three. Use S-R flip-flops in this exercise.

Present State	Next x_1	State x_2
0	0	1
1	0	2, Z
2	0	2

State	Assignments 1	2	3
0	00	00	00
1	01	01	11
2	11	10	01
3	10	11	10

5.8 Carefully examine the flip-flop excitation map shown in Figure E5.8. Then list any map entries (by their $y_3y_2y_1$ values) which are not consistent with the requirements of an actual sequential circuit. Explain your reasoning.

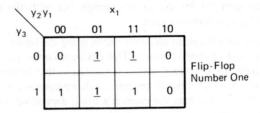

Figure E5.8

5.9 Determine the logic equations for J-K flip-flops, and draw the circuit for a Gray-code counter using the synthesis method. Compare these results with the heuristic design results in the text.

5.10 (a) Using the heuristic method and the fewest possible T flip-flops, design a circuit whose outputs follow the following table. (Assume that the circuit always comes on in the start condition.)

Pulse Number	Level Outputs w_3	w_2	w_1
Start	1	0	0
1	0	1	0
2	0	0	1
3	0	1	0
4	1	0	0
5	0	1	0
6	0	0	1
	etc.		

(b) Repeat part (a) using the formal synthesis method and the specification that $w_i = y_i$, where y_i is the direct output of the flip-flop i.

(c) Discuss and compare the designs.

5.11 Synthesize a practical modulo-5 counter. Take into account the fact that flip-flops have a 0.5 probability of coming on in the 1 state. The T variety of flip-flops should be considered.

5.12 For the state diagram shown in Figure E5.12, synthesize the most economical sequential circuit with assignment 1 (see Exercise 5.7). Assume that flip-flop and gate costs are as given in Section 5.5 in the text, and that the *S-R, T,* and *J-K* devices are available.

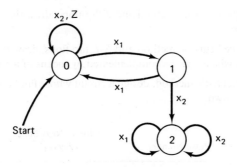

Figure E5.12

5.13 Using the ripple synthesis method, determine the logic circuit for the following flow table.

Present State $(y_3y_2y_1)$	x Input, Next State $(y_3y_2y_1)$
000	000
001	001
010	111
011	000
100	100
101	101
110	011
111	100

5.14 Develop a synthesis technique for the *ripple-carry* method using *T* flip-flops. Explain in detail and verify your work by applying it to the following flow table. (See Figure 5.20 for an example of a ripple-carry circuit.)

Present State $(y_3y_2y_1)$	x input, Next State $(y_3y_2y_1)$
000	000
001	001
010	001
011	110
100	100
101	101
110	101
111	010

5.15 Using the heuristic method, design a modulo-5 ripple-down counter.

5.16 Synthesize a three-bit register (with *S-R* flip-flops) which can be complemented by a pulse on line x_1 and which can be decremented by 1 by a pulse on line x_2.

5.17 Applying the heuristic method and *T* flip-flops, draw the circuit for a four-bit ring counter.

5.18 Applying the heuristic method and *S-R* flip-flops, draw the circuit for a four-bit register which may be complemented by means of a pulse on input line *x*.

5.19 Using the heuristic method, design a circuit whose flip-flop outputs reproduce the table shown.

Pulse Number	*x* Input, Next State $(y_3 y_2 y_1)$
Start	100
1	110
2	111
3	011
4	001
5	000
6	100
etc.	

5.20 A computer having a clock frequency of 5.0 MHz and a 20-ns pulse width requires the waveform shown in Figure E5.20 to implement a particular control function. From heuristic considerations, draw a logic circuit which will satisfy these requirements. (Assume that the voltage levels of all parts of the circuit will be compatible.)

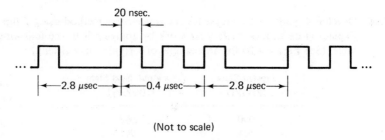

(Not to scale)

Figure E5.20

5.21 Eliminate any redundant states in the following table.

Present State	Next State x_1	x_2
0	3	2
1	1	5
2	0	1, Z
3	4	3
4	3	2
5	0	3

5.22 Repeat Exercise 5.21 for the following table.

Present State	Next State x_1	x_2
0	3	1
1	0	2
2	1	3
3	4	2
4	2, Z	1

5.23 Using the minimum possible (AND, OR, and NOT) gate inputs, convert a T flip-flop to one having the characteristics shown in the table. Only uncomplemented versions of the A and B variables are available.

Flip-flop Input A	B	Next State y_{n+1}
0	0	y_n
0	1	$\bar{y}_n$
1	0	1
1	1	$\bar{y}_n$

6 DIGITAL ELECTRONICS

The discrete-component networks of a few year ago have, to a large extent, been replaced by integrated circuits, ICs. In order to use these new devices effectively, the computer designer needs to have an understanding of basic digital electronics, including the fundamentals of the various IC families and not just a *black-box* concept.

Diodes and transistors will be represented by an approximate circuit model in the discussion that follows because this approach is easily understood and because our present motivation stems more from an interest in the use of electronic building blocks for logic design rather than detailed circuit design itself. On the other hand, a reader whose long-range interests are mainly in computer hardware or in electronic circuits will eventually want to become knowledgeable in the more fundamental approach based on the exact physical behavior of the devices themselves.[5,6,7,10] In Chapter 4 we learned how to construct logic diagrams from AND, OR, NOT, and other types of gates. Here the realization of such gates by means of circuit elements will be presented.

It is assumed that the reader has an elementary acquaintance with electrical networks. This chapter will extend this knowledge into the field of digital electronics in order to briefly explore the circuit level of computer design. We will consider a wide range of topics, from the design of logic gates to flip-flops and from the mechanization with discrete components to integrated circuits.

6.1 DIODE GATES

Although gate circuits constructed exclusively from diodes are no longer extensively used, they exhibit most of the same properties as the more complex, currently popular gate types. In an elementary treatment such as this, it is much easier to derive these properties for diode circuits. Moreover, diodes are the key to understanding transistors and ICs, and diode–transistor combinations are still employed in many digital circuits. For these reasons, we now begin with an elementary consideration of diode physical properties and then devote an extensive discussion to their gate circuits.

Silicon, Si, is the most common semiconductor material from which diodes, transistors, and integrated circuits are constructed. By adding a small amount of an impurity with a valence of three (such as aluminum or indium) to pure silicon, a so-called *p-type* semiconductor is produced. *Holes* are the majority carriers in this type of material, and a memory aid here is to think of *p* as standing for *positive* charge. Conversely, a small amount of an impurity with a valence of five (such as arsenic or phosphorus) may be added to pure silicon to produce an n-type semiconductor. *Electrons* are the majority carriers in this type of material, and the corresponding memory aid is to think of *n* as standing for *negative*. As a first-order approximation, one may view a p-n junction diode, like the one shown in Figure 6.1(a), as consisting of only free positive charge carriers on the p-side of the junction and only free negative charge carriers on the n-side of the junction. With this concept, it can be seen from Figure 6.1(b) that a negative voltage applied to the p-terminal and a positive voltage applied to the n-terminal pull both charge carriers away from the junction. This results in no current flow. On the other hand, the opposite voltages applied to the terminals, as shown in

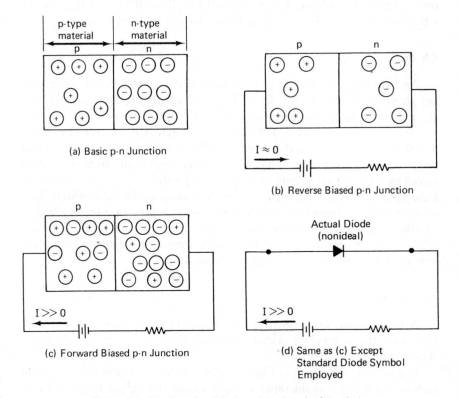

(a) Basic p-n Junction

(b) Reverse Biased p-n Junction

(c) Forward Biased p-n Junction

(d) Same as (c) Except Standard Diode Symbol Employed

Figure 6.1 The Junction Diode and Its Standard Symbol.

Figure 6.1(c), force both charge carriers across the junction, resulting in significant current flow. This simple picture predicts the basic rectifying action of a diode, but it does not account for the fact that silicon diodes require a *cut-in voltage*, V_γ, of about 0.7 volt before appreciable current will flow. In Figure 6.1(d), note the standard diode symbol employed and the fact that the arrow on the symbol points in the direction of easy conventional current flow. (If the battery were reversed, the symbol and current arrows would be in opposition, indicating high diode resistance or nearly zero current flow.)

For analyzing diode networks, the above physical behavior is best represented in terms of the circuit model shown in Figure 6.2(a) and the diode response shown in Figure 6.2(b).

Note the degree to which the model characteristic approaches the actual, and how they both differ from the ideal. The abrupt change in the model diode response at $V_d = V_\gamma$ is due to the fact that the ideal diode, which is an integral part of the model, becomes forward biased. This results in a low input (forward bias) resistance, R_f, instead of the very large (reverse bias) resistance R_r. (For silicon, V_γ typically is 0.7 V, $R_f = 20\ \Omega$, and $R_r > 10^6\ \Omega$, which means R_r will be neglected in the analysis that follows.)

OR Gates

An elementary diode OR gate circuit and the corresponding truth table are shown in Figure 6.3(a). By making the simplifying assumptions that $R_f = 0$ and source resistance $R_s = 0$ we can see that with $x = 5$ V and $y = 0$, D_x is forward biased. Thus, 0.7 V is lost across V_γ and the remainder, $5 - 0.7 = 4.3$, appears at the output, z. Similar reasoning leads to the other three sets of voltages in the table. In producing logic values from the voltage values in the truth table, note that the most positive voltage is assigned the logic value 1 and the most negative voltage is assigned the logic value 0. Such an assignment is known as *positive logic*—more will be said on this subject shortly. Another fact to observe concerning the truth table is that there is a 0.7-V offset between the maximum input voltage and the maximum output voltage, which of course is due to the diode drop.

AND Gates

A very similar arrangement of diodes can be used to produce an AND gate, with the elementary circuit and truth table (using positive logic) as shown in Figure 6.3(b).

In analyzing the circuit in Figure 6.3(b) with $x = 5$ V, $y = 0$ V, we might be tempted to say that the input $x = 5$ V causes the output to go to 5.7 V.

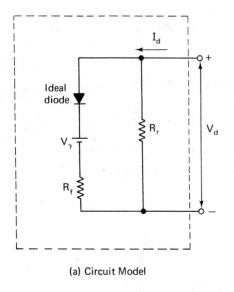

(a) Circuit Model

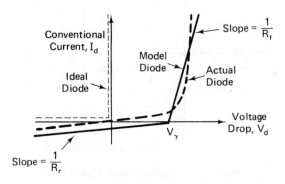

(b) Diode Response

(Note negative I_d scale has been
greatly enlarged for clarity.)

Figure 6.2 The Diode Circuit Model and Its Response.

However, further consideration leads us to the realization that the y diode will now have 5.7 V across it, which will cause it to draw a large amount of current until the output drops to 0.7 V. At this time, the x diode would be reverse biased. Thus, the output is stable at $z = 0.7$ V. Of course, when both inputs reach 5.0 V, the output will be stable at 5.7 V.

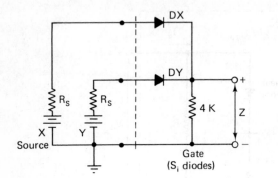

Voltage Values			Logical Values		
X	Y	Z	X	Y	Z
0	0	0	0	0	0
0	5	4.3	0	1	1
5	0	4.3	1	0	1
5	5	4.3	1	1	1

(a) OR Gate and Truth Table

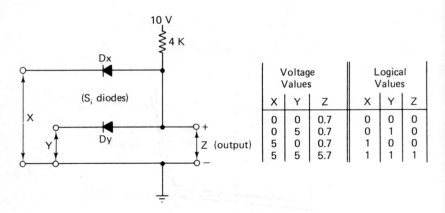

Voltage Values			Logical Values		
X	Y	Z	X	Y	Z
0	0	0.7	0	0	0
0	5	0.7	0	1	0
5	0	0.7	1	0	0
5	5	5.7	1	1	1

(b) AND Gate and Truth Table

Figure 6.3   Simplified Diode Gate Circuits.

Positive vs. Negative Logic

Logic blocks are named according to the function they perform with positive logic. As shown in Figure 6.3, the block name describes the electrical behavior of the device. However, the AND block can be made to produce other logic functions when negative logic is defined on either the input or the output or both. Positive logic has already been defined and *negative logic* is merely the converse, i.e., the most positive level is defined to be a logical 0 and the most negative level is defined to be a logical 1. Table 6.1 shows how the AND block in Figure 6.3(b) can be used to produce the AND, OR, NAND, and NOR functions.

When we draw a logic diagram, the gate symbols will continue to represent the function performed by that gate and not necessarily the electrical

Table 6.1   The Four Functions Produced by the AND Block

Voltage Values			(+) Logic			(−) Logic			(+) Logic	(−) Logic		(−) Logic		(+) Logic
x	y	z	x	y	z	x	y	z	x	y	z	x	y	z
0	0	0.7	0	0	0	1	1	1	0	0	1	1	1	0
0	5	0.7	0	1	0	1	0	1	0	1	1	1	0	0
5	0	0.7	1	0	0	0	1	1	1	0	1	0	1	0
5	5	5.7	1	1	1	0	0	0	1	1	0	0	0	1
			AND Function			OR Function			NAND Function			NOR Function		

performance. The only time an AND function is produced by an AND block is when positive logic is used on both the input and output.

Gate Analysis

Having analyzed simplified diode gates, we will now consider practical gates. Most of the observed effects also carry over to transistor gates, but the principles can be more easily visualized with the simpler diode circuits.

Let us return to the OR gate shown in Figure 6.3(a). The original circuit analysis included V_y, but the voltage drop $I_d R_f$ was neglected as was the voltage drop across the internal resistance of the source $I_d R_s$. Typical values are $R_f = 20 \, \Omega$, $R_s = 1 \, k\Omega$; thus, $R_f \ll R_s$, and R_f can be neglected for the circuits of interest here. A more practical OR circuit, which will now be analyzed, is shown in Figure 6.4. Note that the x input has been tied to zero volts so that diode D_x may be neglected. This has no effect on the conclusions of our analysis, but it does simplify the work. By showing the OR gate input in the form of a Thevenin's equivalent, we can analyze the effect of a general

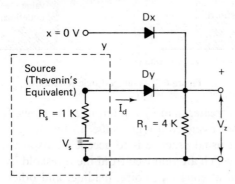

Figure 6.4   OR Circuit to be Analyzed.

source whether the actual source be a pulse generator, an AND gate, or even another OR gate. Now, when $I_d > 0$, the effect of the diode offset voltage V_γ is to make the source appear to be of voltage $V_s - V_\gamma$. Therefore, the output voltage V_z is easily determined from the voltage-divider rule.

$$V_z = \frac{(V_s - V_\gamma)R_1}{R_1 + R_s} \tag{6.1}$$

Substituting the above-mentioned values for the parameters in Equation (6.1), we find that:

$$V_z = \frac{(5 - 0.7)4 \times 10^3}{(1 + 4) \times 10^3} = 3.44 \text{ V} \tag{6.2}$$

The result of Equation (6.2) brings up an interesting question: What logic value should be assigned to an output of 3.44 V? It has been previously stated that with positive logic the most positive voltage level is defined to be a logical 1 and the most negative, a logical 0. While this definition is clear to us, it is too general for use in circuit construction. Gates and other digital circuitry are most easily designed on an absolute basis rather than on the above relative scale. If 0 and 5 volts are the minimum and maximum extremes, the question then becomes: How low can a voltage go (positive logic) before it ceases to be called a logical 1? The actual numbers used in defining logic values are somewhat arbitrary, but the method commonly used in practice employs a voltage margin between the low threshold, V_L, and the high threshold, V_H. Figure 6.5 shows this method for defining both negative and positive logic. For positive logic, any voltage up to V_L is interpreted as a logic

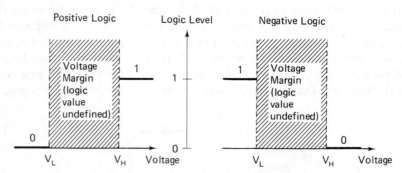

Figure 6.5 Voltage/Logic Relations.

0 and any voltage greater than V_H is interpreted as a logic 1. Voltages between V_L and V_H are not assigned a logic value. In the design gates of and other logic circuits, the usual practice is to accept inputs at the threshold but to develop outputs which are removed from the threshold by a safety factor allowing for circuit attenuation, noise, loading, and other sources of degradation. For example, if values of $V_L = 1$ V and $V_H = 3$ V are employed, the

design goal might typically be to achieve output voltages of 0.5 V and 4 V.

Assuming $V_L = 1$ V and $V_H = 3$ V, and returning to our original question concerning the logic value to be assigned to the result of Equation (6.2), we immediately see that $V_z = 3.44$ V represents a logical 1. Thus, the OR gate in Figure 6.4 generates the correct logical output. On the other hand, it does introduce a considerable loss (from $V_s = 5$ V to $V_z = 3.44$ V). Now Equation (6.1) reveals that all OR gates of this form possess an offset of V_γ and an attenuation of $R_1/(R_1 + R_s)$. Later we will discuss methods for eliminating the offset; but the attenuation is a physical limitation of all passive gates. This is the reason that transistors are introduced in the next section as a means of producing gain to overcome the attenuation. If the numerator and denominator of Equation (6.1) are divided by R_1, a way of minimizing attenuation is suggested:

$$V_z = \frac{(V_s - V_\gamma)}{1 + R_s/R_1} \tag{6.3}$$

If $R_s/R_1 \ll 1$, the attenuation can be neglected. Often the designer does not have control of R_s; thus, from the standpoint of attenuation, it is desirable to make $R_1 \gg R_s$.

Propagation Delay. A large value of R_1, as shown in Figure 6.4, can produce an unacceptable *propagation delay* as the following analysis demonstrates. (Propagation delay is defined as the finite time required for a logic circuit's output to change state in response to a change of state at the input.)

Consider the circuit shown in Figure 6.6. It is the same as in Figure 6.4, except the capacitance C_1 shunting R_1 is explicitly shown. The major contributor to C_1 is stray wiring capacitance, which typically has a value of $C_1 = 50$ pF unless special precautions are taken. If initially $x = 0$ and $V_s = 5$ V, and then later the input changes to $V_s = 0$ V, both diodes will be off and the only discharge path for C_1 is through R_1. (It is assumed that any output load, R_L, satisfies the equation: $R_L \gg R_1$.) From elementary network theory, recall that at time t the parallel resistor-capacitor combination, shown in Figure 6.6, produces the response

$$V_z = V_0 e^{-t/R_1 C_1} \tag{6.4}$$

where V_0 is the initial voltage across the capacitor terminals. If $V_0 = 3.44$ V, which was the value of the circuit output from Equation (6.2), and if $V_z = V_L = 1$ V is substituted into Equation (6.4) along with the values of R_1 and C_1, we find:

$$1 = 3.44 e^{-t/(4 \times 10^3 \times 50 \times 10^{-12})} \tag{6.5}$$

which results in $t = 247$ ns. Therefore, the propagation delay in turning the gate off becomes $Pd_{off} = t = 247$ ns, which is much too long for many digital applications. To reduce Pd_{off}, it is necessary to reduce R_1; but from Equation (6.3), this increases the attenuation problem. Frequently, a satis-

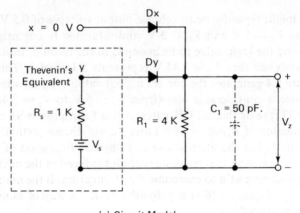

(a) Circuit Model

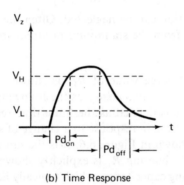

(b) Time Response

Figure 6.6   OR Gate Performance.

factory compromise value for R_1 cannot be achieved, and this is one of the main reasons that transistor logic (discussed in Section 6.2) has replaced diode logic in digital systems.

In the circuit of Figure 6.6(b), Pd_{on} is the delay in turning the gate on. $Pd_{on} < Pd_{off}$, which is a consequence of the fact that a low-resistance charging path, through R_s, is provided for Pd_{on} as opposed to the higher-resistance discharge path through R_1. Thus, Pd_{off} is the major contributor to the delay produced by the diode OR circuit. (Many authors[1] calculate a single propagation delay parameter for a gate or other logic device by averaging Pd_{on} and Pd_{off}.)

AND-OR Circuits

As indicated in Chapter 4, it is very common to develop a logic function in sum-of-products form; thus, we seek a circuit design to mechanize a set of AND gates followed by an OR gate. This situation is represented in Figure 6.7. The dotted sections indicate other portions of the circuit, which are not important for the following *worst-case design*. (This is a conservative design philosophy in which the most pessimistic values possible are assumed for the inputs, the component tolerances, and other circuit parameters.) Consider Figure 6.7 with the V_{si} inputs all at 5 V.

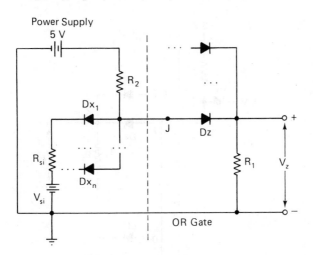

Power Supply 5 V

Figure 6.7 AND-OR Logic.

The circuit to the right of point J can be replaced by an equivalent resistor R_e. Then the power supply, in conjunction with R_e and R_2, forms a voltage divider. Thus, the voltage drop, V_J, across R_e is $V_J \leq 5$ V; and with $V_{si} = 5$ V for each i, all of the D_{xi} diodes are reverse biased and may be omitted from further consideration.

Now consider the voltage divider from the positive side of the power supply, through R_2, D_z, and R_1 and back to the supply. The resulting output voltage becomes:

$$V_z = \frac{(5 - 0.7)R_1}{R_1 + R_2} \tag{6.6}$$

If we continue to use $R_1 = 4$ kΩ and require a nominal value of $V_z = 4$ V, the result from Equation (6.6) is $R_2 = 300$ Ω.

Fan-Out

Having designed the AND-OR circuit with $R_1 = 4$ kΩ and $R_2 = 300$ Ω, we must now determine how many OR gates could be connected to the output of the AND gate, point J. *Fan-out* is a term employed in situations such as this where we desire to have a measure of the number of other circuits a particular circuit can drive. The equivalent circuit for this case is shown in Figure 6.8 where, for worst-case conditions, only one input of each OR gate is assumed to have a nonzero value. Now with every diode D_{zi} on, all points K_i will be at the same voltage and all the resistors R_1 will be effectively in parallel. The resulting equivalent resistor is $R_e = R_1/n$, where n is the fan-out, the number of OR gates being supplied by the AND gate.

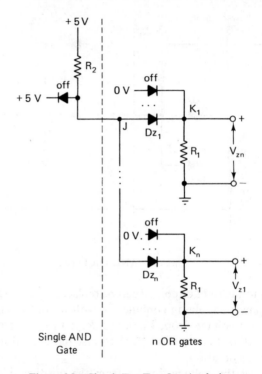

Figure 6.8   Circuit For Fan-Out Analysis.

The general output voltage can now be determined from:

$$V_{zi} = \frac{(5 - 0.7)R_1/n}{(R_1/n) + R_2} \tag{6.7}$$

If we are willing to accept an output value as low as $V_{zi} = V_H = 3$ V, Equation (6.7) may be solved for n (assuming the previous values of $R_1 = 4\,\text{k}\Omega$ and $R_2 = 300\,\Omega$). The result is $n = 5.77$. We round this to the next *lowest* whole number to obtain a fan-out of 5. This means the AND gate can drive five OR gates and the output will still be $V_{zi} \geq V_H$.

Diode Current Constraint

The constraints imposed by the desired frequency response, the voltage margin, and the desired fan-out have already been considered. Now the

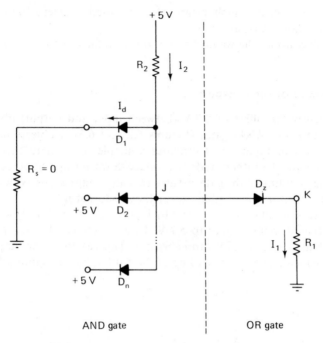

Figure 6.9 Configuration for Maximum Diode Current.

restriction imposed by maximum allowable diode current will be analyzed. Consider the circuit shown in Figure 6.9, with the worst-case conditions of only one gate input at 0 V and negligible source resistance, $R_s = 0$. Under these conditions, diode D_1 will be forward biased and point J will go to 0.7 V. With point K near 0 V and with R_1 of moderate size, I_1 will be negligible; thus, all of I_2 will flow through $D_1 (I_d = I_2)$, resulting in the equation:

$$I_d R_f + I_d R_2 = 5 - 0.7$$

Solving for R_2, we obtain:

$$R_2 = \frac{4.3}{I_d} - R_f \qquad (6.8)$$

For a maximum diode current of $I_d = 10 \text{ mA}$ and the usual value of $R_f = 20 \ \Omega$, Equation (6.8) yields:

$$R_2 = 410 \ \Omega$$

as the minimum allowable resistance.

Employing this R_2, $V_{zi} = V_H = 3 \text{ V}$, and the previous OR gate resistance, i.e., $R_1 = 4 \text{ k}\Omega$, Equation (6.7) now yields:

$$3 = \frac{(4.3)4/n}{4/n + 0.41}$$

Therefore, $n = 4.22$, which means that the diode current limitation has reduced the fan-out from 5 to 4.

We will consider the way other constraints influence diode gate design in the Exercises.

Elimination of Offset Voltage

Frequently the offset of 0.7 V (between input and output) observed in both the OR and AND gates (Figures 6.2 and 6.3) is objectionable. The circuit shown in Figure 6.10 provides a simple solution to this problem through the use of another diode to produce downshifting. In order to gain a basic understanding of the gate, assume that all voltage drops associated with the input circuit are negligible except the diode cut-in voltage, V_y. Then, if 5 V sources are placed at the x and y inputs, and if diodes D_x and D_y are forward biased, point J goes to 5.7 V. Under these conditions, D_z will be forward biased with a 0.7 V drop across it. This results in the output being at 5.0 V, which is what the last line of the truth table in Figure 6.10 shows.

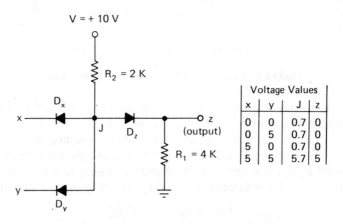

Figure 6.10   Modified AND Gate and Its Truth Table.

The other three lines of that truth table follow in a similar way, with the result that the AND gate is achieved without offset or attenuation. The offset has been eliminated by D_z which, through its back-to-back connection with the other diodes, counters the effect produced by D_x and D_y. This shifting technique is very useful and is employed in several commercial gates, which will be discussed later. On the other hand, the absence of attenuation is simply due to the fact that the source resistance was assumed sufficiently low that the input current produced a negligible voltage drop. (As previously mentioned, this frequently is not a good assumption in practice.)

One of the critical points concerning the proper operation of the modified AND gate is that D_x and D_y always remain forward biased. This can be assured if the open-circuit voltage, V_J, at point J, without D_x and D_y connected, is at a higher potential than V_s. The open-circuit voltage can be calculated from the voltage-divider rule:

$$V_J = \frac{(V - 0.7)R_1}{R_1 + R_2} + 0.7$$

where V is the power supply voltage. Therefore, the requirement on V_s is:

$$V_s < V_J - 0.7$$

or

$$V_s < \frac{(V - 0.7)R_1}{R_1 + R_2} \tag{6.9}$$

For the parameters of Figure 6.10

$$V_s < \frac{(10 - 0.7)4}{4 + 2} = 6.2 \text{ V}$$

which means that the circuit should operate satisfactorily since the source employed was $V_s = 5$ V. (A similar OR circuit which eliminates the offset voltage will be considered in the Exercises.)

‡6.2 TRANSISTOR GATES

In this section, a simple transistor model is developed, and then a number of practical gate circuits are analyzed on the basis of the model.

Transistor Functional Models

Junction transistors, such as those shown in Figure 6.11, are now mass-produced at a very low cost—many common types sell for under 50 cents each. The base-emitter junction, shown in Figure 6-11(a), resembles the diode in Figure 6.1(a) in many ways. It conducts current only when the p-terminal is made positive by at least 0.7 V (in silicon) with respect to the emitter. (Note that the corresponding circuit symbol has an arrow marking the emitter terminal, pointing in the direction of easy current flow.) The collector-base junction also resembles a diode, so that it becomes forward biased when the base is more positive than the collector. (These polarities are exactly reversed for the pnp transistor in Figure 6.11(b).) The major advantage, however, that a transistor has over a diode is that the transistor exhibits *current gain*. A small amount of current flowing in the base-emitter circuit can control a large amount of current in the collector-emitter circuit. This will be explained more fully later.

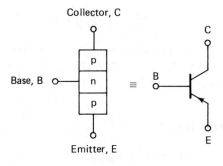

Collector, C

Base, B

Emitter, E

(a) NPN Transistor

Collector, C

Base, B

Emitter, E

(b) PNP Transistor

Figure 6.11   Junction Configurations and Standard Symbols for the Two Transistor Types.

Because of their wide use in digital circuitry and for simplicity, silicon junction transistors in *common-emitter* configuration, as shown in Figure 6.12(a), will be emphasized in our discussion. Moreover, only the *cutoff* and *saturation* regions of the transistor's characteristics are employed in most digital circuits. Cutoff is basically the condition where the transistor is fully off; in silicon, cutoff occurs at $V_{BE} = 0$ volts and $I_C = I_B = I_E = 0$ amperes. Saturation occurs when both the collector junction and the emitter junction are forward biased. In this situation, to a first approximation a transistor can be represented as the two back-to-back diodes shown in Figure 6.13. Since, by definition, both junctions are forward biased, we would expect that $V_{CE,\text{sat}} \approx 0$. For silicon, a better approximation is $V_{CE,\text{sat}} = 0.2$ V, which is the value we will use here.

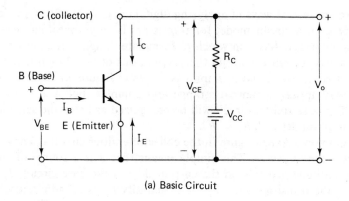

(a) Basic Circuit

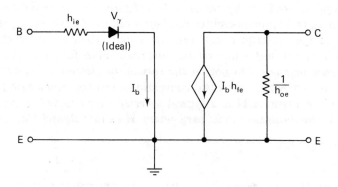

(b) Simplified Common Emitter Model

Figure 6.12   The npn Transistor in a Common-Emitter Configuration.

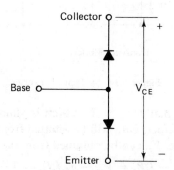

Figure 6.13   Two-Diode Model for an npn Transistor.

There are several ways of analyzing digital circuits; but from an elementary approach, a circuit model for transistors is very appealing because it provides a concrete basis on which to formulate designs. One must exercise care, however, to remain within the assumptions of the chosen model. The simplified model we intend to employ is shown in Figure 6.12(b). At first the moldel may appear somewhat strange and complex, but it is really quite simple. The base-emitter circuit will be recognized as being the same as our diode circuit model with $h_{ie} = R_f$.

The diamond-shaped figure in the collector-emitter circuit is known as a *dependent current source*. This simply means that it is a source of current whose magnitude depends on the current flow in the base circuit, I_b, multiplied by the transistor gain, h_{fe}. The quantity h_{oe} is an admittance; thus $1/h_{oe}$, shown on the diagram, is the corresponding resistance. (The quantities h_{ie}, h_{oe}, and h_{fe} are formally known as *hybrid parameters*, with the e subscript representing the common-emitter configuration, the i subscript representing input, and the subscript o representing output. These standard electronic symbols are employed so that we may find their values for specific transistors in reference manuals. Do not let the formalism create any mystery about their nature—they are simply two conventional resistors and a gain.)

Shown in Figure 6.14 is a typical inverter circuit including reasonable values for the transistor model parameters. Note that R_c and $1/h_{oe}$ are effec-

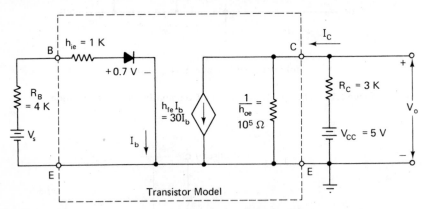

Figure 6.14 A Typical Inverter.

tively in parallel and that $1/h_{oe} \gg R_c$, which is almost always true for our cases of interest. Therefore, $1/h_{oe}$ will be omitted from most future diagrams and the output voltage, V_o, can be obtained from the equation:

$$V_o = V_{cc} - I_c R_c = V_{cc} - h_{fe} I_b R_c = 5 - (30)(3)I_b \qquad (6.10)$$

where all currents are in milliamperes (mA) and the resistance is in thousands of ohms (kΩ).

If $V_o = 0.2$ V when the transistor is in saturation, Equation (6.10) may be solved for the I_b which brings the transistor to the edge of saturation

$$I_b = \frac{0.2 - 5}{-90} = 0.053 \text{ mA}$$

This results in a voltage drop at saturation, $V_{i,\text{sat}}$, across h_{ie} of

$$V_{i,\text{sat}} = (1)(0.053) = 0.053 \text{ V}$$

As the transistor is driven further into saturation, I_b increases causing $V_{i,\text{sat}}$ to increase; and a reasonable approximation is $V_{i,\text{sat}} = 0.1$ V. For this reason, it is common practice in the design of digital circuits to neglect h_{ie}, but to assume $V_{BE,\text{cut-in}} = 0.7$ V and $V_{BE,\text{sat}} = 0.7 + 0.1 = 0.8$ V.

Following the above considerations, the very simple transistor models shown in Figure 6.15 can be constructed.

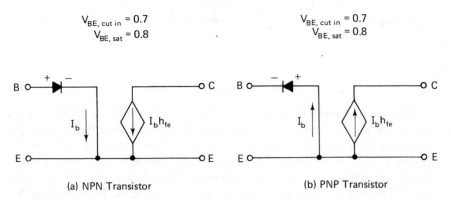

(a) NPN Transistor   (b) PNP Transistor

Figure 6.15 Approximate Models for the Common-Emitter Configuration.

Diode-Transistor Logic (DTL)

Since considerable gain can be obtained from an inverter, a straightforward method for improving the performance of the previously discussed diode gates is to follow each of them with an inverter. The AND gates then become NANDs and the OR gates become NORs. This is not a problem because it has already been proven in Chapter 4 that either of these gates can be employed to produce any desired function. However, our main interest is in gates which can be economically mechanized by integrated circuits, where transistors and diodes are very inexpensive to implement but large capacitors and resistors, $C \geq 100$ pF and $R \geq 10$ kΩ, are expensive. (These conditions are the opposite of those observed in discrete-component circuits.) For this reason, a modified DTL NAND gate, such as that shown in Figure 6.16, is the starting point for IC design.

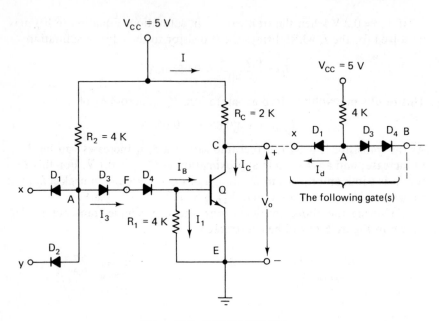

Figure 6.16   DTL NAND Gate.

We will first verify that the circuit is indeed a NAND gate. Then we will determine the operating regions for the gate and calculate the fan-out. In analyzing the circuit of Figure 6.16, we will assume the usual diode cut-in pontetial of 0.7 V and neglect the small voltage drop across R_f. Now, if both inputs are at 0 V, D_1 and D_2 will be forward biased, and point A will reach 0.7 V. This will not be sufficient for D_3 and D_4 to come on. Point B will go to 0 V, Q will be cut off, and V_o will then reach the supply potential of 5 V. These conditions are summarized in the first line of Table 6.2.

Table 6.2   NAND Gate Performance

			Voltage* and State Values					Positive Logic Values		
x	y	D_1	D_2	A	D_3 & D_4	B	Q	V_o	x	y	V_o
0	0	On	On	0.7	Off	0	Cutoff	5	0	0	1
0	5	On	Off	0.7	Off	0	Cutoff	5	0	1	1
5	0	Off	On	0.7	Off	0	Cutoff	5	1	0	1
5	5	Off	Off	2.2	On	0.8	Saturation	0.2	1	1	0

*Ground reference.

On the other hand, if it were thought that 0 V at the inputs would turn Q on, the following conclusion would be true: point B would require 0.7 V;

diodes D_3 and D_4 would need to turn on, requiring point A to be at $B +$ 2(0.7) = 2.1 V. With x at 0 V, point A cannot be at 2.1 V, which means that the assumption of Q turning on has led to a contradiction.

In a similar way, the other three lines of Table 6.2 may be verified. In the fourth line of the table, note that the assumption of Q in saturation leads to point A being at 0.8 + 2(0.7) = 2.2 V, which is consistent with D_1 and D_2 being off. (Here, sufficient current must be provided by the power supply through the resistor R_2 to saturate the transistor. This point will be verified shortly.)

Gate Operating Region

Table 6.2 provides some insight concerning the operation of the DTL gate. But the fourth line indicates $V_o = 0.2$, and with the assumed input values of 0 and 5 V, one might think that some of the offset and perhaps attenuation) problems might be present here as was found in diode gates. Using a method similar to that employed in analyzing Figure 6.16 and ap-plying the same increasing voltage to both inputs, we see that the voltage at point A follows the input but is biased ahead by a diode drop of 0.7 V. Now, in order for Q to reach cut-in, point A must be at 0.7 + 2(0.7) = 2.1 V, which means that the input $x = y = 2.1 - 0.7 = 1.4$ V. Within the model, saturation is reached at 0.1 V higher than cut-in, which leads to $x = y = 1.5$ V. Moreover, all input voltages higher than this will merely drive Q further into saturation; thus, a first-order approximation of the gate's per-formance is given in Figure 6.17.

The operating regions are marked on the diagram to show typical condi-tions rather than absolute limits. Thus, $V_o = 0.2$ indicated in Table 6.2 will easily drive another stage, like itself, into saturation. This means that DTL

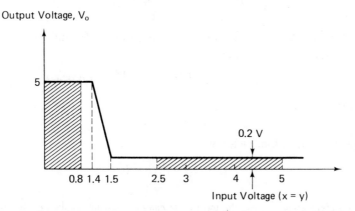

Figure 6.17 DTL Gate Transfer Characteristics (shaded areas show typical operating characteristics).

gates, in contrast to our previous experience with diode gates, may be employed in multilevel logic without detrimental effects. The DTL gate also provides reasonable resistance to noise. Assume that all of the inputs start at 5 V. Then Figure 6.17 indicates that an input could drop to 1.5 V before Q comes out of saturation. This means that a noise spike as large as $5 - 1.5 = 3.5$ V could exist without affecting gate performance. The term *noise margin** is often used to indicate the magnitude of noise required to cause erroneous operation (3.5 V in this case). On the other hand, if at least one input is driven by the output (0.2 V) of a saturated transistor, Figure 6.17 shows that the input could reach 1.4 V before Q reaches cut-in. This means that a noise spike as large as $1.4 - 0.2 = 1.2$ could exist without affecting gate performance.

Fan-out

Now consider the fan-out capabilities of the DTL gate in terms of the number, N, of gates (similar to itself) which it can drive. When Q is cut off, $V_o = 5$ V and the input diode D_1 of the next state will be off. Thus, the next gate draws no current and does not load the gate under study. For these reasons, the loading effects on Q need only be considered when the transistor is in saturation. At that time, the collector current, and hence the base current, must be sufficient to place Q in saturation and to sink the current for the N gates it is driving. An equation describing this situation may be obtained from Figure 6.16:

$$h_{fe}I_B = I_C = I + NI_d \qquad (6.11)$$

where I is the current required to saturate Q and I_d is the current from one of the gates being driven. The value for I is easily determined from:

$$I = \frac{V_{cc} - V_o}{2} = \frac{5 - 0.2}{2} = 2.4 \text{ mA} \qquad (6.12)$$

With $V_o = 0.2$ V driving an identical gate, diodes D_3 and D_4 of the gate being driven will be off; thus,

$$I_d = \frac{V_{cc} - (0.2 + 0.7)}{4} = \frac{5 - 0.9}{4} = 1.025 \text{ mA} \qquad (6.13)$$

From the circuit:

$$I_B = I_3 - I_1 \qquad (6.14a)$$

but Q is in saturation, yielding:

$$I_1 = \frac{V_B}{4} = \frac{0.8}{4} = 0.2$$

*In a more general sense, the term *noise margin* is usually applied to the worst-case conditions expected for a batch of integrated circuits, e.g., the difference between the minimum expected output from one stage and the minimum input to a similar stage which is expected to drive that stage into saturation. For DTL this is typically 0.7 V.

Now with the inputs at 5 V, diodes D_1 and D_2 are cut off, and point A is at 2.2 V, which leads to:

$$I_3 = \frac{V_{cc} - 2.2}{4} = \frac{5 - 2.2}{4} = 0.7 \text{ mA}$$

When these values are substituted into Equation (6.14a), the result is:

$$I_B = 0.7 - 0.2 = 0.5 \text{ mA} \qquad (6.14b)$$

After substituting Equations (6.12), (6.13), and (6.14b) into Equation (6.11), we find:

$$(30)(0.5) = 2.4 + N(1.025)$$

Therefore, $N = 12.6/1.025 = 12.3$, which means a fan-out of 12 may be achieved.

Direct-Coupled Transistor Logic (DCTL)

In the cutoff/saturated mode that has been assumed here, the transistor acts much like a conventional switch. Thus, the question arises whether we can directly stack transistors in series or parallel to do logic operations, as can be done with ordinary switches. A NOR circuit following this concept is shown in Figure 6.18. Since it is assumed that the gate inputs come from the outputs of similar gates, values of 0.2 V and 3 V are employed in the truth table in Figure 6.18(b). From the logic portion of the table:

$$V_o = \bar{x}\bar{y} = \overline{x + y}$$

which verifies that a NOR gate is realized. One of the major disadvantages of this type of circuit is that the noise margin is low. Transistor cut-in is at 0.7 V and an input low is 0.2 V; thus, a noise spike of $0.7 - 0.2 = 0.5$ V will cause erroneous operation. In addition, there is danger of *current hogging* with

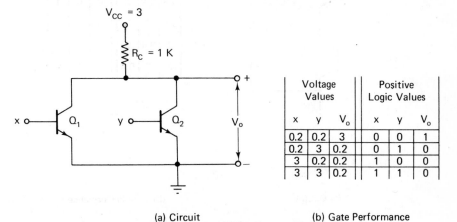

Voltage Values			Positive Logic Values		
x	y	V_o	x	y	V_o
0.2	0.2	3	0	0	1
0.2	3	0.2	0	1	0
3	0.2	0.2	1	0	0
3	3	0.2	1	1	0

(a) Circuit   (b) Gate Performance

Figure 6.18 DCTL NOR Gate.

parallel-coupled DCTL. This condition occurs in multilevel logic when one gate on a lower level tries to drive two separate gates on another level. If the two transistors being driven have different values for $V_{BE,\text{sat}}$, the one that has the lowest value will tend to clamp the common input voltage, and the second transistor may remain near its cut-in value. Thus, the second transistor may not get sufficient current to reach saturation, leading to $V_o \gg 0.2$ V.

A DCTL series circuit is shown in Figure 6.19. In the truth table portion of the figure, we see that the $V_{CE,\text{sat}}$ voltages add when both transistors are in saturation, resulting in $V_o = 0.4$ V; thus, input voltages of 0.4 V and 3 V are assumed. From the logic portion of the table, we see that

$$\bar{V}_o = xy \quad \text{or} \quad V_o = \overline{xy}$$

which verifies that a NAND gate is realized.

The noise margin of this gate is even worse than that for the NOR gate in that a noise spike of $0.7 - 0.4 = 0.3$ V will cause erroneous operation. Moreover, if we try to stack more transistors in series to produce a three- or four-input gate, the $V_{CE,\text{sat}}$ buildup becomes intolerable. Another disadvantage of the series gate configuration is the fact that the transistor farthest

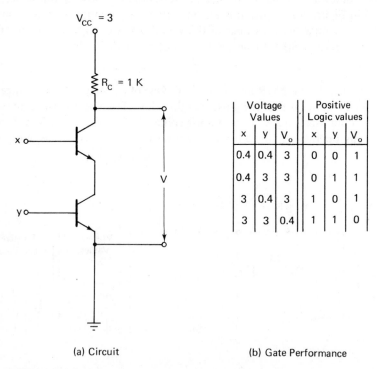

| (a) Circuit | (b) Gate Performance |

Figure 6.19   DCTL NAND Gate.

from ground requires a larger base-to-ground voltage to drive it into saturation. As a result of this, a multilevel logic circuit having a cutoff gate on a lower level supplying two different gates on another level would experience the problem of current hogging if the two transistors being supplied were not the same position above ground. Thus, both the series and the parallel DCTL bipolar configurations have serious limitations which prevent them from being employed in standard integrated circuits. (Later in the chapter we will find that the configuration is much more satisfactory with unipolar transistors.)

Resistor-Transistor Logic (RTL)

Having observed in the previous section that DCTL is not usable as a standard bipolar configuration, we will now attempt to eliminate the disadvantages through a simple modification. By placing a resistor in series with each input, we can eliminate the problem of current hogging; in trying to draw a large share of current, the transistor will also produce a voltage drop across its base resistor, causing a larger voltage to appear at the other transistors.

The main advantage of the RTL family of gates is low cost. Of all the standard bipolar configurations employed in integrated-circuit mechanization, it requires the smallest space on the silicon wafer.

A typical medium-power RTL gate employed in practical integrated circuits is shown in Figure 6.20. Note that both inputs at 0 V cut off the transistors and cause the unloaded output to reach the power supply potential of 3 V. On the other hand, one or more of the inputs at 3 V cause that transistor(s) to go into saturation, resulting in $V_o = 0.2$ V. Thus, Figure 6.20(b) shows that $V_o = \bar{x}\bar{y} = \overline{x + y}$, which means that a NOR gate has been realized.

One of the problems with the RTL circuit is that V_o falls quite rapidly as it is loaded with succeeding RTL gates. Figure 6.20(c) shows the equivalent circuit for a gate having a high output and a load of N succeeding gates. From the figure, V_o can be calculated:

$$V_o = 0.8 + \frac{2.2(450/N)}{640 + 450/N} = 0.8 + \frac{(2.2)(450)}{640N + 450} \qquad (6.15)$$

As expected, for no loading ($N = 0$), $V_o = 0.8 + 2.2 = 3.0$; and as N increases, V_o approaches 0.8 V. A graph of this function is shown in Figure 6.21. In practice, the maximum fan-out for this circuit is 5, which means that $V_o \approx 1$ V (actually 1.07 V). Since 0.8 V is the minimum signal which will drive the succeeding gate into saturation, a fairly small noise spike (1.07 − 0.8 = 0.27 V) will cause erroneous circuit operation.

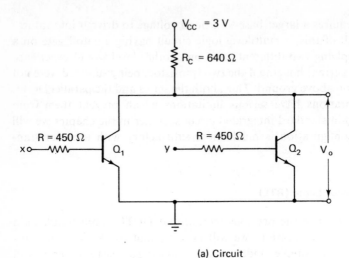

(a) Circuit

Voltage Values			Positive Logic values		
x	y	V_o	x	y	V_o
0	0	3	0	0	1
0	3	0.2	0	1	0
3	0	0.2	1	0	0
3	3	0.2	1	1	0

(b) Gate Performance

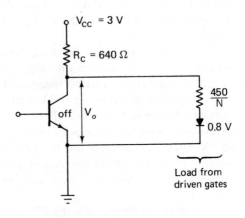

Load from driven gates

(c) Equivalent Circuit for Fan-Out Calculations

Figure 6.20   RTL NOR Gate.

Using the calculated input potential of 1.07 V and the usual value of $V_{BE,sat} = 0.8$, we obtain the required base current:

$$I_B = \frac{1.07 - 0.8}{0.45} = 0.6 \text{ mA}$$

But

$$I_C = \frac{3.0 - 0.2}{0.64} = 3.38 \text{ mA}$$

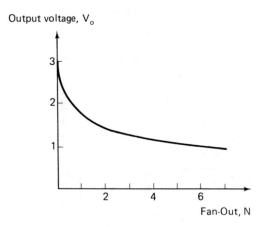

Figure 6.21   Fan-Out Characteristics of a Typical RTL Gate.

therefore, the required transistor gain is:

$$h_{fe,\min} \geq \frac{I_C}{I_B} = \frac{3.38}{0.6} = 5.63$$

Transistor-Transistor Logic (TTL or T^2L)[3,8,9]

With integrated circuit technology, it is easy to produce multiple emitters on a single transistor. Configurations of this type are called TTL and have the fastest response of the saturating logic group. A basic circuit for the TTL family is shown in Figure 6.22. Considering the two-diode model of the transistor, we see that Figure 6.22 has the same labeling (points A, F, X, etc.) and the same topology as Figure 6.16, where Q_1 replaces D_1, D_2, and D_3. The truth tables for the two circuits are also similar, as a comparison between Figure 6.22(b) and Table 6.2 shows. In Figure 6.22(b), note that with both inputs low the two emitter junctions behave like a pair of forward-biased diodes; hence point A is a diode drop above the input (i.e., $0.2 + 0.7 = 0.9$ V). In order for Q_2 to reach the cut-in voltage, however, it requires that the B-C junction of Q_1, D_4, and the B-E junction of Q_2 be on. This, in turn, requires $3(0.7) = 2.1$ V at point A, which is considerably more than the 0.9 V available. Thus, Q_2 remains off and the unloaded output reaches the power supply potential of 5 V. It is easy to see from the figure that the above conclusions continue to hold as long as at least one input is at a low value.

When all inputs are at 5 V, Figure 6.22(b) shows that the B-E junction of Q_1 will be off (reverse biased). Under these conditions, point A will rise toward the supply voltage, turning Q_2, D_4, and the B-C junction of Q_1 on. Thus, point A will reach $0.7 + 0.7 + 0.8 = 2.2$ V, and with Q_2 in saturation,

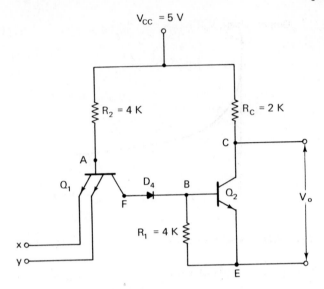

(a) Elementary Circuit

		Voltage and state values							Positive Logic values		
x	y	B-E Junction	B-C Junction	A	B	Q_2	V_o		x	y	V_o
0.2	0.2	on	off	0.9	0	off	5		0	0	1
0.2	5	on	off	0.9	0	off	5		0	1	1
5	0.2	on	off	0.9	0	off	5		1	0	1
5	5	off	on	2.2	0.8	sat.	0.2		1	1	0

(b) Gate Performance

Figure 6.22 TTL Gate Principles. (The voltages shown for V_0 are unloaded values.)

$V_o = 0.2$ V. It is easily seen from the table that the function realized is a NAND ($\bar{V}_o = xy$ or $V_o = \overline{xy}$).

Passive vs. Active Pull-Up

A circuit such as the one in Figure 6.22 is said to possess *passive pull-up* because the transistor does not carry current when V_o goes high. (V_o is determined only by the passive part of the circuit.) If a circuit containing an "on" transistor is employed to achieve V_o high, *active pull-up* is said to exist. With the same limitations on transistor current and other parameters, active pull-up is able to dirve heavire loads (from both transistor states); it leads to faster circuit response; and it requires less power.

In order to achieve basic understanding of these two types of output circuits, let us consider a switch model for the transistors, i.e., an on transistor will be considered to have zero resistance (between collector and emitter) and an off transistor will be considered to have infinite resistance. The circuits for passive and active pull-up are shown in Figure 6.23(a) and (b), respectively. Assume that the transistors have a current limit of 10 mA and it is desired to obtain a nominal high output state of $V_o = 4$ V. Let us determine how heavy a load (or how large a fan-out) the circuit can supply, i.e., find the smallest possible value for R_L.

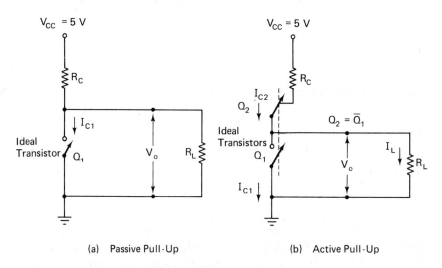

(a)   Passive Pull-Up   (b)   Active Pull-Up

Figure 6.23 Switch Model for Two Types of Output Stages. (Q_1 and Q_2 represent logic states of the corresponding transistors.)

From the circuit in Figure 6.23(a), the $Q_1 = 1$ (transistor on or switch closed) condition for maximum allowable transistor current leads to

$$\frac{5}{R_C} = I_{C1} \leq 10 \text{ mA}$$

(It is assumed that the transistor is not required to sink any current from the load.) Therefore,

$$R_C \geq 0.5 \text{ k}\Omega \tag{6.16}$$

For $Q_1 = 0$:

$$V_o = \frac{5R_L}{R_L + R_C} = \frac{5}{1 + R_C/R_L} = 4 \, V$$

therefore,

$$R_L = 4R_C \tag{6.17}$$

Substituting Equation (6.16) into (6.17):

$$R_L \geq 4(0.5) = 2 \text{ k}\Omega \tag{6.18}$$

In contrast to the above results, consider the simplified model for active pull-up shown in Figure 6.23(b). With $Q_1 = 0$ ($Q_2 = \bar{Q}_1 = 1$) and recalling the transistor current limit of 10 mA:

$$\frac{5}{R_C + R_L} = I_{C2} \leq 10 \text{ mA}$$

therefore,

$$R_C + R_L \geq 0.5 \text{ k}\Omega \tag{6.19}$$

From the requirement of a nominal upper value for V_o of 4 V:

$$V_o = \frac{5R_L}{R_L + R_C} = \frac{5}{1 + R_C/R_L} = 4 \text{ V}$$

therefore,

$$R_L = 4R_C \tag{6.20}$$

Substituting Equation (6.20) into (6.19):

$$R_C + 4R_C \geq 0.5 \text{ k}\Omega$$

thus,

$$R_C \geq 0.1 \text{ k}\Omega \tag{6.21}$$

and from Equation (6.20):

$$R_L \geq 0.4 \text{ k}\Omega \tag{6.22}$$

Comparing Equations (6.18) and (6.22), we see that the active circuit can handle a load which is five times as heavy as passive pull-up. In addition, the active circuit can sink 10 mA when delivering the lower value for V_o (here $Q_1 = 1$ and $Q_2 = 0$). If the passive circuit is required to sink any current from the load under similar conditions, R_C must be increased and the load must be even lighter than specified by Equation (6.17), i.e., $R_L > 2 \text{ k}\Omega$.

A Practical TTL Circuit

A practical TTL circuit employing active pull-up is shown in Figure 6.24. Here Q_1 has the same role as it did in the elementary circuit of Figure 6.22(a); but the diode D_4 is replaced by the transistor Q_3, which also assumes the task of providing an inverted input to Q_4. (The requirements of supplying an input to Q_2 and the inverse to Q_4 gives the circuit associated with Q_3 the name *phase splitter*.) Active pull-up is employed in the output circuit with one transistor stacked upon a second, resulting in the term *totem-pole output*.

Voltage and State Values

x	y	A	Q_2	Q_3	B	D	Q_4	V_o
0.2	0.2	0.9	off	off	0	≈ 5	on	≈ 3.6
0.2	$\geqslant 1.6$	0.9	off	off	0	≈ 5	on	≈ 3.6
$\geqslant 1.6$	0.2	0.9	off	off	0	≈ 5	on	≈ 3.6
$\geqslant 1.6$	$\leqslant 1.6$	2.3	sat.	sat.	0.8	1	off	0.2

(a) Circuit Performance
(The voltages shown for V_o are unloaded values.)

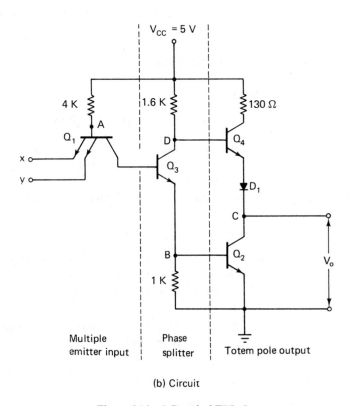

(b) Circuit

Figure 6.24 A Practical TTL Gate.

When comparing the performance tables for the elementary and practical TTL circuits, we see that the voltage values for points A and B are the same,* as is the performance of Q_1. Moreover, the differences observed between the

*The voltage at point A shown in the table for the practical gate is 2.3 V instead of 2.2 V because the saturated transistor Q_3 yields approximately 0.1 V greater drop than diode D_4.

tables are a matter of degree rather than fundamental nature. For example, the totem-pole output circuit leads to the unloaded high output being $V_o \approx$ 3.6 V instead of 5 V; but of course, the gate still realizes the NAND function. When transistor Q_3 is off, point D rises toward the supply voltage, turning Q_4 and D_1 on. With the output lightly loaded, Q_4 and D_1 are near their cut-in values, and the very small base current produces negligble drop across the 1.6 kΩ resistor. This results in $V_o \approx 5 - 0.7 - 0.7 \approx 3.6$ V. When transistor Q_3 is on, there is sufficient base current to drive it into saturation. Taking into account that Q_2 is also in saturation, we obtain:

$$D = 0.8 + 0.2 = 1.0 \text{ V}$$

Since point A must be at 2.3 V in order to saturate Q_2 and Q_3, the inputs must be no more than one diode drop less; thus, the table shows values of $\geq$1.6 V. Of course, to allow for noise and other practical considerations, the inputs should actually be considerably larger.

Emitter-Coupled Logic (ECL)[4]

Emitter-coupled logic is the highest-speed circuitry in common use for digital data processing. The main reason for the improved performance is that the delay enocuntered by driving transistors into saturation, as is done with TTL and other common types of logic, is avoided. Gate delay in the neighborhood of 1 ns can be obtained.

An elementary ECL gate circuit is shown in Figure 6.25. The basic principles employed in ECL are:

1. Transistors are never allowed to reach saturation.

2. The current through a common-emitter resistor, R_E, is supplied either entirely by the reference transistor, Q_3, or by one (or both) of the input transistors, Q_1 or Q_2. (The family name, emitter-coupled logic, comes from this fact.)

If the inputs are low, say -1.58 V, Q_1 and Q_2 will be cut off, resulting in Q_3 being in the active region. We will asume that the resulting base-emitter voltage drop for an active transistor is 0.75 V; thus,

$$V_E = V_{\text{ref}} - 0.75 = -1.18 - 0.75 = -1.93 \text{ V}$$

From this current, I_E, is found to be:

$$I_E = \frac{V_E - V_{EE}}{R_E} = \frac{-1.93 - (-5.2)}{1.18 \times 10^3} = 2.76 \text{ mA}$$

Since the other two transistors are cut off, all of I_E must pass through R_{C2}, resulting in:

$$V_{02} = -(0.3)(2.76) = -0.83 \text{ V}$$

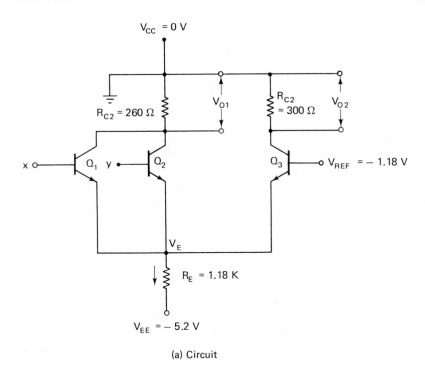

(a) Circuit

								Positive Logic values			
		Voltage and State Values									
x	y	V_E	Q_1	Q_2	Q_3	V_{O1}	V_{O2}	x	y	V_{O1}	V_{O2}
− 1.58	− 1.58	− 1.93	off	off	active	0	−0.82	0	0	1	0
− 1.58	− 0.75	− 1.50	off	active	off	− 0.82	0	0	1	0	1
− 0.75	− 1.58	− 1.50	active	off	off	− 0.82	0	1	0	0	1
− 0.75	− 0.75	− 1.50	active	active	off	− 0.82	0	1	1	0	1

(b) Gate Performance

Figure 6.25 Elementary ECL Gate.

Also, the cut-off transistors prevent current from flowing in R_{C1}, resulting in $V_{01} = 0$ V.

Now, if at least one of the inputs goes high, say -0.75 V, its transistor will go into the active region, resulting in:

$$V_E = \text{(Input Voltage)} - V_{BE} = -0.75 - 0.75 = -1.5 \text{ V}$$

Then:

$$I_E = \frac{V_E - V_{EE}}{R_E} = \frac{-1.5 - (-5.2)}{1.18 \times 10^3} = 3.14 \text{ mA}$$

Now, all of I_E flows through R_{C1}, resulting in:

$$V_{01} = -(0.26)(3.14) = -0.82 \text{ V}$$

Since Q_3 is cut off, no current flows through R_{C2}, resulting in $V_{02} = 0$ V.

The above facts are summarized in the table shown in Figure 6.25(b). Now the circuit actually realizes two functions: $V_{01} = \bar{x}\bar{y} = \overline{x + y}$, which is a NOR; and $\bar{V}_{02} = xy$ or $V_{02} = \overline{\bar{x}\bar{y}} = x + y$, which is an OR. Thus, the gate contains two outputs that are complements of each other, and the gate symbol is represented by:

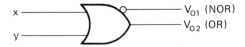

A practical version of Figure 6.25 is shown in Figure 6.26. Although the new figure seems much more complicated, it is exactly the same as the previous one except for a stage (Q_5, D_1, and D_2) which implements the voltage V_{ref} and the emitter-follower output stages (Q_4 for the V_{01} output and Q_6

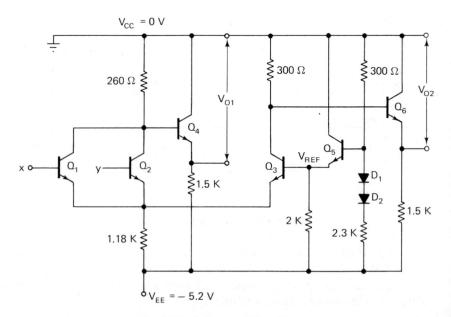

Figure 6.26 Practical ECL Gate.

for the V_{02} output). While V_{ref} could be generated simply by a pair of resistors forming a voltage divider, the diodes provide a measure of compensation for temperature variations. The emitter followers provide a low output impedance; and since they operate in the active region, the previous outputs [from Figure 6.25(b)] are merely shifted down by -0.75 V. Thus, the outputs from the practical circuit are $0 - 0.75 = -0.75$ V for the higher value and $-0.83 - 0.75 = -1.58$ V for the lower value. With these shifts, the gate inputs and outputs are compatible, which means that multilevel logic, with the output of one ECL gate feeding another ECL gate, can be implemented. Moreover, since the emitter followers only shift the output levels, the same functions are mechanized as shown in Figure 6.25(b).

Wired Logic

From an economic standpoint and because of the simplifications produced it is frequently desirable to directly connect the outputs of certain gates as shown in Figure 6.27. In this case the wired OR function is indicated; but wired AND is also very common. On the other hand, wired logic cannot be implemented by certain types of circuitry. Thus, in practical design situations one must first ask whether wired logic is possible and then ask what resulting function is produced. We now ask these questions concerning the DTL gate shown in Figure 6.16.

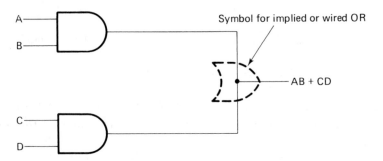

Figure 6.27   Wired Logic.

It appears that the gate can be divided into a diode stage (including D_1 through D_4) and an inverter stage (including Q and R_C). This scheme is schown in Figure 6.28 for the wired logic connection of two gates, where the dashed outline indicates each actual DTL NAND gate. The circuit reveals that the direct connection of the two gates effectively places the collector resistors in parallel, which means that more power is dissipated when only one transistor is on; other than that, no detrimental effects occur. (Some DTL gates are constructed with open collectors so that a single collector resistor of proper size can be added during wiring.) The table in Figure 6.28(b) demonstrates the wired logic performance and results in $V_o = \overline{x^*y^*}$. But the logic values at the gate terminals are $x = \overline{x^*}$ and $y = \overline{y^*}$; thus, $V_o = xy$,

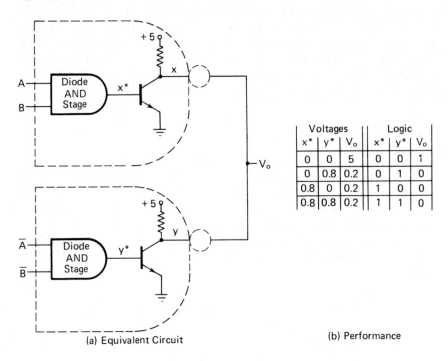

Voltages			Logic		
x^*	y^*	V_o	x^*	y^*	V_o
0	0	5	0	0	1
0	0.8	0.2	0	1	0
0.8	0	0.2	1	0	0
0.8	0.8	0.2	1	1	0

(a) Equivalent Circuit   (b) Performance

Figure 6.28   DTL Wired Logic Connection.

which means that *wired logic produces AND for DTL gates.* Going one step further, for the circuit shown, we see that $x = \overline{AB}$ and $y = \overline{\bar{A}\bar{B}}$, which means:

$$V_o = (\overline{AB})(\overline{\bar{A}\bar{B}}) = (\bar{A} + \bar{B})(A + B) = \bar{A}B + A\bar{B} = A \oplus B$$

Thus, it is easy to produce the EXCLUSIVE OR function with only two DTL gates.

Wired logic yields the OR function with ECL gates and the AND function for some RTL gates. On the other hand, a TTL gate in the high state with its output directly wired to the output of a TTL gate in the low state results in a large current flow between the two gates, as shown in Figure 6.29. Thus, wired logic is not feasible for TTL.

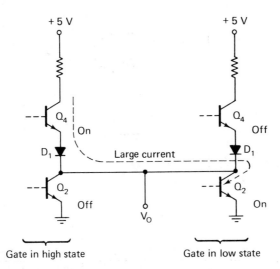

Figure 6.29 Difficulty with Wired Logic in TTL Gates. (Note that only totem-pole stages are shown.)

6.3 FLIP-FLOP REALIZATION METHODS

Flip-flops can be designed directly from transistors,[10] but the logic principles upon which various types of flip-flops are based can be best understood if gates are used as building blocks. Moreover, in many practical integrated circuits, gate flip-flops are used. Figure 6.30 demonstrates how an elementary *S-R* circuit can be constructed from two NAND gates and two inverters. The following facts should be observed concerning the logic diagram and the associated truth table.

1. The negated input OR form for each NAND is used to make it easier to visualize circuit action.

2. Pairs of inverters in the diagram have a slash drawn through them to emphasize that they effectively cancel each other.

3. The variables *a*, *b*, *c*, and *d* at the gate inputs represent logic values after inversion.

4. In the truth table, the last column indicates the sequence followed as the logic variables change in response to a particular input.

5. The flip-flop starts in the 0 state, and during time interval 2 the switching of *S** from 0 to 1 causes the variables ($S^*ayc\bar{y}b$) to change in a sequence, which eventually terminates with the flip-flop in the 1-state.

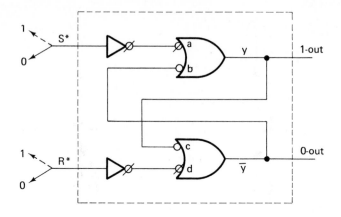

Time Interval	S*	R*	a	b	c	d	y	ȳ	Operation Sequence
1	0	0	0	0	1	0	0	1	(F.F. initially in 0-state)
2	1	0	1	1	0	0	1	0	S* aycȳb
3	0	0	0	1	0	0	1	0	S*a
4	0	1	0	0	1	1	0	1	R* dȳbyc
5	0	0	0	0	1	0	0	1	R* d

Figure 6.30   Elementary *S-R* Flip-Flop Logic Diagram and Truth Table.

6. When S^* returns to 0, only the variable a changes, and the flip-flop remains in the 1 state.

7. Similarly, the switching of R^* from 0 to 1 causes the variables ($R^*d\bar{y}byc$) to change in sequence, ending with the flip-flop in the 0 state.

8. This is called an *elementary flip-flop*† because it changes state when the input changes, as opposed to the previous circuit which changed on the trailing edge. The symbol * is used on the S and R inputs in Figure 6.30 to denote that an elementary flip-flop is considered.

If the inverters in Figure 6.30 are replaced with NAND gates, it is easy to convert the device into the clocked elementary *S-R* flip-flops shown in Figure 6.31. Because of the inverter canceling, we see that an $S^* = 1$ input immediately sets the device, $y^* = 1$, as long as the clock input, C^*, is high. The one major disadvantage of this configuration, however, is that the flip-flop outputs do not change exclusively on the leading edge of the clock pulse or on the trailing edge. On the contrary, there is no real limit on the number of times it may change state, as long as the clock is high and the S^* and R^*

†The particular form of circuit shown in Figure 6.30 is known as a *latch*, but our term *elementary flip-flop* is somewhat more general.

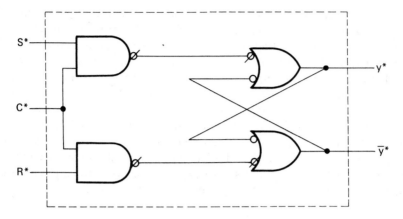

Figure 6.31 Elementary Clocked *S-R* Flip-Flop.

inputs keep changing. These properties are not allowed in many digital circuits. For example, a counter constructed from them would likely fail because the operation of one flip-flop depends on the signals from the others being constant during each clock pulse.

As we saw in Chapter 5, one way to eliminate the above problem is to employ flip-flops whose inputs respond to signals only while the clock pulse is present, but whose outputs change *only* a short time after the clock pulse has passed. These may be constructed by means of a cascade connection of two elementary clocked *S-R*'s in the so-called *master-slave* arrangement shown in Figure 6.32.

In this circuit, the presence of the clock pulse allows the master to go toward the state dictated by its inputs. The occurrence of the clock pulse also immediately disconnects the slave from the master (through the inverter at the slave's *C**-input) so that the flip-flop's output holds the previous state until the clock pulse goes low. At that time, the slave's inputs are opened, the master's are closed, and the master's state is transferred to the slave.

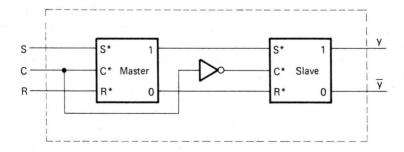

Figure 6.32 Master-Slave *S-R* Flip-Flop.

Another improvement that can be made to the elementary flip-flop is to provide *asynchronous*† preset, P_r^*, and clear, C_r^*, terminals. In this way, complicated external gating is not required when it is desired to directly load particular bits with ones or zeros. This action can be accomplished by the addition of another output to each side of the flip-flop as shown in Figure 6.33. Because of the inversion at the input of each OR, note that the non-actuating state of these asynchronous inputs is $P_r^* = C_r^* = 1$. The device is immediately cleared when $P_r^* = 1$ and $C_r^* = 0$, and it is immediately preset when $P_r^* = 0$ and $C_r^* = 1$. The symbol for the complete elementary flip-flop is shown in Figure 6.33(b). The inversion symbols have been pulled out of the logic diagram and shown explicitly in order to emphasize that asynchronous action is initiated by a low, i.e., logical zero, input.

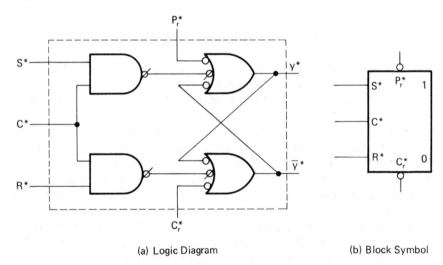

(a) Logic Diagram   (b) Block Symbol

Figure 6.33   Elementary Clocked Flip-Flop with Asynchronous Inputs.

How may elementary flip-flops be used to construct a master-slave *J-K* rather than an *S-R*? Recall that the major difference in performance between the two units is that the *J-K* flip-flop allows $J = K = 1$. This is easily arranged by feeding cross-coupled signals from the output to each input AND gate as shown in Figure 6.34. The concept here is that an input is allowed to pass on to the master flip-flop only if the opposite side of the slave is currently in the 1 state. Of course, such a constraint is consistent with $J = 1, K = 0$ for setting the device; $J = 0, K = 1$ for resetting; and $J = 1, K = 1$ for complementing. As in the master-slave *S-R* unit in Figure 6.32, the clock pulse is fed to the slave through an inverter so that the slave does not receive

†These are called asynchronous because the flip-flop is made to change state independently of the clock pulse.

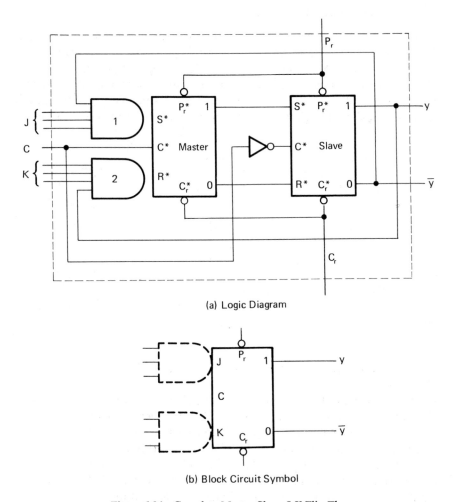

(a) Logic Diagram

(b) Block Circuit Symbol

Figure 6.34 Complete Master-Slave *J-K* Flip-Flop.

the input from the master until the negative edge of the clock pulse has passed. It should be observed that the cross-coupled feedback required to create the *J-K* characteristics comes from the slave output, since that output is the one which determines the true state of the flip-flop. Note that extra terminals are provided at both the *J* and the *K* inputs. These permit an AND operation at the *J* and *K* inputs without an external gate. (Any unused inputs should be tied to used ones so that the input gates will function properly.) Figure 6.34(b) shows the block diagram symbol employed for this unit, including the implicit AND gate terminals. (The AND gates are shown dashed because they are physically inside the flip-flop itself.)

Leading-Edge-Triggered Flip-Flops

One way to insure determinant flip-flop operations is to use master-slave devices. Another solution, which is sometimes preferable, is edge triggering. In essence, this method employs the (positive- or negative-going) edge of the clock pulse to transmit the original input state to an elementary flip-flop and to simultaneously lock out any input changes. A simplified* positive-edge-triggered (positive-going) flip-flop is shown in Figure 6.35. Gates 3 and 6

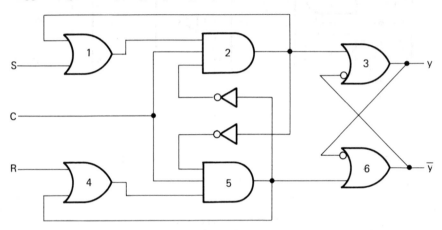

Figure 6.35   A Simplified Positive-Edge-Triggered *S-R* Flip-Flop.

form an elementary flip-flop. As long as the clock is low, gates 2 and 5 have low outputs. With the inputs at $S = 1$, $R = 0$, when the clock goes high, the outputs of gates 1 and 2 will go high, causing the flip-flop to be set. Simultaneously, the output of gate 2 will feed a high level back around the loop to gate 1, locking both of them to a high state independent of the S input. Also, the inverted output of gate 2 locks gate 5 into a low state. Thus, further inputs are ineffective and the entire circuit is static until the clock returns to a low state. Upon transition to clock low, the elementary flip-flop continues to hold its state, but the remainder of the circuit unlocks—gates 2 and 5 return to low outputs, and the outputs of gates 1 and 4 are determined only by their external inputs. The flip-flop is now ready to change state as soon as the leading edge of another clock pulse occurs.

It is easy to convert the gating within the above *S-R* flip-flop to the more popular *J-K* type. (This will be the subject of some of the exercises at the end of the chapter.) On the other hand, it is not difficult to use external gating to also achieve the conversion of one type of flip-flop to another. Figure 6.36

---

*It is termed a simplified flip-flop because, for tutorial reasons, its lock-out capability omits one case, which will be considered in the Exercises.

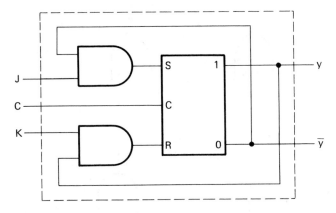

Figure 6.36 Conversion of an *S-R* to a *J-K* Flip-Flop.

demonstrates how an *S-R* may be converted to a *J-K* flip-flop. Note that the cross-coupling between the outputs and the input gates allows a high input to pass only if the flip-flop is in the opposite state to the one initiated by the input.

Noise Considerations in Flip-Flop Selection

Consider the flip-flop of Figure 6.34. One disadvantage of this master-slave concept is that the *J* and *K* inputs must not be allowed to change while the clock pulse is high; otherwise an erroneous output can be developed. (For example: with $J = 1$, $K = 0$, $y = 1$, the occurrence of the clock pulse should not change the state of the flip-flop. While the clock pulse is high, however, a momentary noise spike causing $K = 1$ would result in the master changing state and would eventually result in the erroneous flip-flop state $y = 0$.) A complete edge-triggered *J-K* flip-flop requires the inputs to be constant for only a short time (a few nanoseconds) before and after the triggering edge of the clock pulse passes and not during other phases of the clock signal. Thus, noise which occurs during the bulk of the time that the clock pulse is high has no effect on the state of the flip-flop.

A *J-K* flip-flop of this type is shown in Figure 6.37. Comparing it with the *S-R* device in Figure 6.35, we see that gates 1 through 6 serve exactly the same purpose in both units. (For example, both employ gate loops 1–2 and 4–5 to lock the flip-flop into the state existing at the positive edge of the clock pulse.) Gates 7, 8, and 9 in Figure 6.37 provide the *J-K* coupling and supply (through gate 8) a high to either the *y* or *ȳ* channel of the flip-flop so that, for *all* input combinations, the onset of the positive edge will lock the unit in one state. The reader should check several input combinations to verify this. In particular, note that $J = K = 0$, $y = 0$ yields $G_7 = G_8 = G_9 = 0$, and $\bar{G}_9 = 1$ (where $G_i \equiv$ output of gate *i*); therefore $G_4 = 1$ which locks $G_5 =$

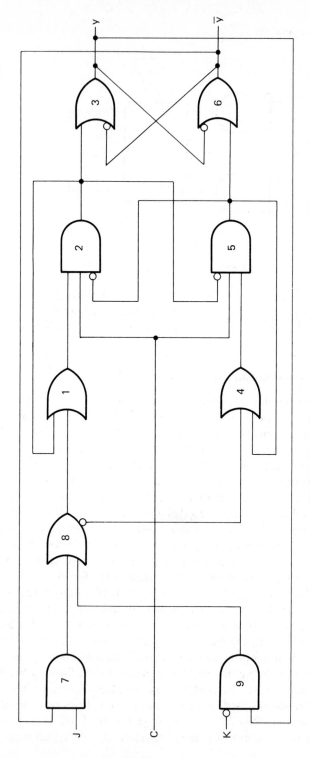

Figure 6.37 Positive-Edge-Triggered *J-K* Flip-Flop.

1 and $G_2 = 0$, preventing input changes from having an effect until the clock goes low. (If the flip-flop remained unclocked for any combination of inputs, it would be vulnerable to any noise occurring while the clock input was high.) Gate methods may be applied to the design of other types of logical devices; some of these (e.g., D-type flip-flops) will be considered in the Exercises at the end of this chapter.

‡6.4 MOSFET CIRCUITS[1,7]

A special type of semiconductor device called the *metal-oxide semiconductor field-effect transistor*, or MOSFET for short, is becoming increasingly popular for large shift registers, memories, and other digital applications because of its efficient use of silicon chip area and its ease of manufacture. Thus far, we have considered only bipolar transistors, which employ both holes and electrons in their operation. The MOSFET, however, employs only one current carrier, either holes alone or electrons alone. For this reason, it is known as a *unipolar* device. Although it has a formidable name, the principles of operation of this type of transistor are fairly simple. As shown in Figure 6.38, two separate p-regions are produced on an n-type substrate.

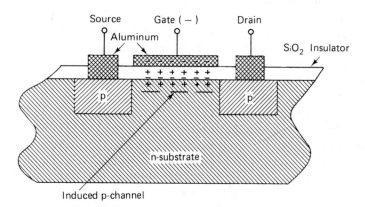

Figure 6.38 p-Channel MOSFET.

A silicon-oxide insulator layer is formed above them, and an isolated metal layer called a *gate*, is formed between the p-regions but above the oxide. (The name MOS comes from the three layers just mentioned.) Since there is n-material between the two p-regions, back-to-back diodes are effectively produced; and with zero potential on the gate, a very high resistance exists from source to drain. It should be observed, however, that the gate, the oxide, and the substrate form a sandwich similar to a parallel-place capacitor. Thus, the application of a negative voltage to the gate (with respect to the substrate) causes a positive charge to be induced on the upper surface of the substrate as shown in Figure 6.38. This layer of positive charge effectively forms a conducting *p-channel* between the p-regions. The resistance between source and drain, which originally was many megohms, has been reduced to a value in the neighborhood of a few hundred ohms by application of the negative gate voltage. Since the change in resistance covers so many orders of magnitude, the above action is often compared to that of a mechanical switch. Another similarity to a switch is the fact that the device will conduct current equally well in either direction between source and drain.

The MOSFET in Figure 6.38 is classified as an *enhancement* type because the current flow increases, is enhanced, by an increasing negative voltage applied to the gate. (A somewhat different structure leads to a MOSFET in which an increase in gate voltage leads to a reduced current flow; hence it is called a *depletion* type. Since the depletion MOSFET is not as important as the enhancement type, it will not be considered here.) By analogy with Figure 6.38, it is easy to see that interchanging the n- and p-type materials leads to an n-channel enhancement-type MOSFET. Of course, the gate voltage must be positive to turn on an n-channel device. Since the p-channel device is actuated by a negative signal between the gate and substrate, its action is similar to a pnp transistor; furthermore, an n-channel device is similar to an npn transistor.

Because a very good insulator is present between the gate and the other elements, except for a small capacitance, the input impedance of the gate is very high—actually as much as 10^{14} ohms in some cases. Incidentally, as we view the action of the gate, it is evident where the name *field-effect transistor* originated.

Circuit symbols for MOSFETs are not as well-standardized as they are are for ordinary transistors; but a common set of symbols is shown in Figure 6.39. By analogy with the structure of pnp transistors, the p-channel MOS-FET, P-MOS, has the arrow pointing into the transistor at the source terminal. The n-channel device, by analogy with an npn transistor, has the arrow pointing out at the source terminal. Since an external substrate terminal is not shown in parts (a) and (b) of Figure 6.39, *it is assumed that the substrate is internally connected to the source.* When it is necessary to make a different connection to the substrate, however, the symbol shown in part (c) of the figure will be employed.

MOSFET Gates

Direct-coupled transistor logic is used in the construction of MOSFET gates; but there is no possibility of current hogging since the gate terminal has such a high input impedance. Another interesting feature of MOSFET circuits is that considerable room on the silicon chip is saved by using a properly biased transistor to simulate the load resistor. (A MOS transistor

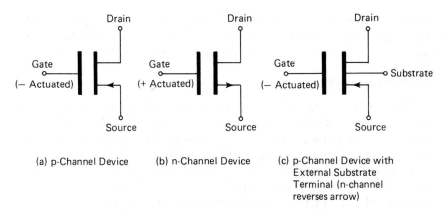

(a) p-Channel Device   (b) n-Channel Device   (c) p-Channel Device with
                                                            External Substrate
                                                            Terminal (n-channel
                                                            reverses arrow)

Figure 6.39   Circuit Symbols for MOSFETs.

requires approximately 1 mil² on the chip, while a corresponding resistor requires more than 10 times that amount.)

A typical MOSFET circuit is shown in Figure 6.40. Since the gate of Q_3 is tied to the negative supply, the transistor is always on; thus, it serves as a load resistance. The other two transistors serve in their usual capacity as switching transistors. Part (b) of the figure is easily verified if we consider the p-channel gate as a switch, which is closed when a large negative voltage is applied and open when 0 V is applied. Using positive logic, we see from the table that the function realized is:

$$V_o = \bar{A}\bar{B} = \overline{A + B}$$

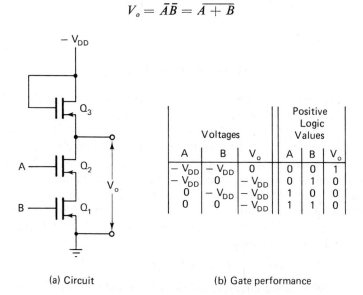

	Voltages			Positive Logic Values	
A	B	V_o	A	B	V_o
$-V_{DD}$	$-V_{DD}$	0	0	0	1
$-V_{DD}$	0	$-V_{DD}$	0	1	0
0	$-V_{DD}$	$-V_{DD}$	1	0	0
0	0	$-V_{DD}$	1	1	0

(a) Circuit   (b) Gate performance

Figure 6.40   A p-Channel NOR Gate.

which is a NOR. In the Exercises at the end of the chapter, a parallel rather than a series structure will be shown to yield a NAND gate, and other slightly more complicated structures will be shown to yield multilevel logic.

Complementary MOS

By employing both p-channel and n-channel MOS transistors in the same circuit, a family of gates called *complementary MOS* (CMOS), having very low power requirements, may be constructed. The inverter in Figure 6.41 demonstrates the basic principles of the CMOS family. Note that the substrate of Q_2 is shown explicitly connected to the drain terminal. This allows zero volts on input A to place a positive voltage between the gate and substrate, turning the FET on. The remaining aspects of circuit performance are summarized in the table, where it is verified that inverter action is achieved.

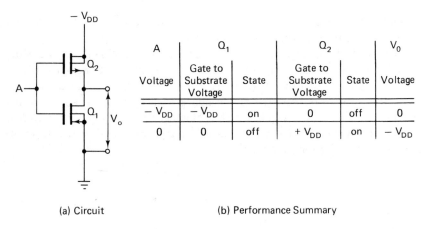

A	Q_1			Q_2		V_0
Voltage	Gate to Substrate Voltage	State		Gate to Substrate Voltage	State	Voltage
$-V_{DD}$	$-V_{DD}$	on		0	off	0
0	0	off		$+V_{DD}$	on	$-V_{DD}$

(a) Circuit   (b) Performance Summary

Figure 6.41   CMOS Inverter.

It is evident that essentially zero static current flows through the unloaded gate when it is in either state. Moreover, connection of the output to the high impedance gate terminals of other circuits provides little additional power dissipation. Thus, the CMOS circuit produces very low power drain.

CMOS devices may, of course, be assembled into NAND and NOR gates as well as into other useful logic elements. A typical CMOS gate is shown in Figure 6.42. The reader should verify that the circuit shown in part (a) of the figure does indeed produce the results shown in part (b). Since positive logic represents the most positive level as a logical 1, then $0 \text{ V} =$ logic 1 and $-V_{DD} = $ logic 0;

$$V_o = \bar{A}\bar{B} = \overline{A + B}$$

which verifies that the circuit is indeed a NOR gate. Other types of CMOS gates will be considered in the Exercises.

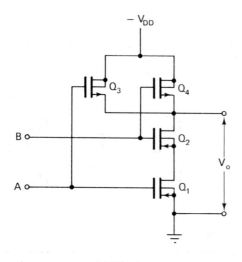

(a) Circuit

Voltage and State Values						
A	B	Q_1	Q_2	Q_3	Q_4	V_o
$-V_{DD}$	$-V_{DD}$	on	on	off	off	0
$-V_{DD}$	0	on	off	off	on	$-V_{DD}$
0	$-V_{DD}$	off	on	on	off	$-V_{DD}$
0	0	off	off	on	on	$-V_{DD}$

(b) Performance Summary

Figure 6.42   CMOS NOR Gate.

MOSFET Flip-Flops

There is, of course, no difficulty in constructing conventional gate type flip flops from MOSFET devices, and this is sometimes done; but other more interesting possibilities exist. Since the input impedance to the gate of a MOSFET is so high, 10^{14} Ω, it is practical to employ the very small gate-substrate capitance, C_{gs}, to temporarily store the desired state of the flip-flop. Thus, master-slave units can be constructed with much smaller area on the IC chip than with bipolar methods.

The above principle is demonstrated in Figure 6.43 for a so-called *dynamic* flip-flop, which is one that must receive a new input at about a 1-kHz rate to refresh the signal stored on C_{gs}. Note that clock pulse C_{p1} controls the master flip-flop, Q_1 through Q_3; and clock pulse C_{p2} controls the slave, Q_4 through Q_5. The timing diagram for the circuit is shown in Figure 6.43(b), and the

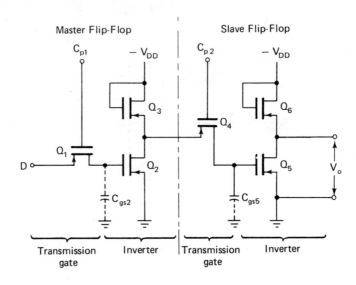

(a) Circuit

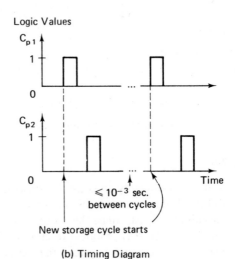

(b) Timing Diagram

Figure 6.43 Dynamic Master-Slave D Flip-Flop.

basic principles of operation are as follows:

1. $C_{p1} = 1$ turns Q_1 on and allows the input to charge the gate-substrate capacitance, C_{gs2}, of Q_2 to the potential of the input signal. The notation $C_{p1} = 1$ means that a *logical-1* signal exists at the C_{p1} input,

which in turn closes the circuit between the source and drain of the *transmission gate* Q_1. (A transmission gate is merely a MOSFET employed as a switch between two separate sections of a circuit.)

2. The signal is inverted by the Q_2–Q_3 circuit but is blocked by Q_4 because $C_{L2} = 0$. At this point the master has been set to the new state, but the slave still holds the old state.

3. C_{p1} returns to zero. Later $C_{p2} = 1$ and the signal is allowed to pass from the master to the slave and to be stored in the gate substrate capacitance, C_{gs5} of Q_5.

4. The input signal, doubly inverted, now appears at the output V_o, and the master slave action is complete.

Of course, the signal stored on C_{gs5} will eventually decay due to the finite resistance in the circuit; this is why it must be refreshed periodically. Many different techniques could be employed to restore the signal, but one natural situation occurs when the flip flop is part of a dynamic shift register. There, many flip-flops, like that shown in Figure 6.43, are connected in a series loop so that data can be circulated and usually picked off, or changed, one bit at a time at a common point.

Another type of MOSFET flip-flop is shown in Figure 6.44. This is a static-type device since refreshing is not necessary; but to keep the signal from degrading, a feedback path is provided from the output to the input. Note that the transmission gate, Q_7, is the only feature added to the previous figure in order to construct the new flip-flop. Figure 6.44(b) shows the timing sequence required for proper circuit operation. Observe how the following storage cycle is executed:

1. A new cycle begins when transmission gates Q_4 and Q_7 open, i.e., the slave is disconnected from the master and the feedback path is interrupted. The current flip-flop state is still held by C_{gs5}; thus, the output is not disturbed.

2. A short time later $C_{p1} = 1$, and the next state is transferred to C_{gs2}. (Note that C_{p1} returns to zero before any of the other inputs change.)

3. Then $C_{p2} = 1$ and the master-slave transfer occurs.

4. Finally, a short time later, the feedback path is closed by $C_{p3} = 1$. With both Q_4 and Q_7 actuated, the flip-flop is in what can be called steady state. As long as power is supplied through $-V_{DD}$, the circuit will properly hold the given state without any changes in the clock signals. The broken sections in Figure 6.44(b) are employed to represent a steady-state condition of arbitrary length. The MOSFET flip-flop are employed in shift registers (discussed in Section 6.5), in random-access memories (discussed in Chapter 9), and in many other

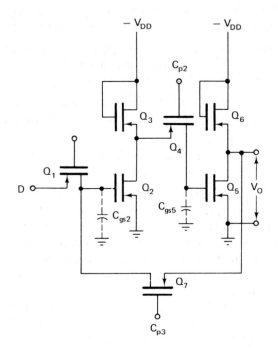

(a) Circuit

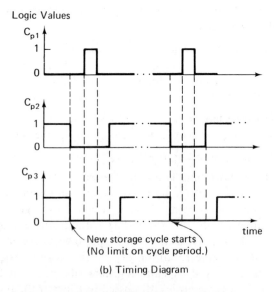

(b) Timing Diagram

Figure 6.44 Static Master-Slave D Flip-Flop.

types of sequential circuits. Both the static and dynamic flip-flops have similar applications; but the static circuit has the advantage that a refresh operation is not required. On the other hand, the dynamic circuit requires somewhat less chip area and less average power.

6.5 INTEGRATED CIRCUITS[1,4,8]

Integrated circuits may be classified in several different ways. One way is according to families such as ECL or TTL, which will be compared in a table shortly. A completely different approach is to place them in one of three groups according to the number of equivalent gates in one package or on a single silicon chip. Here, a circuit is termed *small-scale integration*, SSI, if it contains the equivalent of nine gates or less in one package. A circuit is called *medium-scale integration*, MSI, if it contains the equivalent of between 10 and 100 gates in one package. Finally, a circuit is called *large-scale integration*. LSI, if it contains the equivalent of over 100 gates in a single package.

SSI, MSI, and LSI

A typical SSI configuration is the 74H11 TTL triple three-input AND, which, including the output terminals, requires $3 \times 4 = 12$ pins for the gates plus the ground and V_{cc}, yielding a total of 14 pins. Thus, it conveniently fits into a standard 14-pin dual in-line package illustrated by the logic diagram in Figure 6.45. (Incidentally, the H in the part number indicates that it is a high-speed circuit—about a factor or two faster than standard TTL.)

A typical MSI circuit is the 8202 TTL 10-bit buffer register, which is available in the 24-pin configuration shown in Figure 6.46(a). Part (b) of the figure shows a photomicrograph of the actual circuit. It employs *D*-type

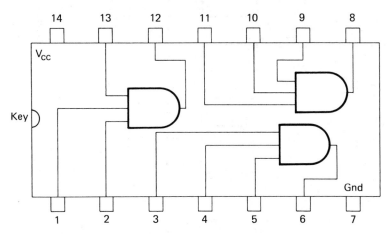

Figure 6.45 Type 74H11, SSI Triple Three-Input AND.

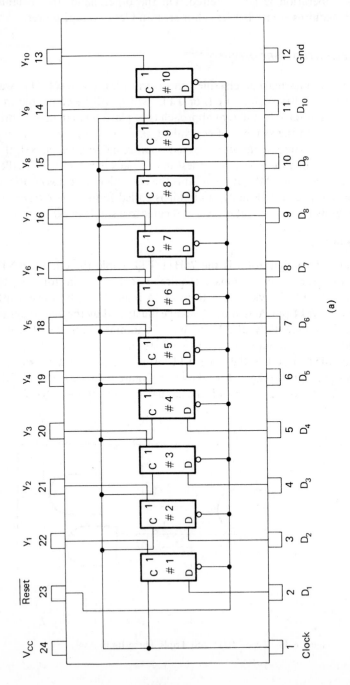

Figure 6.46(a) Type 8202, MSI 10-Bit Buffer Register.

(a)

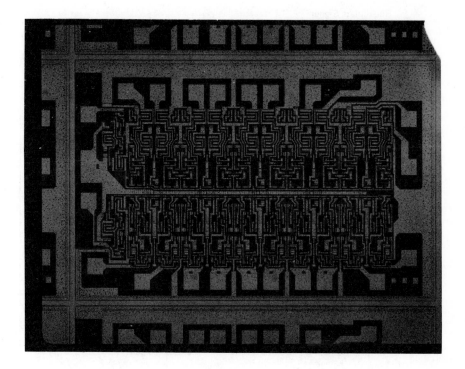

(b)

Figure 6.46(b) Photomicrograph of Integrated Circuit of Fig. 6.46(a).

flip-flops having a direct reset. (When $\overline{\text{reset}} = 0$, all flip-flops are cleared; when $\overline{\text{reset}} = 1$, all flip-flops are enabled.) Since all 10 D-inputs as well as all true outputs are available, many different serial and parallel register operations can be performed.

A typical LSI circuit is the 2533 MOS 1024-bit static shift register, which is available in the 8-pin configuration shown in Figure 6.47. This circuit has a large number of possible applications such as a CRT refresh circuit, a delay-line replacement, and a low-cost sequential-access memory. When the clock ϕ_{in}, is at logical 1, the stream-select input allows one of two signals to enter the register; later, when the clock goes low, the register is shifted. As shown by the broken lines in the figure, frequently one of the input channels is employed to provide a recirculate path so that data reaching the output of the shift register are reentered. On the other hand, new data can be inserted, where desired, through the other channel by changing the SS input. Comparing Figures 6.46 and 6.47, an interesting point is observed—the number of pins on an IC package is not necessarily an indication of circuit complexity. Here the 8-pin package contains a circuit which is many times more complex

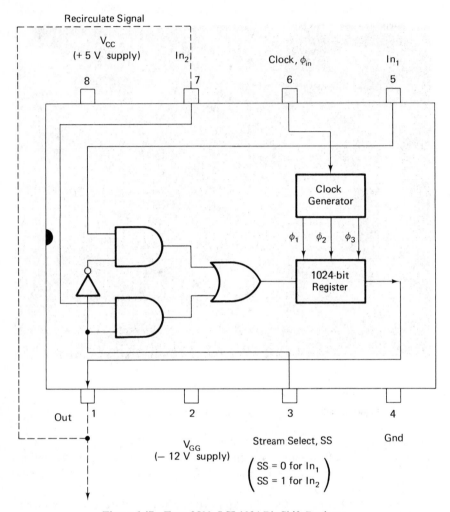

Figure 6.47 Type 2533, LSI 1024-Bit Shift Register.

than that in the 24-pin package. But the fact that the input and output leads
for each bit are available means that a much larger variety of circuits may be
constructed with the 24-pin package.

Integrated Circuit Families

In previous sections we have considered several digital logic families.
It is appropriate at this point to make a detailed comparison of them so that
the reader can see why a certain family is chosen for a particular application.

Table 6.3 serves as the main medium in the comparison. Note that the

Table 6.3 Characteristics of IC Digital Logic Families[4]

Parameter	DTL	ECL*	CMOS	P-MOS	RTL	TTL
1. Positive logic function of gate	NAND	NOR/OR	NOR or NAND	NAND	NOR	NAND
2. Wired positive logic function	AND	OR	None	None	AND (some functions)	None
3. Propagation delay per gate, ns.	30	1	70	300	12	6
4. Fan-out	8	10	≥ 50	20	5	10
5. Immunity to external noise	Good	Good	Very good	Fair	Fair	Very good
6. Cost per function	Low	High	Medium	Medium	Low	Low
7. Typical power dissipation per gate	10 mW	50 mW plus load	0.01 mW static 1 mW at 1 MHz	0.2 to 10 mW	12 mW	15 mW

*Slower gates having greater fan-out and lower cost are available.

rows in the table indicate nine different parameters and the columns list the six logic families. Of course, the manufacturer's catalogs should be consulted for the actual parameters of a specific circuit. Moreover, subfamilies frequently exist within a given IC type, and these may have different parameter values than the typical ones listed in the table. (This is particularly true for propagation delays, fan-out, and power dissipation.) Also, the catalogs will often carry other important information not considered here. On the other hand, the table does give a fairly good general view of the most popular IC families as they currently exist.

Most of the parameters are self-explanatory, but a few should be discussed:

Row 1—While the positive logic function listed is the basic one implemented, catalogs should be consulted for other functions which may be available.

Row 3—The propagation delay listed is the expected (not worst-case) value for a gate not heavily loaded. Also, when a range of delays exists for different family members, the value given tends toward the short end of the scale. Moreover, wiring delays, which average 1 ns per 15 cm must be added to the table values.

Row 4—The values of fan-out given account for worst case conditions of output driving capacity and input loading.

6.6 BIBLIOGRAPHY

1 CARR, WILLIAM N., and JACK P. MIZE, *MOS/LSI Design and Applications*. New York: McGraw-Hill Book Company, 1972.

2 GARRETT, L. S., "Integrated-Circuit Digital Logic Families, Part I—RTL, DTL, and HTL Devices." *IEEE Spectrum*, Vol. 7, October, 1970, pp. 46–58.

3 GARRETT, L. S., "Integrated-Circuit Digital Logic Families, Part II—TTL Devices." *IEEE Spectrum*, Vol. 7, November, 1970, pp. 63–72.

4 GARRETT, L. S., "Integrated-Circuit Digital Logic Families, Part III—ECL and MOS Devices." *IEEE Spectrum*, Vol. 7, December, 1970, pp. 30–42.

5 GRAY, P. E., and C. L. SEARLE, *Electronic Principles: Physics, Models, and Circuits*. New York: John Wiley & Sons, Inc., 1969.

6 KEYES, ROBERT W., "Physical Limits in Digital Electronics." *Proc. IEEE*, Vol. 63, 1975, pp. 740–67.

7 MILLMAN, JACOB, and CHRISTOS C. HALKIAS, *Integrated Electronics: Analog and Digital Circuits and Systems*. New York: McGraw-Hill Book Company, 1972.

8 MORRIS, ROBERT L., and JOHN R. MILLER (Eds.), *Designing with TTL Integrated Circuits*. New York: McGraw-Hill Book Company, 1971.

9 SCARLETT, J. A., *Transistor-transistor Logic and Its Interconnections.* New York: Van Nostrand Reinhold Company, 1972.

10 STRAUSS, LEONARD, *Wave Generation and Shaping* (2nd ed.). New York: Mc-Graw-Hill Book Company, 1970.

6.7 EXERCISES

6.1 What function is mechanized by the circuit shown in Figure E6.1 (positive logic)? Assume: Inputs = 0 V or −3 V.

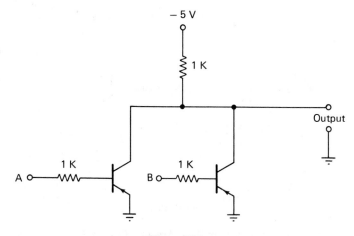

Figure E6.1

6.2 Repeat Exercise 6.1 for Figure E6.2. Also determine what voltages point *P*

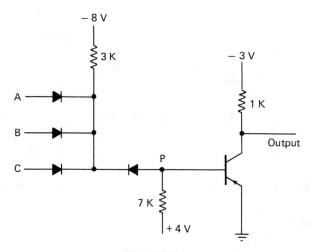

Figure E6.2

can assume (with respect to ground). Use: V_{EB} (forward bias) $= 0.5$ V, Diode (forward bias) $= 0.7$ V, V_{CE} (sat.) $= 0$ V.

6.3 Repeat Exercise 6.1 for Figure E6.3. Also, the inputs are either $+5$ V or -1 V.

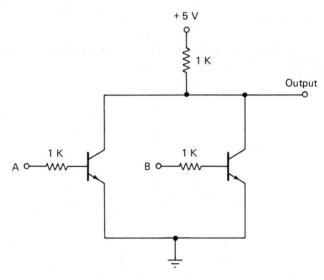

Figure E6.3

6.4 Refer to the circuit in Figure E6.4. Assume: Positive Logic (A & $B = +5$ V, $\bar{A}$ & $\bar{B} = -5$ V), V_{diode} (forward bias) $= 1$ V, V_{BE} (forward bias) $= 1$ V. Write the voltage at each node.

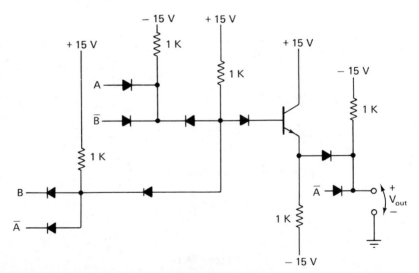

Figure E6.4

6.5 If two RTL IC gates have their outputs connected together, what function is produced by the composite gate? Compare wired logic for RTL with that for DTL.

6.6 (a) For the gate circuit shown in Figure E6.6, determine the purpose of the input diodes and the function mechanized.

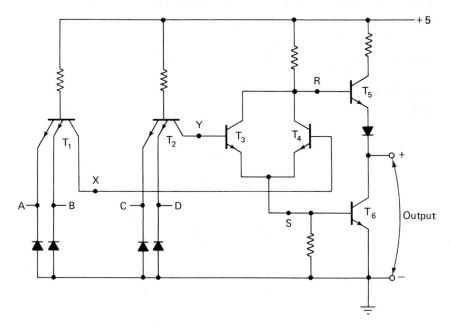

Figure E6.6

(b) Propose an expander for use with this gate. Where would it be connected?

6.7 What type of logic block is required, with negative logic on the input and positive on the output, to produce the NOR function?

6.8 Design a diode OR gate (input voltage levels 0 and 3 volts) which does not show an offset in either the voltage equivalent to a logical 1 or the voltage equivalent to a logical 0. What is the disadvantage of this gate?

6.9 Consider the AND-OR logic of Figure 6.7 and determine the fan-out for a circuit which has the following worst-case limitations:

$$T_f = 200 \text{ ns}, \qquad I_{d \max} = 8 \text{ mA}$$

$$V_H = 3 \text{ V}, \qquad V_L = 1 \text{ V}$$

Stray wiring capacitance $= 50 \text{ pF}$

6.10 In a DTL gate, the noise margin from the nominal input values of 0.2 V and 5 V is smaller for the lower input voltage. What can be suggested to more nearly equalize this, and what is the new performance?

6.11 Is it possible to construct a two-state device with only inverters? Explain.

6.12 Draw a gating circuit, similar to Figure 6.35, for a positive-edge clocked *J-K* flip-flop with asynchronous inputs.

6.13 Redraw Figure 6.35 with NAND gates only.

6.14 Redraw Figure 6.35 for a trailing-edge (negative-going) trigger.

6.15 Draw a gating circuit for a leading-edge clocked type-*D* flip-flop.

6.16 Draw the gating, similar to Figure 6.36, to convert a clocked *J-K* block to a clocked *S-R* block.

6.17 Using p-channel MOSFETs, draw the circuit, similar to Figure 6.40, for a three-input NAND gate.

6.18 Repeat Exercise 6.17 for an n-channel NAND gate.

6.19 Find the logic block realized by the structure shown in Figure E6.19.

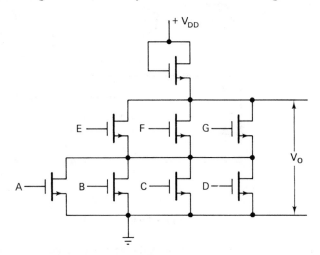

Figure E6.19

6.20 Make a logic diagram showing the additional gating necessary to make the lock-out capability of Figure 6.35 complete.

6.21 Using the fewest possible NAND gates (all in common form), draw the diagram for a positive-edge-triggered *D* flip-flop.

7 | COMPUTER ARCHITECTURE

Part I
Register Transfer Logic, Organization, and Simulation

Three major levels of computer design can be identified. We have already considered the first level, electronics, in Chapter 6, where, transistors are the primitives which are assembled into gates, flip-flops, and other useful circuits. The second level is logical design, considered mainly in Chapters 4 and 5; there, the resulting circuits (e.g., gates) of the previous stage become the primitives of the next stage, and the resulting products are registers, counters, etc. Finally, the third level is reached, which is the subject of this chapter. Here, registers, memories, and similar subsystems are primitives, and the resulting products are digital processors. This highest level of design is called *computer architecture*. It has existed informally since the beginning of computers but is only now starting to receive the systematic development it deserves. By definition, architecture covers a wide range of design topics: from decisions concerning whether a small special-purpose computing system should process data (bits) in series or parallel, to decisions concerning the optimum method for connecting many general-purpose digital computers into a large information-processing network.

We begin the discussion of architecture with the development of additional design tools—tools that allow not only computers but all types of digital circuitry to be described on the register level. These tools are then applied to several practical situations, including a detailed explanation of computer organization. Finally, two elementary digital simulation models, based on FORTRAN, are developed. Examples are then presented for the simulation of some basic arithmetic operations and for the simulation of a simple computer.

In order to properly establish the fundamentals, our approach to computer organization is more from the standpoint of a minicomputer, such as the Digital Equipment Corporation PDP-8, rather than from the standpoint of a large complicated system, such as the Control Data 6600. This approach does not sacrifice generality because, once the fundamentals have been firmly established, the reader is then equipped to grasp more complicated structures. Many of these are discussed in Chapter 10, and others can be gleaned from the Bibliography.

From a single chapter, one cannot hope to become an expert; but sufficient knowledge should be gained here so that the literature and advanced textbooks can be read much more easily. In addition, the reader's design capabilities should be strengthened and his horizons expanded.

7.1 REGISTER TRANSFER LOGIC

When considering problems in computer architecture, we frequently need a concise language to describe the information flow and processing between registers, much as Boolean Algebra is available for describing the flow and processing of separate bits. Register Transfer Logic, RTL,* is currently an *informal method* which is being developed to satisfy this requirement. The simplest operation that may be represented in RTL is $A \leftarrow B$. This type of expression is called a *production*, and it indicates the transfer of data from register B to register A. Although this is a very simple production, other more complicated ones will soon be introduced, and sequential combinations of these can be employed to describe complex arithmetic and logic operations in a very general way. More powerful and general RTL methods exist;[1,3,4,10] but the one employed here, which emphasizes familiar engineering notation, is thought to be advantageous—in a first book—for understanding the basic principles. The following list will make our definition of RTL more precise:

1. The *contents* of the registers will be denoted by one or more letters, with the first always being in upper case.
2. Parallel (not serial) transfers will occur between registers.
3. All bits will transfer at the same time, i.e., a synchronous rather than asynchronous case will be represented.
4. The bits of each register are numbered from right to left, with the least significant bit (LSB) labeled zero, e.g., A_0 is the LSB of register A.

Table 7.1 further expands the notation and presents examples for clarification. The variety of ways bits can be transferred should be studied, e.g., attention should be paid to the difference between A_{1-3} and $A_{1,3}$. Also, observe how the subset of bits, E, is labeled in the last example.

Figure 7.1 presents the logic diagram for a typical member of Table 7.1. Note that solid lines are employed for data and broken lines for the transfer signal. The labeling for the bits of register A is standard; but BAC_{E3}, for

*The name *register transfer language* is also commonly employed, but there should be no confusion between the abbreviation and the resistor transistor logic described in Chapter 6.

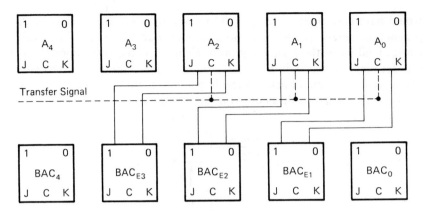

Figure 7.1   Logic Diagram for $A \leftarrow BAC_E$.

Table 7.1   Basic Register Transfer Operations

Operation	Meaning	Example (Before transfer; After transfer)
$A \leftarrow B$	The contents of B are transferred to A.	$A = $ -, $B = 110$; $A = 110, B = 110$
$A_2 \leftarrow B_3$	Transfer of bit 3 register B to bit 2 Register A.	$A = 1010, B = 1011$; $A = 1110, B = 1011$
$A_2 \leftarrow A_3$	Transfer bit 3 in A to bit 2 in A.	$A = 1010; A = 1110$
$A_{1-3} \leftarrow B_{1-3}$	Transfer bits 1 through 3 from B to A.	$A = 1$ --- $0, B = 01011$; $A = 11010, B = 01011$
$A_{1,3} \leftarrow B_{1,3}$	Transfer bits 1 and 3 from B to A.	$A = 1$-0-$0, B = 01011$; $A = 11010, B = 01011$
$A \leftarrow BAC_E$	Transfer the group of bits called E from BAC to A with right justification.	$A = 10$--- , $BAC = \underset{E}{\underbrace{01011}}$; $A = 10101, BAC = 01011$

Notes:
1. Whenever the symbol "-" appears in the example column, that bit or group of bits is a don't-care.
2. Unless explicitly stated, the registers on the right side of the production are not changed by the transfer operation.
3. In the example column the information preceding the semicolon represents conditions before the transfer and that following represents conditions after the transfer.

example, represents a bit from the subset of *BAC*, which is named *E*, and this happens to be bit 3 of the *BAC* register. Table 7.1 specifies a right-justified transfer in which the least significant bit of BAC_E becomes the least significant bit of *A*. Finally, Figure 7.1 shows a jam transfer between the registers. If a clear and then transfer had been desired, a sequence of two productions would have been needed: $A_{0-2} \leftarrow 0$ followed by $A \leftarrow BAC_E$. Here the first production merely specifies that bits 0 through 2 are set to zero, i.e., cleared. Corresponding to the logic diagram in Figure 7.1, the very simple register transfer diagram shown in Figure 7.2 can be drawn. It should be recognized that a certain amount of detail is lost in going to Figure 7.2; this may or may not be a disadvantage, depending on the purpose of the diagram.

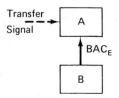

Figure 7.2   Register Transfer Diagram for $A \leftarrow BAC_E$.

The clearing of a register is the simplest arithmetic operation that can be performed, but several other possibilities are also shown in Table 7.2. Multiply, divide, and other more complicated processes could also be included in the table, but we will concentrate on defining primitive register transfer operations and will build more complicated ones from combinations of the simpler types. Figure 7.3 represents a typical member of Table 7.2—that of the parallel addition circuit.

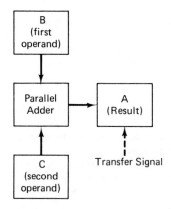

Figure 7.3   Register Transfer Diagram for Parallel Addition: $A \leftarrow B + C$.

Besides the transfer operations already defined, we will find that the logic operations between registers, defined in Table 7.3, are also very useful. A word of explanation is necessary at this point concerning the logic symbols

Table 7.2 Arithmetic Operations

Operation	Meaning	Example
$A \leftarrow 0$	Clear register A.	$A = - ; A = 00 \ldots 0$
$A \leftarrow B + C$	Transfer the arithmetic* sum of B and C into A.	$A = -, B = 011, C = 001$; $A = 100, B = 011, C = 001$
$A \leftarrow B - C$	Subtract C from B and transfer the result to A.	$A = -, B = 101, C = 010$; $A = 011, B = 101, C = 010$
$A \leftarrow A + 1$	Increment A by 1.	$A = 0111 ; A = 1000$
$A \leftarrow 1346_8$	Load 1346_8 into A with right justification.	$A = ----- ; A = 01346_8$

*Whether this is 1's or 2's complement arithmetic must be made evident from the context.

Table 7.3 Logic Operations*

Operation	Meaning	Example
$A \leftarrow B \lor C$	Transfer to register A the OR combination of the corresponding bits of B with those of C.	$A = -, B = 101, C = 100$; $A = 101, B = 101, C = 100$
$A \leftarrow B \land C$	Transfer to register A the AND combination of the corresponding bits of B with those of C.	$A = -, B = 101, C = 100$; $A = 100, B = 101, C = 100$
$A \leftarrow B \land C_2$	Transfer to register A the AND combination of bit 2 of C with the individual bits of B.	$A = -, B = 101, C = 100$; $A = 101, B = 101, C = 100$
$A \leftarrow B \land C_1$	Transfer to register A the AND combination of bit 1 of C with the individual bits of B.	$A = -, B = 101, C = 100$; $A = 000, B = 101, C = 100$
$A \leftarrow B \lor C_2$	Transfer to register A the OR combination of bit 2 of C with the individual bits of B.	$A = -, B = 101, C = 100$; $A = 111, B = 101, C = 100$
$A \leftarrow \bar{A}$	Complement the individual bits of A.	$A = 1011 ; A = 0100$

*The AND and OR operations are undefined unless the two registers contain the same number of bits or one of them specifies a single bit.

employed: $\lor \equiv$ OR and $\land \equiv$ AND. These are the symbols commonly used in mathematical literature; but they will be employed in this text to represent logic operations between registers in order to avoid confusion between the symbol for arithmetic addition, $+$, and the logic OR symbol $\lor$. The realization of a typical member of Table 7.3 is shown in Figure 7.4.

Shift, rotate, and scale manipulations constitute major primitives from which multiply, divide, and some other common operations are formed. These are summarized in Table 7.4.

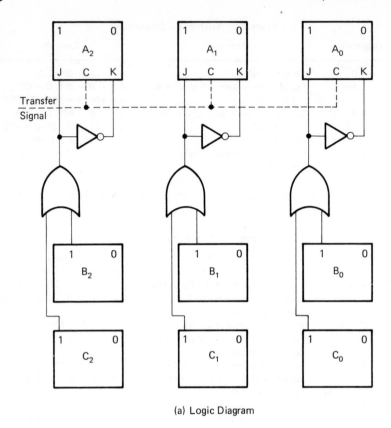

(a) Logic Diagram

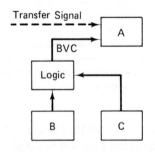

(b) Register Transfer Diagram

Figure 7.4 Realization of $A \leftarrow BVC$.

A typical logic diagram for a member of this group is shown in Figure 7.5. The circuit is very similar to the shift registers that were discussed in Chapter 5; also the circuitry required to mechanize the other productions in Table 7.4 should be obvious.

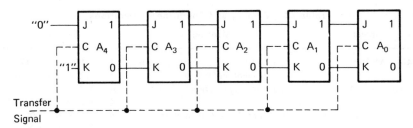

Figure 7.5 Logic Diagram for $A \leftarrow srA$.

Table 7.4 The Shift, Rotate, and Scale Operations

Operation	Meaning	Example
$A \leftarrow srA$	Shift register A right one bit.	$A = 10001 ; A = 01000$
$A \leftarrow slA$	Shift register A left one bit.	$A = 10001 ; A = 00010$
$A \leftarrow sl3A$	Shift register A left three bits.	$A = 10001 ; A = 01000$
$A \leftarrow rrA$	Rotate register A right one bit.	$A = 10001 ; A = 11000$
$A \leftarrow rlA$	Rotate register A left one bit.	$A = 10001 ; A = 00011$
$A \leftarrow rr2A$	Rotate register A right two bits.	$A = 10001 ; A = 01100$
$A \leftarrow scA$	Scale register A right one bit.	$A = 10000 ; A = 11000$
$A,B \leftarrow slA,B$	The concatenated register A,B is shifted left one bit.	$A,B = 1001 ; A,B = 0010$ $\underbrace{\quad}_{A}\underbrace{\quad}_{B}$

Notes:

1. A group of two lowercase letters represents operations; except for scale, which is only defined for one direction, the first letter denotes the operation type and the second denotes the direction. (If the number of bits for the operation is greater than one, it is indicated by a digit following the two letters.)
   Type: r ≡ rotate, s ≡ shift, sc ≡ scale
   Direction: r ≡ right, l ≡ left
   Example: rr ≡ rotate right
2. In the rotate operation, the least and most significant bits are adjacent to each other as far as bit transfer is concerned.
3. In the shift operation, a zero is moved into the vacated bit.
4. Scale is the same as shift right except that the sign (bit $n - 1$) is left unchanged instead of reading a zero into that position.
5. A concatenated register is the result of joining two (or more) registers in a string so that the LSB of the first-mentioned register is one bit left of the MSB of the second. (The concatenation operator is a comma and the symbol(s) for the next register follows without a space.)

The final RTL operations to be introduced are shown in Table 7.5. These two groups are very important in that they give the designer the ability to do hardware branching, and they provide him with the opportunity to communicate with memory. Instead of working with a fixed memory address, as shown in the table, if it is desired, for example, to load register B from a memory location under program control, the production $B \leftarrow M(A)$ is

Table 7.5 Memory and Conditional Operations

Operation	Meaning	Example
$A \leftarrow M(203)$	Load A from the contents of memory location 203.	$A = -, M(203) = 312;$ $A = 312, M(203) = 312^*$
$M(203) \leftarrow A$	Load memory location 203 from A.	$A = 312; M(203) = -;$ $A = 312; M(203) = 312$
IF $(A_3 = 1)$ $B \leftarrow 0$	If bit 3 of A is a 1, then clear B.	$A = 01001, B = 001;$ $A = 01001, B = 000$
IF $(A \neq 0)$ $B \leftarrow 0$ ELSE $B_2 \leftarrow 1$	If A is not 0, then clear B; otherwise set bit 2 of B to 1.	$A = 01001, B = 011;$ $A = 01001, B = 000$
IF $(A \geq 1)$ $B \leftarrow 0,$ $C_1 \leftarrow 1$ See note 1.	If $A \geq 1$, then clear B and set bit 1 of C to 1.	$A = 01, B - , C = -0;$ $A = 01, B = 00, C = 10$
IF $(A, B_2 = 0, 1)$ $C \leftarrow 0$ See note 2.	If A is cleared and if bit 2 of B is a 1 then clear C.	$A = 00, B = 100, C = -;$ $A = 00, B = 100, C = 00$
IF $((B_1 = 1$ OR $C = 0)$ AND $D = 0)$ $A_1 \leftarrow 1$ See note 3.	If bit 1 of B is a 1 or C is cleared and if D is cleared, then set bit 1 of A to 1.	$A = 01, B = 00, C = 0,$ $D = 0; A = 11, B = 0,$ $C = 0, D = 0;$

Notes:
1. This production demonstrates that multiple consequences, separated by commas, are allowed with a single IF statement.
2. Registers or parts of registers may be concatenated to form the premise portion of a conditional operation. [The production is a concise from of IF $(A = 0$ AND $B_2 = 1)$ $C \leftarrow 0.$]
3. Parentheses may be nested, as required, to clarify complicated conditional relations.

---

*Some memory types clear the memory location during the read process, destructive readout. These will be thoroughly discussed in Chapter 9.

employed, where the value of A has previously been determined. Figure 7.6 has been constructed corresponding to the second conditional transfer shown in the table. Note how the register transfer diagram employs standard software flow-chart symbols when possible, and how much more concise it is than the logic diagram.

7.2 APPLICATIONS OF RTL

Up to this point, only a single production has been employed to represent an RTL process, e.g., Figure 7.6(b). However, most practical processors need a more complicated representation; they require a sequence of productions, each of which is initiated by a transfer signal. Once this ability has been added to our RTL repertoire, a very powerful method will have been created. In fact, we will have the ability to represent any digital system from small special-purpose digital interface units to full-scale general-purpose computers. Often the transfer signal is a clock pulse or is produced from the

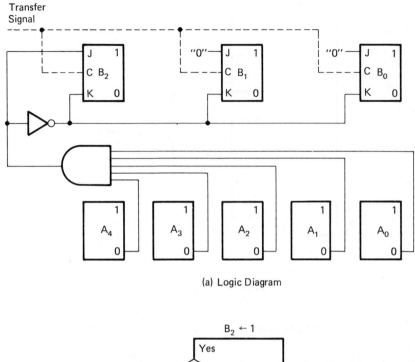

Transfer
Signal

(a) Logic Diagram

(b) Register Transfer Diagram

Figure 7.6 Realization of IF $(A = 0)$ $B_2 \leftarrow 1$ ELSE $B \leftarrow 0$.

clock by means of frequency division; however, with some extensions, RTL can be employed for asynchronous systems.

Example: A Data Storage and Comparison Unit

Frequently in practice (e.g., in the arithmetic section), data are brought in from one unit and processed, and the results are transmitted to the next unit. Figure 7.7 shows an RTL program for the case where one number is

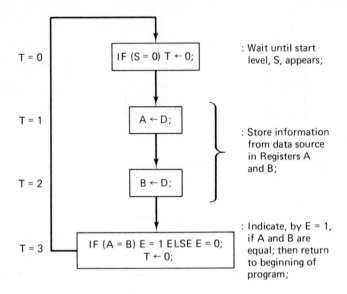

(Note: Except where stated otherwise the production T ← (T + 1) is
implied at each state, T_i.)

Figure 7.7   RTL Program for the Data Storage and Comparison Unit.

read into register *A* from data source *D*,* and another number is later read
into register *B*. The numbers are checked for equality, and the process is
repeated with new data, as long as the start level, *S*, remains true. In the
figure, the column down the left side indicates the sequencing. To be more
specific, each number there denotes the presence of a *level*, one clock period
long, which uniquely determines when a particular step in the RTL process
is to occur. (Our definition of a level is a signal which remains in a particular
logic state, either a 1 or 0, for a length of time at least as long as the period of
the clock pulse.)

For this example, the states† T_0 through T_3 occur in sequence, and then
the circuit returns to state T_0. In general, the order can be interrupted and
various states repeated or skipped, as is done with the assembly language
programs discussed in Chapter 3. As a matter of fact, we could call Figure
7.7 a hardware program since it realizes an algorithm for hardware much
as algorithms were previously realized for software. Contrary to most software programs, more than one RTL operation can occur simultaneously
(while the processor is in a particular state). The RTL program represents
this by means of additional statements having the same T_i value but separated

---

*New data reaches *D* after each clock pulse.
†The abbreviation T_i will frequently be employed to represent the state $T = i$.

by semicolons, as is done at T_3 in Figure 7.7. (observe that a semicolon also terminates the last production of each state, and a colon is placed before each comment.)

Figure 7.8(a) shows the block diagram for the circuit that realizes the program of Figure 7.7. The reader should now carefully study these two figures and visualize the way the data and control signals, specified by the program, are implemented by the hardware. To further clarify the process, a *timing diagram*, employing a simplified case with two bits for each of the A, B, and D registers, is presented in Figure 7.8(b). We have previously employed these diagrams in Chapters 5 and 6; but in this application, their full value as a design tool for indicating time sequences and interrelationships between data and control signals will be realized. Since we have been emphasizing clocked J-K flip-flops, these are implied in the timing diagram; and because it is convenient with this type of flip-flop, the information is communicated in a level mode with the actual transfer being completed a few nanoseconds following the trailing edge of the clock pulse. The clock pulse signal, C_p, is the key component in Figure 7.8(b). (The pulses are numbered below the time base for easy reference.) The other signals are synchronized with respect to it; and the state changes, for all signals, occur a few nanoseconds following its trailing edge. In agreement with the program of Figure 7.7, the T_i states in the timing diagram occur in sequence, with a reset to T_0 after T_3. Level T_1 furnishes the ready signal to register A, and clock pulse 3 (also later $C_p = 7$) initiates the actual transfer from the data source* (a similar transfer to register B occurs at $C_p = 4$ and $C_p = 8$). Only when the two registers are equal does the output signal, E, go to the logical-1 state (after $C_p = 8$).† It should be noted that the data source is synchronized with the clock, which is evident from Figure 7.8(a) and (b). However, the start level, S, is not synchronized since it occurs between clock pulses 1 and 2.

In order to provide more details concerning the operation of the data storage and comparison unit, the logic circuits for register A, the state generator, and the comparator are presented in Figures 7.9, 7.10, and 7.11, respectively. (The register B circuit is the same as that for register A.) The following facts should be observed concerning these figures:

1. The terminals for the individual logic diagrams are all arranged in the same relative positions as they appear in the block diagram of Figure 7.8(a). (This should be helpful in following the information paths between the various circuits.)

---

*The fact that the clock simultaneously advances the state generator to T_2 should not be allowed to hide the fact that it is the T_1 level that really opens register A to the incoming data. The register circuit logic details will be discussed shortly to prove this point.

†Since the comparator is a combinational circuit, to be discussed in detail shortly, Figure 7.8(b) correctly shows that, when $A = B$, $E = 1$ during the entire T_3 period.

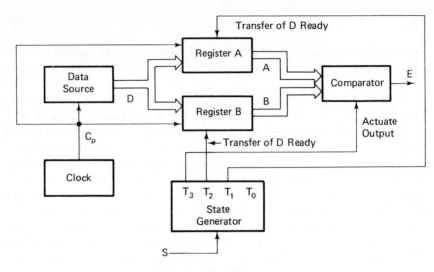

Transfer of D Ready

Register A

A

Data
Source

D

Comparator

E

B

Register B

Clock

Cp

Transfer of D Ready

Actuate
Output

T_3 T_2 T_1 T_0
State
Generator

S

(a) System Block Diagram

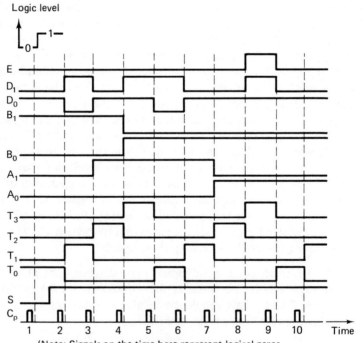

Logic level

1
0

E
D_1
D_0
B_1

B_0
A_1

A_0
T_3

T_2
T_1
T_0

S
C_p

1 2 3 4 5 6 7 8 9 10   Time

(Note: Signals on the time base represent logical zeros,
those above the base represent logical ones.)

(b) Timing Diagram

Figure 7.8 Data Storage and Comparison Unit.

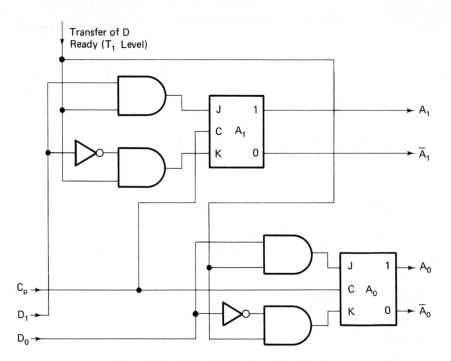

Figure 7.9 A Two-Bit Version of Register *A*.

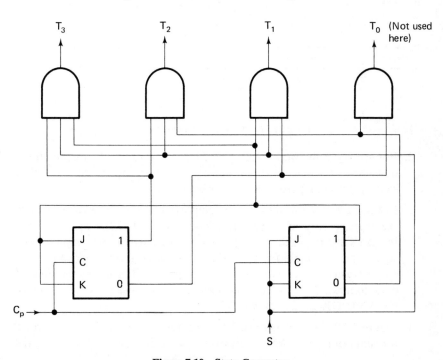

Figure 7.10 State Generator.

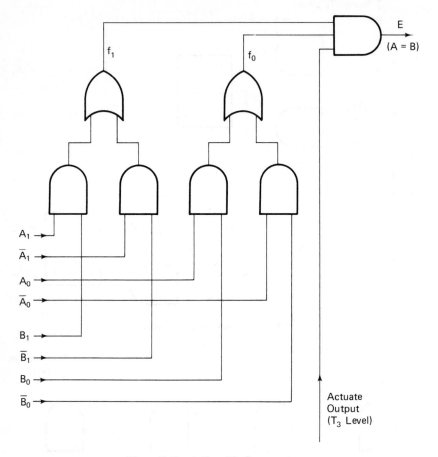

Figure 7.11 A Two-Bit Comparator.

2. For brevity, only two-bit versions of Figures 7.9 and 7.11 are presented; however, extensions of these circuits to as many bits as desired follow exactly the same pattern as shown here.

3. In Figure 7.9, the AND gates are employed to allow new data to reach the register only when the transfer-ready level ($T_1 = 1$) exists. The inverters are required to generate $\bar{D}_i$ since only the uncomplemented signals are available from the data source. A general version of the data transfer circuit is shown in Figure 7.12. There, any number of data sources, (1 to N) can be individually transferred by means of the AND-OR gating shown.

4. Figure 7.10 shows a standard up-counter which requires the level S for actuation. The AND gates serve as a decoder to transform the

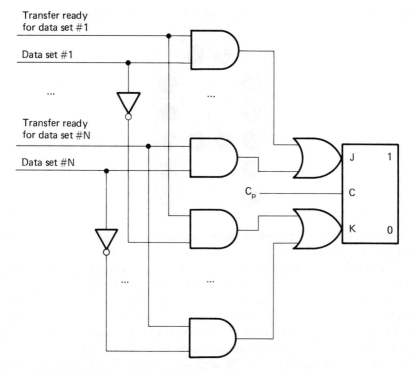

Figure 7.12 General Circuit for the Transfer of Data into a Clocked *J-K* Flip-Flop.

counter outputs into the required T_i states. Note that the S level also controls whether or not the decoder's output is allowed to appear.

5. Figure 7.11 shows a comparator; and more specifically, it is an EQUALITY or EXNOR circuit. To determine whether, for example, the least significant bits of registers A and B are equal, the EXNOR $f_0 = A_0 B_0 + \bar{A}_0 \bar{B}_0$ is evaluated. This output is then ANDed with similar circuit outputs for all the other bits to determine whether the registers agree. In this case, the evaluation is valid only at state T_3; thus, the T_3 level is employed to actuate the output AND gate.

Example: Character Generator for a CRT Display

Let us now consider another practical example, that of generating alphanumeric characters for a cathode ray tube (CRT) display. There are several ways of accomplishing this, but one of the common methods is to intensify certain positions in a 4-by-6 array, much as is done with small lights in a time-and-temperature sign. For example, the letter A can be represented as:

where the filled areas represent intensified points and the remainder represents points not intensified.

The system, shown in Figure 7.13(a), operates essentially as follows. The 24 bits (4 by 6) which form a character are stored in the memory, then transferred into a register in the logic unit, and finally read onto the CRT one bit at a time. The Z signal controls the beam of the CRT, with a $Z = 1$ turning the beam on and $Z = 0$ turning the beam off. The position of the individual bits on the face of the CRT is controlled by the X and Y deflection signals, which in turn are produced by the digital-to-analog (D/A) converters. The D/A converter is a special electronic circuit that develops an analog voltage proportional to the input digital number. For simplicity, we will assume that the CRT in Figure 7.13 has a deflection system such that the entire display is in the first quadrant, with the origin in the lower left corner. To emphasize principles, we have also made a number of other simplifying assumptions, all of which can easily be removed in practice, such as the fact that the appearance of the character will not be objectionable if some points are displayed somewhat longer than others.

Because it is the value employed by several minicomputers and because it is desirable to indicate some of the problems in producing a CRT display, we will assume that the system memory contains 12-bit words. Thus, two words will be required to get sufficient bits for a 4-by-6 grid. Figure 7.13(b) shows the assumed structure of each character, including bit numbering.

Now that the basic system has been described, let us examine Figure 7.14, which shows the algorithm used in RTL notation. Again, the comments adjacent to each block should help in understanding the process. Notice that three loops are present in the algorithm and each of these employs a different counter: C, which accumulates the number of points completed in each column of the character; D, which accumulates the number of bits completed in each 12-bit word; and E, which accumulates the number of

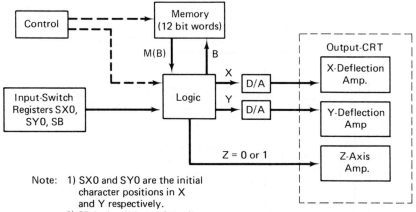

Note: 1) SX0 and SY0 are the initial
character positions in X
and Y respectively.
2) SB is the address of the first
word in memory containing the
character pattern.
3) A switch register is a row of miniature toggle switches, which
represents a binary number. (The up position represents a
logical one.)

(a) System Block Diagram

First   Second
word   word

5	11	5	11
4	10	4	10
3	9	3	9
2	8	2	8
1	7	1	7
0	6	0	6

(b) 4 by 6 Character

Figure 7.13   Character Generator and CRT Display.

points completed in the 24-bit character. Examination of Figure 7.13(b)
will reveal the physical reasons for these three loops.

Figure 7.14 indicates that once the circuit starts, the character is displayed
over and over again, which is necessary to make it visible (unless a storage

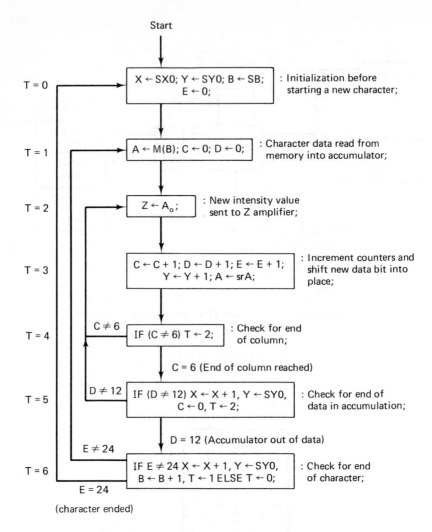

Start

T = 0 — X ← SX0; Y ← SY0; B ← SB; E ← 0; : Initialization before starting a new character;

T = 1 — A ← M(B); C ← 0; D ← 0; : Character data read from memory into accumulator;

T = 2 — Z ← A$_o$; : New intensity value sent to Z amplifier;

T = 3 — C ← C + 1; D ← D + 1; E ← E + 1; Y ← Y + 1; A ← srA; : Increment counters and shift new data bit into place;

C ≠ 6

T = 4 — IF (C ≠ 6) T ← 2; : Check for end of column;

C = 6 (End of column reached)

D ≠ 12

T = 5 — IF (D ≠ 12) X ← X + 1, Y ← SY0, C ← 0, T ← 2; : Check for end of data in accumulation;

D = 12 (Accumulator out of data)

E ≠ 24

T = 6 — IF E ≠ 24 X ← X + 1, Y ← SY0, B ← B + 1, T ← 1 ELSE T ← 0; : Check for end of character;

E = 24

(character ended)

Figure 7.14 RTL for a 4-by-6 Character Generator.

CRT is used). Also, the fact that there is a return to $T = 0$ after the completion of point 24 means that the input switches can be effective at that time in changing the character's position and even in selecting a different character by changing the SB switch. (Of course, the memory must be preloaded with the proper alphanumeric patterns.)

Methods for Resetting the State Generator

In the previous example, there were four branch points where the state generator's normal sequence was broken, and these must be accounted for in the hardware design. There is, however, no unique solution to the state generator reset problem—each particular case has its own requirements and its own optimum solution. On the other hand, Figure 7.15 presents a general

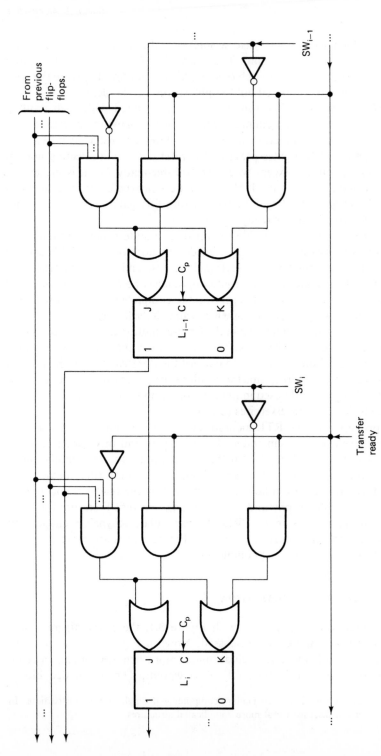

Figure 7.15 A General-State Generator Configuration.

approach which will always yield a solution, although for any particular case it may be not the optimum. The figure shows two typical bits, L_i and L_{i-1}, from the generator. A switch register, SW, is employed to store the reset state, but frequently some circuit simplification can be achieved if a fixed source of data is used. The top AND gate associated with each flip-flop forms part of a parallel up-counter, with the accompanying inverter serving to disable the counter function when a new state is transferred in. Finally, the lower pair of AND gates prevents the reset data from affecting normal up-counting operations except when the transfer-ready signal is present. If other reset states are required, they may be constructed from additional pairs of AND gates feeding into the existing OR gates. (Each reset state requires a separate transfer-ready signal.)

Thus, it is technically possible to construct a state generator which can be reset to as many different states as needed by adding pairs of AND gates to all bits for each reset state. Nevertheless, cost and other practical considerations often dictate that alternate approaches be considered. One way to repeat certain states without repeating the complete sequence is to break the process into two parts, either of which can call itself or call the other part. This is the general approach taken in the fetch-execution cycle to be described in the next section. Another possibility is to interrupt the state generator and allow a special down-counter to control the iterative part of the process until it is complete, at which time the regular state generator resumes control. This method is frequently used in multiply operations. Still other approaches will be considered in the Exercises (Section 7.7).

The principles of the RTL method have been developed, and examples have been given. In the remainder of this chapter and in Chapter 8, many applications of the RTL method to the design, analysis, and simulation of digital systems will be presented. Since RTL is still an informal method that is in the formative stages, following this discussion the reader is encouraged to consult the Bibliography and to examine the current literature for more advanced developments in the field. In particular, Rhyne's[12] chapter on controller design (the control logic of our Figure 3.1) is an interesting supplement to the discussion just completed.

7.3 COMPUTER ORGANIZATION

In Chapter 3, in conjunction with Figure 3.1, the block diagram of a general-purpose computer was discussed. Now, in Figure 7.16, we want to expand the control unit and include much more detail, with emphasis on computer organization* from an RTL viewpoint. The nonshaded blocks

*Here we are considering an elementary computer with single address instructions. In Chapter 10 we will discuss several more complicated structures.

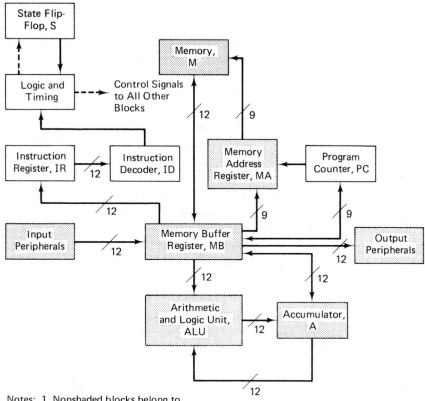

Notes: 1. Nonshaded blocks belong to
control unit.
2. Solid lines are for data and
dotted lines are for control.
3. The numbers on the data paths,

e.g. $\frac{}{9}$, represent the

number of bits transmitted.

Figure 7.16 Organization of an Elementary Computer System.

shown in the figure belong to the control section; all of these are registers
except the logic and timing block and the instruction decoder block.

Before describing the system as a whole, the individual blocks will be
explained:

1. *Memory buffer register*—As the name implies, this register serves as
an interface between the memory and other units. Thus, all data and
instructions entering or leaving the memory must pass through the
memory buffer.

2. *Memory address register*—The location in memory being employed as a source or receiver of data is determined by this register.

3. *Instruction register*—As indicated in Chapter 3, a typical machine language instruction is composed of an operator and an address. The instruction register receives those bits of the current instruction which convey operator information.

4. *Instruction decoder*—The contents of the instruction register are decoded here by means of a combinational circuit having as many outputs as there are instructions in the computer's repertoire. The single output possessing a logical 1 indicates which instruction is being executed at a particular time.

5. *Program counter*—This is the register which keeps track of the address in memory where the next instruction is to be found. Normally it is incremented by 1 each time so that the next instruction in sequence will be available for execution; but branch instructions are an exception. For example, in the case of JMP, the program counter has a new location transferred to it from the address section of the memory buffer.

6. *Accumulator*—The accumulator, which is actually part of the arithmetic unit, is the most prominent register in the machine. It always holds one operand for each arithmetic or logic operation, and the result is always placed there.

7. *Logic and timing circuit*—This unit furnishes all of the control signals which allow the computer to execute a program. The master clock, the state generator, and an extensive logic circuit are present in order to initiate transfers between the various registers at the proper times.

8. *State*—In order to keep track of the mode in which the computer is operating, the control unit employs a one-bit register called the *state* flip-flop. (This flip-flop is shown explicityly in the figure to indicate its relationship to the other units; but it is really part of the logic and timing unit and can be thought of as a means to break the state generator's sequence into two sections.)

Now that we have some concept of the various circuits contained in a computer, several typical machine language instructions will be explained in order to give the reader a deeper understanding of this aspect of computer architecture. First, it must be understood that a typical machine language instruction requires several clock pulses to complete because there are many suboperations that the machine must accomplish sequentially* in obtaining

---

*In Section 7.4 we will consider both the restrictions on performing two or more RTL operations concurrently and those imposed by typical hardware configurations.

an instruction from memory, decoding it, obtaining the data from memory, performing the operation, and preparing to obtain the next instruction.

The process of carrying out an instruction will be divided into two modes called the *fetch* (or *instruction*) cycle and the *execution* cycle. The fetch cycle is concerned with obtaining the instruction from memory and preparing for execution, while the purpose of the execution cycle is to obtain the data and carry out the operation. Some other functions of both cycles will become obvious later in this section. As might be expected, the fetch cycle is common to all instructions, but the execution cycle is different.

An RTL diagram for the fetch cycle in a simple computer is shown in Figure 7.17. The reader should study this figure in conjunction with Figure 7.16, completely visualize the process that is taking place, and agree that by the completion of $T = 2$ sufficient information has been transferred so that the execution cycle can be initiated. Note that the convention for the machine state is $S = 1$ for execute and $S = 0$ for fetch; also, the symbols used for memory address register, program counter, etc. are the same ones used in Figure 7.16. At $T = 2$ it should be observed that several transfers are occurring simultaneously, which is perfectly feasible and certainly efficient.

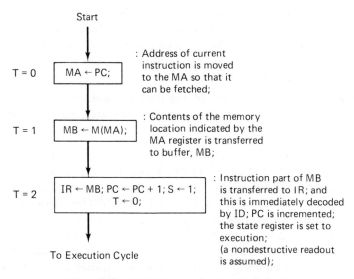

Figure 7.17 RTL Diagram for the Fetch Cycle.

Having discussed the fetch cycle, we are now ready to consider the execution cycle. These will be different for each instruction; but several have been gathered together in Table 7.6 for comparison purposes. First note the following format differences between this program and previous ones. For simplicity, RTL statements are written in more conventional programming form rather than in flow-chart form. Moreover, the production $T \leftarrow (T + 1)$

Table 7.6   RTL Programs for the Execution Cycle of Several Instructions

	ADD	COM	JMP	STC	ROR
$T=0$	$MA \leftarrow MB_{0-8};$	$A \leftarrow \bar{A};$	$PC \leftarrow MB_{0-8};$	$MA \leftarrow MB_{0-8};$	$A \leftarrow rrA;$ $MB_{0-5} \leftarrow MB_{0-5}-1;$
$T=1$	$MB \leftarrow M(MA);$	- - -	- - -	$MB \leftarrow A; A \leftarrow 0;$	IF $(MB_{0-5} \neq 0)$ $T \leftarrow 0;$
$T=2$	$A \leftarrow A+MB; S \leftarrow 0;$ $T \leftarrow 0;$	$S \leftarrow 0;$ $T \leftarrow 0;$	$S \leftarrow 0;$ $T \leftarrow 0;$	$M(MA) \leftarrow MB;$ $S \leftarrow 0;$ $T \leftarrow 0;$	$S \leftarrow 0; T \leftarrow 0;$

Note: A detailed description of each of the instructions is found in Table 3.2.

is implied at each state unless an explicit branching statement for T is given. Now let us make detailed comments concerning the RTL statements required by the STC instruction in the table:

$T = 0$:The address section of the memory buffer is transferred to the memory address register in order to establish the storage location required by the STC instruction;

$T = 1$:The number to be stored is transferred from the accumulator to the memory buffer register and simultaneously the accumulator is cleared. (As previously mentioned, there is no conflict in having a simultaneous transfer because of the type of flip-flops employed.);

$T = 2$:The actual transfer of the data to memory is accomplished. The state flip-flop and state generator are reset so that another fetch cycle can begin;

As an exercise, carefully study the composition of the entire set of instructions in the table. Again, Figure 7.11 will be of considerable aid in visualizing the details of the data flow. At the end of each execution cycle, note that preparation is made to start the fetch cycle for the next instruction by means of the $S \leftarrow 0$ and $T \leftarrow 0$ transfers. Although certain instructions like COM and JMP require only two clock pulses for completion, the full three clock pulses have been employed for uniformity, and nothing is accomplished during $T_1 = 1$. The philosophy of making most execution cycles of constant length is fairly common in practice since it simplifies the timing circuitry; but of course, there may be cases where the extra cost and complication could be justified by the time saved. On the other hand, the ROR instruction is an interesting one in that the number of clock pulses composing the execution cycle is not fixed at three, as it is with the other four instructions shown; but it is reasonable to make it depend on the number of bits the accumulator is rotated right.

Of course, part of the art of computer architecture is to find more effective ways to compose machine instructions within the host of practical constraints that exist, e.g. in Chapter 9, we will learn that a core memory requires approximately one microsecond to complete a memory cycle. With these increasing requirements, the number of clock pulses within the instruction cycles may increase; but this will be partially compensated by more sophisticated instruction design.

Since we are interested in the logical design of computers, let us now consider how the operations indicated in Table 7.6 can be realized. Figure 7.18

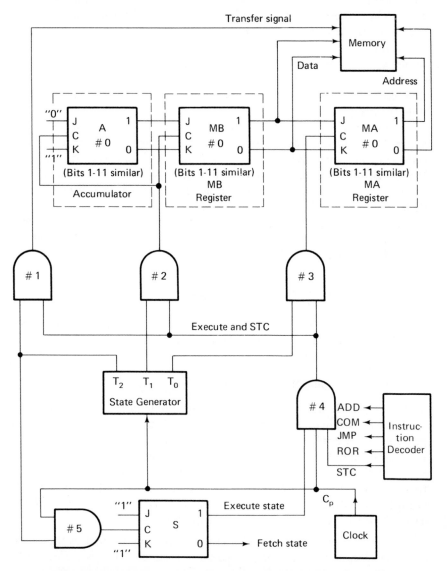

Figure 7.18 Execution Cycle Logic for the STC Instruction in an Elementary Computer.

shows the logic diagram of the STC instruction in the execution cycle. There the accumulator and the other registers have their individual bits arranged in a vertical line. Since the others follow the same pattern as bit 0, only that one is shown for each register. Note that the instruction decoder indicates which instruction is being executed by providing a logical-1 level output at the proper terminal. A careful study of this figure will reveal several interesting features. For example, gate 4 generates a signal which indicates that the STC instruction is being executed, the machine is in the execute state, and a clock pulse is present. In this simplified case, only three gate inputs are saved over running the STC, $S = 1$, and clock signals directly to gates 1 through 3; but it demonstrates a principle which is frequently employed in practice. In agreement with Figure 7.17 and Table 7.6, note that the T_2 signal from the timing generator goes directly to the S register, causing it to change state each time $T_2 = 1$ regardless of the instruction being executed or the state the machine is in.

In Figures 7.9, 7.10, and 7.12, the clock signal is permanently connected to the C terminal of each flip-flop and the transfer-ready signal from the state generator controls when each flip-flop receives new data. An alternate, but less general, approach is shown in Figure 7.18. Here, it is convenient to employ the state generator to gate the clock signal so that clock pulses reach each flip-flop only when data is to be received.

Several additional concepts concerning computer organization, the composition of machine instructions, and logic will be considered in the Exercises and in Chapter 10.

7.4 FURTHER DEVELOPMENT OF RTL

What technique can be employed to make RTL programs more concise? Are all possible RTL statements allowed? It is the purpose of this section to answer practical questions such as these.

Criterion for Grouping RTL Statements

In Chapter 5, we found that state tables sometimes contain redundant states whose elimination can result in significant savings. A similar reduction process can be applied to simplify an RTL program, either by eliminating both a state and its production(s) entirely or by eliminating a state by grouping its production(s) with those of another state. In either situation, the criterion is similar to that in Chapter 5: assume that two RTL states may be grouped (or one eliminated) unless prevented by a *proven incompatibility*. Two major sources of proven incompatibilities are:

1. Any specific limitations on circuit construction dictated by the system specifications, e.g., there may be a requirement concerning which registers must be employed.

2. Constraints established by the desired sequence of circuits actions, e.g., when one register is loaded from another, the first must receive the data at least one clock pulse before the second register is loaded.

As an example of the criterion's use, consider the RTL program in Figure 7.7. Because the data comparison unit requires that a sequential pair of samples from source D be transferred into registers A and B, an attempt to group the production from T_1 and T_2 into a single state results in a *proven* incompatibility. Thus, it may seem that four is the minimum number of states for the RTL program. The specification, however, did *not* explicitly require that the checking of level S and the transfer $A \leftarrow D$ be done serially, which means that the productions in T_0 and T_1 may be combined in a single state.

What about also eliminating state T_2 by removing register B and making a direct comparison between A and D as the following production indicates?

$$\text{IF } (A = D) \, E = 1 \text{ ELSE } E = 0;$$

This production contains a proven incompatibility, however, because the original specification explicitly stated that the second number is to be read into register B before the comparison. Therefore, the simplified RTL program is the three-state version shown in Figure 7.19. Although register A is loaded from D each time T_0 is executed, only the value contained in A when $S = 1$ is actually compared with the contents of register B. (Because A is loaded at T_0 in the simplified program, compared to being loaded at T_1 in the original program, two different sequential pairs from D are compared in the two programs. Nevertheless, the original specification is sufficiently broad to permit both solutions.)

```
T = 0  IF  (S = 0)  T ←— 0;  A ←— D;
T = 1  B ←— D;
T = 2  IF  (A = B)  E = 1  ELSE  E = 0;  T ←— 0;
```

Figure 7.19 Simplified Version of the RTL Program in Figure 7.7.

Limitations on RTL Statements

There are always hardware constraints whenever one constructs electronic systems. Moreover, these constraints are not absolute. Sometimes they are influenced by the environment in which the circuit must operate; they are subject to change as technology improves; and sometimes they may exist because of established conventions or practical considerations such as the desire for a finite inventory of circuit modules types. Hardware constraints, as well as certain practical rules which have already been mentioned, cause limitations on RTL production formats. Several typical productions which are forbidden for the purposes of *all RTL programs* in this text are shown in Table 7.7 along with a brief reason for the limitation.

In addition to the general limitations considered above, there are some specific limitations for digital computers with the classical organization shown in Figure 7.16. These limitations are listed in Table 7.8 and will be assumed to hold for *all digital computers* discussed in this text unless explicitly indicated. (For digital logic not part of a computer, however, the Table 7.8 restrictions do not apply; but those of Table 7.7 are still valid.) Neither of the tables is exhaustive—the reader, however, should be able to visualize many variations of the examples shown. Moreover, as a new RTL program is being checked, the table should serve as a model for validating any questionable productions. Just like pure hardware constraints, our limitations on productions are not absolute since a different environment or a different technology could add to or subtract from the lists; but for purposes of this text, we will agree to employ the limitations of Tables 7.7 and 7.8.

Table 7.7   General RTL Statement Limitations

Statement Type	Reason for Limitation
1. $M(C) \leftarrow A$; $B \leftarrow M(D)$;⎫ 2. $M(B) \leftarrow M(A)$; ⎬	A conventional, single, unified memory only allows one location to be operated on at a time.
3. $A \leftarrow A + B + C$;	A conventional adder permits only two operands.
4. $A \leftarrow A \times B$;	Multiply is not an RTL primitive.
5. $A \leftarrow A \div B$;	Divide is not an RTL primitive.
6. $B \leftarrow B + M(A)$;	A special circuit would be required.
7. $M(A) \leftarrow B$; $B \leftarrow M(A)$;	In contrast to the legal simultaneous register exchange $A \leftarrow B$; $B \leftarrow A$; we assume memory may not be read and written simultaneously even if the same address is employed.

Table 7.8   RTL Statement Limitation
(for the System of Figure 7.16)

Statement Type	Reason for Limitation
1. $MA \leftarrow srMA$	Although the *accumulator and memory buffer* do have circuitry for shifting and other logical operations, the other registers do not.
2. $T = 0$ $MB \leftarrow M(MA)$; $T = 1$ $MC \leftarrow M(MA)$;	Memory can not be read directly into two different buffer registers because only one buffer is assumed to exist. $MC \equiv$ a second buffer register.
3. $T = 0$ $M(MA) \leftarrow MB$; $MA \leftarrow MA + 1$; $T = 1$ $M(MA) \leftarrow MB$;	For *destructive readout only* (see Chapter 9), it is assumed that a particular address must be read before writing can take place.
4. $M(MA) \leftarrow A$; $PC \leftarrow A$;	Data paths do not exist for these separate productions.
5. $MB \leftarrow MA$; $MB \leftarrow IR$;	Although data paths exist between these registers, the arrow on each path indicates that information may not flow in the direction specified by the RTL statement. (Note the production $MB \leftarrow PC$ is allowed because it is needed in Chapter 11, particularly Figure 11.4.)

‡7.5 SIMULATION OF DIGITAL SYSTEMS

One of the main reasons for employing digital simulation models is to test new computer designs, algorithms, and concepts before going to the trouble of actually building a hardware prototype. This does not mean that simulation is a panacea, since there are cases where a model representing the system to the degree of detail required would actually be more complicated than the hardware itself. On the other hand, simulation is a valuable design tool, and it is assuming a very prominent role in computer engineering.

There are a number of good digital simulation language packages.[4, 5, 10, 14] From the standpoint of an introduction to computer design, however, these are usually time-consuming to learn, and the student has a tendency only to become well versed in a particular procedure (black-box approach) rather than also gaining insight about the simulator's internal design principles. Although the following is an elementary discussion, it attempts to overcome these disadvantages through the following features: modular program construction (which facilitates system reconfiguration and experimentation), availability of complete programming details, and use of FORTRAN[13] (which is widely available and is popular for engineering applications).

Two simulation approaches, the *arithmetic model* and the *logic model*, are discussed here. The first approach is so named because it employs FORTRAN arithmetic statements to represent RTL expressions, and the second approach employs FORTRAN logic statements. The arithmetic model achieves conciseness at the expense of certain hardware details. On the other hand, the logic model permits considerable details, but it is not very concise. (Although the principles developed here apply to simulation of computer system probabilistic models, it is beyond our scope to consider this topic.)

The Logic Model

The simulation of a six-bit, 2's complement, serial adder employing FORTRAN logic statements is shown below in Figure 7.20. The heart of the program is found in lines 11 and 12, where the Boolean expressions for the sum, S, and carry, C, outputs are realized. The following facts should be observed concerning the program:

1. Line 1 indicates that the augend, addend, and sum are each stored as a six-element logical array with each element occupying one byte.

2. To avoid unwieldy FORTRAN for the input/output of arrays, the first element is the most significant bit, the sign, for all numbers; e.g., $S(1)$ is the sign bit of the sum.* Lines 6 and 7 of the DO loop are consistent with this and begin by adding the least significant bits.

---

*The FORTRAN convention is different from the bit numbering of Table 7.6 and other parts of the text, but if one exercises care, it should cause no difficulty.

```
          C*** SIMULATION OF A SERIAL ADDER
0001           LOGICAL*1  C,   S(6),  X(6),Y(6),P,Q,D
0002           WRITE (6,8)
0003      5    READ (5, 10, END=50) X,Y
0004           WRITE (6,11) X, Y
0005           C=.FALSE.
0006           DO 1 J=1,6
0007           I=7-J
0008           P=.NOT. X(I)
0009           Q= .NOT. Y(I)
0010           D=.NOT. C
0011           S(I) = (P.AND. Q .AND. C) .OR. (P.AND. Y(I)
               $    .AND.  D)
               $    .OR. (X (I).AND. Q .AND. D)  .OR.  (X(I)
               $    .AND. Y(I) .AND. C)
0012      1    C = (X(I) .AND.C ) .OR. (X(I) .AND. Y(I) )
               $    .OR. (Y(I) .AND. C)
0013           WRITE (6,13 )  S
0014           GO TO 5
0015      50   STOP
0016      8    FORMAT  ('1',15X'SIMULATION OF A SERIAL ADDER'
               $    /26X  'S = X+Y')
0017      10   FORMAT ( 6L2/6L2 )
0018      11   FORMAT   ('0',22X'X = ', 6L2/ 23X'Y = ',6L2)
0019      13   FORMAT   (23X 'S = ', 6L2)
0020           END
```

Figure 7.20   Simulation of a Six-Bit, 2's Complement Serial Adder.

3. Line 3 indicates that the program will continue to read sets of data, compute the sum, and print the data until it runs out of cards.

4. The adder will correctly produce the 2's complement result for all sums between $\pm 37_8$. (Note: No end-around carry is required.)

Input numbers and results, all in octal, for $S = 26 + 5 = 33$, $S = 35 + (-17) = 16$, and $S = (-17) + (-12) = 47 = -(30 + 1)$ are shown in Figure 7.21. Note that instead of the usual binary representation FORTRAN employs $T = 1$ and $F = 0$; however, this difference should cause no difficulty and the sums are seen to be correct.

Lines 5 through 12 of the program we have just discussed can easily be made into a two's complement subroutine which realizes the RTL statement $S \leftarrow X + Y$. The resulting FORTRAN call is: CALL TADD(X, Y, S, C, N) where N is the bits contained in each number and C is the carryout.* Similarly it is convenient to produce a subroutine for most of the other RTL statements. Then, one can simulate complicated systems by means of a few easy-to-remember subroutine calls and a few other supporting FORTRAN statements. Table 7.9† shows a representative set of RTL statements and their FORTRAN counterparts. (The actual programs for the subroutines are

---

*The two's complement subroutine is called in constructing the programs for other subroutines, e.g. 1's complement add, and it is here that the carryout is needed.

†This table is far from exhaustive particularly in the nonsubroutine cases; but the reader should have no trouble in achieving other needed functions.

Table 7.9   Logic Model Equivalents for Typical RTL Statements

RTL	FORTRAN*	Remarks
1. $S \leftarrow X + Y$	1. CALL ADD (X,Y,S,N)	1's complement addition.
2. $A \leftarrow B \wedge C$	2. CALL AND (B,C,A,N)	
3. $B \leftarrow \bar{A}$	3. CALL COM (A,B,N)	1's complement.
4. $A \leftarrow B$	4. CALL EQ (B,A,N)	
5. $A_{J-K} \leftarrow B_{L-M}$	5. CALL EQUAL (B,L,M,A,J,K)	For one bit transfer: $J=K$, $L=M$. ($J,K,L,$ and M are integer, decimal format.)
6. $A \leftarrow 0$	6. CALL FALSE (A,N)	
7. Fill Memory Array $M(L,N)$ (from cards)	7. CALL FILMEN (M,L,N)	$L\equiv$Locations filled. (L in decimal)
8. $A \leftarrow M(B)$	8. CALL LODA (M,B,A,L,N) ⎱	Symbols same as
9. $M(B) \leftarrow A$	9. CALL LODM (A,M,B,L,N) ⎰	line 7.
10. T-F/Octal conversion	10. CALL LOTO (L,O,N)†	LOgical To Octal
11. T-F/Decimal conversion	11. CALL LOTD (L,D,N)†	LOgical To Decimal
12. $A \leftarrow B \vee C$	12. CALL OR (B,C,A,N)	
13. Octal/T-F conversion	13. CALL OTOL (O,L,N)†	
14. Print Memory Array, M	14. CALL PRTMEM (M,L,N)	Symbols same as line 7.
15. $B \leftarrow$ sl$A$	15. CALL SL (A,B,N)	
16. $B \leftarrow$ sr$A$	16. CALL SR (A,B,N)	
17. $S \leftarrow X + Y$	17. CALL TADD (X,Y,S,C,N)	2's comp. $C\equiv$carry-out.
18. $A \leftarrow \bar{A} + 1$	18. CALL TCOM (A,N)	2's complement.
19. IF $(A=B)$ $C \leftarrow 1$	19. CALL TESTEQ (A,B,C,N)	C contains one bit.
20. IF $(A=0)$ $B \leftarrow 1$	20. CALL TESTZE (A,B,N)	B contains one bit.
21. IF $(A_3=0)$ $B_4 \leftarrow 0$	21. IF $(\cdot\text{NOT}\cdot A(3))B(4) = \cdot\text{FALSE}\cdot$	
22. IF $(A_3=1)$ $T \leftarrow 5$	22. IF $(A(3))$ GO TO 5	
23. $A_1 \leftarrow (\bar{B_1} \vee C_2) \wedge D_4$	23. $A(1)=(\cdot\text{NOT}\cdot B(1)\cdot\text{OR}\cdot C(2))\cdot\text{AND}\cdot D(4)$	
24. $D \leftarrow (A \wedge B) \vee C$	24. CALL AND (A,B,D,N) CALL OR (D,C,D,N)	

*The meaning of the symbols A, B, C, etc. are described by the corresponding RTL statements; all are N-bit arrays, in FORTRAN logical format except where noted. $N \equiv$ bits/register, in decimal. A complete listing of the subroutines is found in Appendix A.

†Here $O \equiv$ octal number with sign; $L \equiv$ an N-bit logical array, i.e., T-F symbols, in two's complement form. $D \equiv$ decimal number with sign.

given in Appendix A.) Lines 10 and 13 in the table, of course, do not have a RTL counterparts; but they are very convenient routines to convert between octal and true-false logic. In this way the user can handle input/output data in the more convenient octal numbers rather than the form of logic which FORTRAN must ultimately employ. (The Fill Memory and Print Memory subroutines, on Lines 7 and 14 will be discussed later.) Note that, on Lines 21 through 23 of the table, the RTL equivalents are simply individual FORTRAN statements.

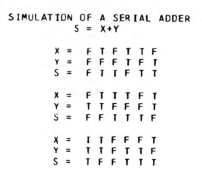

```
SIMULATION OF A SERIAL ADDER
            S = X+Y

        X =   F  T  F  T  T  F
        Y =   F  F  F  T  F  T
        S =   F  T  T  F  T  T

        X =   F  T  T  T  F  T
        Y =   T  T  F  F  F  T
        S =   F  F  T  T  T  F

        X =   T  T  F  F  F  T
        Y =   T  T  F  T  T  F
        S =   T  F  F  T  T  T
```

Figure 7.21   Serial Adder Results.

As an example of the application of the logic model, consider the problem of multiplying two 6-bit positive numbers and obtaining the 12-bit product. Figure 7.22 shows the block diagram and RTL statements required to solve the problem. In the logic diagram, note that the multiplier basically consists

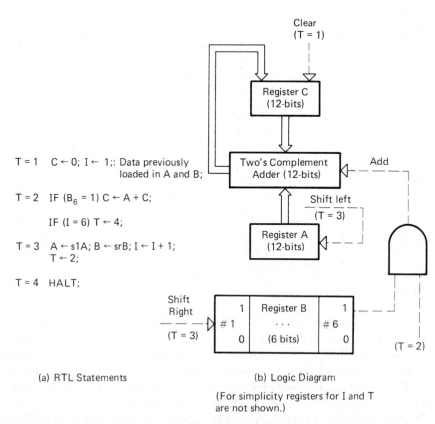

(a) RTL Statements   (b) Logic Diagram

(For simplicity registers for I and T are not shown.)

Figure 7.22   Multiply Algorithm ($C = B \times A$).

of two shift registers, A^* and B, for the two input numbers, a result register, C, and a 12-bit full adder. The timing signals are indicated by means of quantities within parentheses.

The FORTRAN program to realize the solution of this problem is shown in Figure 7.23. Examine the program carefully and observe how the various subroutines are called in correspondence to the RTL statements and the logic diagram. The convention employed is that FORTRAN statement numbers between 1 and 99 are placed in columns 4 and 5 to represent the T-states and statement numbers above 99 are placed in columns 2 through 4 to serve auxiliary portions of the simulation, such as numbering of format statements. In particular, note how the octal numbers are converted to T-F form at lines 7 and 8 and how the least significant bit of B is tested, at line 11, in order to decide whether to add the contents of A to C. The results obtained with the multiply program for several sets of input numbers are shown in Figure

```
      C***  SIMULATION OF A MULTIPLY UNIT. C=AXB
 1          IMPLICIT INTEGER (3)
 2          LOGICAL*1 A(12),B(6), C(12), D, L(12)
 3          WRITE (6,130)
 4          WRITE (6,140)
 5          WRITE (6,150)
 6     600  READ (5, 100, END= 5CC) 01, 02
      C     REGISTERS A AND B ARE FILLED FROM OCTAL INPUTS.
 7          CALL OTOL (01,  A,12)
 8          CALL OTOL (02,  B,6)
      C     THE RTL PORTION OF THE SIMULATION BEGINS
      C     HERE ANC ENDS BEFORE LINE 1E.
      C  T           ( RTL STATES, T, ARE IN COLUMNS 4&5.)
 9     1 CALL FALSE (C,12)
10          I=1
11     2 IF (B(6)) CALL TADD (A,C,C,D,12)
12          IF (I.EQ.6) GO TO 4
13     3 CALL SL (A,A,12)
14          CALL SR (B,B,6)
15          I=I+1
16          GO TO 2
17     4 CONTINUE
      C  4A HALT
      C     OUTPUT OF THE RESULTS NEXT OCCURS (LTNES 18 & 19).
      C     THEN A CHECK IS MADE TO SEE IF ANOTHER
      C     SIMULATION IS REQUIRED (LINE 2C).
18          CALL LOTO (C,0,12)
19          WRITE (6,170) 01,02,0
20          GO TO 600
21     500  STOP
22     100  FORMAT (2I4 )
23     130  FORMAT ('1',15X,'SIMULATION OF A MULTIPLY UNIT ' )
24     140  FORMAT ( 26X,'A X B = C ')
25     150  FORMAT ( '0',14X,'A',14X,'B',16X,'C' / '0' )
26     170  FORMAT ( 12X, I4,11X,I4, 13X, I6 )
27          END
```

Figure 7.23 Multiply Unit.

*Since it must be shifted left without losing significant bits and since it operates with a 12-bit adder, observe that register A contains 12 bits.

```
SIMULATION OF A MULTIPLY UNIT
        A X B = C

   A              B              C


   42             10             420
    6              5              36
   25             34             1114
   32             73             2776
```

Figure 7.24   Results of Simulation (all numbers in octal).

7.24. When checking these results, remember that all the numbers are in octal.

Associative Memory Circuit

In order to become familiar with more of the Logic Model Statements in Table 7.9, particularly, FILMEM, PRTMEM, LODA and LODM, consider the following problem: We want to add an associative memory circuit to the elementary computer of Figure 7.16. The purpose of this circuit is to search the memory until a match is found between the word in Memory Location One and any later word. The address of the first matching word found is to be placed in the last memory location, and if none is found, the last memory location is to be cleared. This circuit has practical value, e.g., in performing table lookup, and could easily be added to our elementary computer. In contrast to a program constructed from conventional machine language instructions, this "hardware routine" could be made to run very fast since once it is started the circuit does not require any fetch cycles until the task is completed.

Figure 7.25 shows the associative memory circuit and the components common with our elementary computer. Actually, the only major blocks added to our original computer is the register C and the comparator. The register C stores the contents of the first memory location, and the comparator indicates when the circuit has been successful in finding a matching word in memory. The blocks common with our elementary computer perform the same functions they did previously; e.g. the Memory Address Register points to the location in memory currently being processed. (One small change, however, is the fact that data can be transferred directly from MA into MB.)

Let us now consider the RTL program required to implement the circuit. To make this program more specific and to tie it to the simulation program to follow, however, assume that 11_8 is the last location in memory. The RTL program is shown in Figure 7.26. Note that T-states 3, 4 and 5 form a loop which checks one memory location at a time until either a match is found or the end of memory is reached. State 3 has the dual function of filling register C (first statement) and detecting end of memory (last statement). Since the

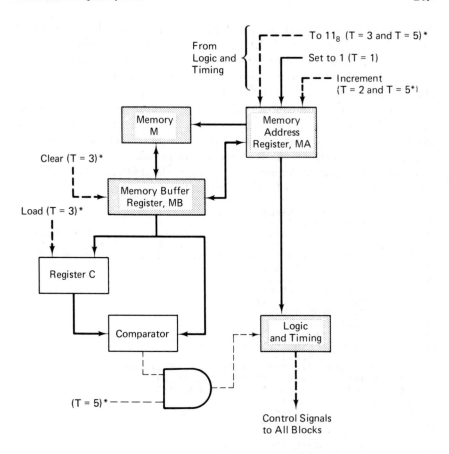

Figure 7.25 The Associative Memory Circuit.

$T = 1$ $MA \leftarrow 1$;
$T = 2$ $MB \leftarrow M(MA)$; $MA \leftarrow MA + 1$;
$T = 3$ IF $(MA = 2)$ $C \leftarrow MB$; IF $(MA = 12_8)$ $MB \leftarrow 0$, $MA \leftarrow 11_8$, $T \leftarrow 6$;
$T = 4$ $MB \leftarrow M(MA)$;
$T = 5$ IF $(MB = C)$ $MB \leftarrow MA$, $MA \leftarrow 11_8$ ELSE $MA \leftarrow MA + 1$, $T \leftarrow 3$;
$T = 6$ $M(MA) \leftarrow MB$;
$T = 7$ HALT;

Figure 7.26 RTL Program for the Associative Memory Circuit.

Memory Address Register is incremented, in both cases, before reaching State 3, it at first appears that MA is one number ahead of where it should be.

The simulation routine to realize the RTL program is shown in Figure 7.27. Observe that the program is divided into three sections: Statements 1 through 11 initialize the simulation, fill the memory from cards, and then print the initial memory contents; Statements 12 through 28 implement the formal RTL portion of the program; finally, Statements 29 through 36 print the final memory contents and serve miscellaneous functions e.g., format operations. The details of the RTL section, in particular, should be studied thoroughly and the correspondence between the FORTRAN statements and the RTL statements of Figure 7.26 should be checked.

```
     C*** ASSOCIATIVE MEMORY CIRCUIT
1         IMPLICIT INTEGER (O)
2         LOGICAL*1 A(6),B(6),C(6),D,E(6),G(6),M(9,6),MA(6),MB(6)
     C  READ IN CONSTANTS OA=1, OB=2, OE=11, OG=12 ALL IN OCTAL
     C  WITH A,B,E, AND G THE CORRESPONDING NUMBERS IN T-F FORM
3         READ (5,900) OA, OB, OE, OG
4         CALL OTOL (OA,A,6)
5         CALL OTOL (OB,B,6)
6         CALL OTOL (OE,E,6)
7         CALL OTOL (OG,G,6)
8         WRITE (6,840)
9         WRITE (6,860)
10        CALL FILMEM (M,9,6)
11        CALL PRTMEM (M,9,6)
     C  THE ACTUAL RTL PORTION OF THE SIMULATION
     C  BEGINS HERE AND ENDS BEFORE LINE 29
     C    T    (RTL STATES, T, ARE IN COLUMNS 4 AND 5 ONLY.)
12        1 CALL EQ (A,MA,6)
13        2 CALL LODA (M,MA,MB,9,6)
14          CALL ADD (MA,A,MA,6)
15        3 CALL TESTEQ (MA,B,D,6)
16          IF (D) CALL EQ (MB,C,6)
17          CALL TESTEQ (MA,G,D,6)
18          IF (D) CALL FALSE (MB,6)
19          IF (D) CALL EQ (E,MA,6)
20          IF (D) GO TO 6
21        4 CALL LODA (M,MA,MB,9,6)
22        5 CALL TESTEQ (MB,C,D,6)
23          IF (D) CALL EQ (MA,MB,6)
24          IF (D) CALL EQ (E,MA,6)
25          IF (.NOT.C) CALL ADD (MA,A,MA,6)
26          IF (.NOT.D) GO TO 3
27        6 CALL LODM (MB, M, MA, 9, 6)
28        7 CONTINUE
     C    7A HALT
     C        FINISH BY WRITING OUT THE RESULT.
29        WRITE (6,870)
30        CALL PRTMEM (M,9,6)
31        STOP
32   840  FORMAT ('1')
33   860  FORMAT (25X,'INITIAL')
34   870  FORMAT (26X,'FINAL')
35   900  FORMAT (5I10)
36        END
```

Figure 7.27 Simulation for the Associative Memory Circuit.

```
           INITIAL
       MEMORY CONTENTS
       ------------------

       1  F  T  F  F  T  F
       2  F  F  T  T  F  T
       3  F  T  T  F  T  T
       4  F  F  F  T  F  T
       5  T  F  F  T  T  T
       6  F  T  F  F  T  T
       7  F  T  F  F  F  T
       8  F  T  F  F  T  F
       9  T  F  F  T  T  F


            FINAL
       MEMORY CONTENTS
       ------------------

       1  F  T  F  F  T  F
       2  F  F  T  T  F  T
       3  F  T  T  F  T  T
       4  F  F  F  T  F  T
       5  T  F  F  T  T  T
       6  F  T  F  F  T  T
       7  F  T  F  F  F  T
       8  F  T  F  F  T  F
       9  F  F  T  F  F  F
```

Figure 7.28   Results of the Associative Memory Simulation.

The results of the associative memory simulation are shown in Figure 7.28. (Note that the memory contents are printed with the locations given in decimal down the left column and the contents of each location occupies a separate line.) In the initial memory contents, it is seen that the first match between the contents of Memory Location One and the contents of a later location occurs at Location Eight. The final memory contents, at Location Nine, verifies that the circuit has indeed found this match but that no other location in memory was disturbed.

The Arithmetic Model

In contrast to the logic model, where individual bits are processed and one concentrates on the detailed operations of each computer subsystem, there are many occasions when one would be willing to sacrifice the bit-by-bit details in order to simulate an entire computer system by means of a relatively short program. The arithmetic model is well suited to this need. In essence, it employs a FORTRAN arithmetic expression to represent each

RTL statement, and the data are represented by decimal digits in the integer mode rather than individual bits in the T-F mode.

Table 7.10 shows the correspondence between statements in the arithmetic model and those in RTL. The present model is mainly useful for simulation at a fairly high level, e.g., the shift family of instructions does not even appear explicitly. As mentioned above, our model operates on decimal numbers; thus, the add operations at statements 4 and 5 are really decimal. In contrast, many computers, such as the LINC, actually operate in the octal system. The above properties of the arithmetic model must be correctly interpreted for any digital simulation; but when properly used, the model can produce very helpful and concise simulation programs.

In Section 7.3 we described the organization of a computer by employing the RTL notation. Figure 7.29 shows the FORTRAN simulation, arithmetic model, for a simple version of that computer. The fetch cycle in the program should be compared with the RTL diagram in Figure 7.17, and the execution cycle for the ADD, STC, and other instructions should be compared with Table 7.6. (Note that the variables MA, PC, etc. have the same meanings that they had in Figure 7.16 and that there are comments to indicate the beginning of each main section of the program.)

In analyzing the beginning of the program, the reader should observe that the computer memory is represented by a 100-element array, $M(1)$ through $M(100)$, and that for purposes of memory management:

$ML \equiv$ the lower memory limit actually printed

$MH \equiv$ the upper (highest) memory limit printed

$N \equiv$ the highest memory location into which instructions and data are initially entered.

Table 7.10   Arithmetic Model Equivalents for Typical RTL Statements

RTL	FORTRAN Equivalent	Remarks
1. $A \leftarrow B$	1. $A = B$	
2. $A_2 \leftarrow B_3$	2. $A(2) = B(3)$	A and B are arrays.
3. $A \leftarrow 0$	3. $A = 0$	
4. $A \leftarrow B + C$	4. $A = B + C$	
5. $A \leftarrow A + 1$	5. $A = A + 1$	
6. $A \leftarrow \overline{A}$	6. $A = -A$	
7. $A \leftarrow M(B)$	7. $A = M(B)$	$M(B)$ = contents of memory location B. $M(B)$ is an array.
8. $M(B) \leftarrow A$	8. $M(B) = A$	
9. IF $(A = 1)$ $B \leftarrow 0$	9. IF $(A \cdot EQ \cdot 1)$ $B = 0$	
10. IF $(A > 1)$ $T \leftarrow 5$	10. IF $(A \cdot GT \cdot 1)$ GO TO 5	T is the state of the system.
11. $T \leftarrow 5$	11. GO TO 5	

```
                C****    SIMPLE COMPUTER SIMULATION
      0001                DIMENSION  M(100)
      0002                INTEGER PC,A
                C****    MEMORY CLEAR, READ IN, AND PRINT
      0003      2         DO 3 I= 1,100
      0004      3         M(I) =0.
      0005                WRITE (6,80 )
      0006                READ (5,72, END=70 )PC,N, ML, MH
      0007                READ (5,79) (M(I), I=20,N )
      0008                WRITE (6,73) PC,N, ML, MH
      0009                WRITE (6,74)
      0010                WRITE (6,75 ) (M(I), I=ML, MH )
                C****    FETCH CYCLE
      0011      5         MA =PC
      0012                MB= M(MA)
      0013                PC =PC +1
                C****    DECODER PART OF FETCH
      0014                IF (MB.LT. 2000. ) GO TO 10
      0015                IF (MB.LT. 4000. ) GO TO 20
      0016                IF (MB.LT. 6000. ) GO TO 30
      0017                IF (MB.LT. 8000. ) GO TO 40
      0018      10        IF (MB .EQ. 0.  ) GO TO 50
      0019                IF (MB .EQ. 11.) GO TO 12
      0020                IF (MB .EQ. 17.) GO TO 14
      0021                IF (MB .EQ.451.) GO TO 15
      0022                GO TO 2
                C****    ACD INSTRUCTION
      0023      20        MA=MB-2000.
      0024                MB = M(MA)
      0025                A=A +MB
      0026                WRITE (6,77) PC,MA,A
      0027                GO TO 5
                C****    STC INSTRUCTION
      0028      30        MA= MB-4000.
      0029                MB=A
      0030                A=0
      0031                M(MA) =MB
      0032                WRITE (6,77) PC,MA,A
      0033                GO TO 5
                C****    JMP INSTR.
      0034      40        PC= MB-6C00.
      0035                WRITE (6,77) PC,MA,A
      0036                GO TO 5
                C****    CLR INSTR.
      0037      12        A=0
      0038                WRITE (6,77) PC,MA,A
      0039                GO TO 5
                C****    COM INSTR.
```

Figure 7.29 Computer Simulation—FORTRAN Model.

Note that the read statement on line 6 allows several programs to be run in sequence. The statement on line 22 causes any undecoded instruction, a NOP, to cause the current simulation to terminate and a new program to be read in.

Figure 7.30 presents the results after the problem shown in Table 3.4 is run with our simulator. Observe the following facts about the results:

1. The first line shows the starting value of PC along with *N*, *ML*, and *MH*. Contents of the memory before the program ran are printed

```
0040            14      A =-A
0041                    WRITE (6,77) PC,MA,A
0042                    GO TO 5
                C****   APO INSTR.
0043            15      IF (A .LT. C. ) GO TO 16
0044                    PC =PC+1
0045            16      WRITE (6,77) PC,MA,A
0046                    GO TO 5
                C****   HLT INSTR.
0047            50      WRITE (6,77 ) PC,MA,A
0048                    WRITE (6,81)
0049                    WRITE (6,75 ) (M(I), I=ML, MH )
0050                    GO TO  2
0051            70      STOP
0052            72      FORMAT (413)
0053            73      FORMAT ( 31X,' PC=',  I3, '    N=', I3,
                $      '   ML=',  I3, '   MH=', I3)
0054            74      FORMAT('0',35X,'INITIAL MEMORY CONTENTS ')
0055            75      FORMAT ( 20X, 5I10)
0056            77      FORMAT ( 35X, ' PC=',  I3,3X,'MA=',I3,5X,'A=',I4)
0057            79      FORMAT (           5I10)
0058            80      FORMAT ('0')
0059            81      FORMAT ('0',35X,' FINAL MEMORY CONTENTS ')
0060                    END
```

Figure 7.29—*Cont.*


```
              PC= 20    N= 28    ML= 15    MH= 50

                    INITIAL MEMORY CONTENTS
        0              0              0              0              0
       11           2027           2025           4025           2028
     4030              0              5              1              0
        0              0              0              0              0
        0              0              0              0              0
        0              0              0              0              0
        0              0              0              0              0
        0
                    PC= 21    MA= 20    A=    0
                    PC= 22    MA= 27    A=    5
                    PC= 23    MA= 25    A=4035
                    PC= 24    MA= 25    A=    0
                    PC= 25    MA= 28    A=    1
                    PC= 26    MA= 35    A=    0
                    PC= 27    MA= 26    A=    0

                    FINAL MEMORY CONTENTS
        0              0              0              0              0
       11           2027           2025           4025           2028
     4035              0              5              1              0
        0              0              0              0              0
        1              0              0              0              0
        0              0              0              0              0
        0              0              0              0              0
        0
```

Figure 7.30   Results From (Fig. 7.29) Simulation.

with $ML = 15$, indicating that the first memory location, column 1 row 1, is address 15.*

2. Each time an instruction is completed, the values for PC, MA, and A are printed.

3. Finally, after the program has halted, the memory contents are again printed, and the changes to the various memory locations can be noted.

4. For convenience, some small changes were made in the memory locations employed in the simulation as compared with Table 3.4.

Since the simulation is for a decimal machine and since Table 3.4 is actually an octal program, some minor differences appear, e.g., a memory address 28 exists. This and other possible differences could be eliminated; but it is felt that the added program complexity is not warranted. The arithmetic model will be further developed in the Exercises for this chapter.

7.6 BIBLIOGRAPHY

1 BARBACCI, M. R., "A Comparison of Register Transfer Languages for Describing Computers and Digital Systems." *IEEE Trans. Comput.*, Vol. C-24, 1975, pp. 133–50.

2 BELL, C. G., J. GRASON, and A. NEWELL, *Designing Computers and Digital Systems*. Maynard, Mass.: Digital Press, 1972.

3 BELL, C. G., J. L. EGGERT, J. GRASON, and P. WILLIAMS, "The Description and Use of Register Transfer Modules (RTM's)." *IEEE Trans. Comput.*, Vol. C-21, 1972, pp. 495–500.

4 BREUER, MELVIN (ed.), *Digital System Design Automation: Languages, Simulation and Data Base*. Woodland Hills, California: Computer Science Press, Inc., 1975.

5 CHU, YAOHAN, *Computer Organization and Microprogramming*. Englewood Cliffs, N.J.: Prentice-Hall, Inc., 1972.

6 CHU, YAOHAN (ed.), "Hardware Description Languages." *Computer* (IEEE Computer Society), Vol. 7, No. 12, 1974, pp. 18–66.

7 FOSTER, CAXTON C., *Computer Architecture*. New York: Van Nostrand Reinhold Company, 1970.

8 FRIEDMAN, T. D., and S. C. YANG, "Methods Used in Automatic Logic Design Generator (ALERT)." *IEEE Trans. Computers*, Vol. C-18, 1969, pp. 593–614.

*Each row contains the contents of five memory locations in sequence.

9   GREENBERG, STANLEY, *GPSS Primer.* New York: John Wiley & Sons, Inc., 1972.

10 HILL, FREDRICK J., and GERALD R. PETERSON, *Digital Systems: Hardware Organization and Design.* New York: John Wiley & Sons, Inc., 1973.

11 MANO, M. MORRIS, *Computer Logic Design.* Englewood Cliffs, N.J.: Prentice-Hall, Inc., 1972.

12 RHYNE, V. THOMAS, *Fundamentals of Digital Systems Design.* Englewood Cliffs, N.J.: Prentice-Hall, Inc., 1973.

13 STUART, FREDRIC, *WATFOR-WATFIV FORTRAN Programming.* New York: John Wiley & Sons, Inc., 1971.

14 TEICHEROW, D., and J. F. LUBLIN, "Computer Simulation Discussion of the Technique and Comparison of Language." *Comm. ACM,* Vol. 9, 1966, pp. 723–41.

7.7 EXERCISES

7.1 The initial contents of a group of registers are: $A = 10$, $B = 00$, and $C = 01$. List their contents after execution of the following operation:

$$\text{IF } ((A_1 = 1 \text{ AND } B_0 = 0) \text{ OR } C_1 = 1) \ A_0 \leftarrow 1$$

7.2 The initial contents of a group of registers are: $A = 101$, $B = 0001$, $C = 0010$, and $D = 101$. List their contents after execution of the following operation:

$$\text{IF } (A_1, B = 0, 1) \ C \leftarrow \text{rl2}D$$

7.3 By employing three-bit, *J-K*, registers, draw the logic diagram, similar to Figure 7.1, for the production:

$$\bar{B} \leftarrow \bar{A} \wedge B$$

7.4 After the RTL program below halts, determine the following:
(a) The contents of registers A, B, and C.
(b) The number of times the production at $T = 2$ was executed.
(c) Where the program halts.

$T = 0$ $A \leftarrow 0011$; $B \leftarrow 1011$; $C \leftarrow 0010$;: Only positive numbers allowed;

$T = 1$ IF $(A_1, B_2 = 1, 0) \ C_2 \leftarrow 1$ ELSE $C \leftarrow \bar{C}, T \leftarrow 6$;

$T = 2$ $C \leftarrow C + 1$;

$T = 3$ IF $(C < B) \ T \leftarrow 2$;

$T = 4$ $B \leftarrow 0$;

$T = 5$ HALT;

$T = 6$ HALT;

7.5 By employing an algorithm based on repeated subtraction of the divisor from the dividend, write an RTL division program which places the quotient in register Q and the remainder in register R. Assume:

(a) Initially the divisor is in register DR and the dividend is in register DD.

(b) The given values of DR and DD are always positive.

(c) The fewest possible states, T, are to appear in the program.

7.6 It is desired to add a new instruction called S̲kip on P̲ositive M̲emory to Table 3.2. The mnemonic is SPM XXX and the code is 7XXX. The purpose of the instruction is to test whether the content of memory location XXX is positive. If it is, the next instruction is skipped; otherwise, the next instruction in sequence is performed. Write an RTL program for this instruction similar to what is done in Table 7.6.

7.7 The LINC employs an instruction whose mnemonic is LDA, code 1000, with meaning L̲oaD̲ A̲ccumulator with the number from memory location Y. The *address* of Y is in $p + 1$. Take the next instruction from $p + 2$. (p is the location of the LDA instruction.) Write an RTL program for this instruction and comment on the number of clock pulses required.

7.8 Write the RTL programs for two new pseudo-LINC instructions: jump to subroutine, JMS, and return from subroutine, RFS. JMS is an instruction located at address p; location $(p + 1)$ contains the address where the start of the subroutine may be found; and after the subroutine is completed the next instruction is taken from $(p + 2)$. Employ the smallest possible number of clock cycles and, within this limit, the smallest number of RTL statements. Any extra registers needed may be labeled RA, RB, etc. (Assume that a subroutine may not be called from within another subroutine.)

7.9 Add the NOP instruction (see Table 3.2) and the LDA instruction (see Exercise 7.7) to your arithmetic simulation deck (Figure 7.25). Next, employ a program similar to that in Table 3.5 to sum the numbers in locations 40 through 47. (Place a 1 in each data location.) Make the first instruction in your program a NOP, and replace the CLR and ADD instructions at the beginning of the program with LDA. Run your program and comment on the results.

7.10 Write a FORTRAN subroutine which is the equivalent of the RTL statement:

$$B \leftarrow \text{rr}A$$

(Any of the routines listed in Table 7.9 may be called as part of this subroutine.)

7.11 Write the FORTRAN arithmetic model equivalent to the RTL statement:

$$\text{IF } ((C \geq 2 \text{ OR } B = 1) \text{ AND } D = 0) \ A \leftarrow 1$$

7.12 By employing the logical model, write a FORTRAN program to simulate the division algorithm mentioned in Exercise 7.5. Test your program, using six-bit numbers for DD and DR, on the following problems: 54/6, 57/16.

7.13 (a) Using clocked *J-K* flip-flops, draw the logic circuit for the state generator and the IF statement part of the RTL program shown below:

$$T = 1 \quad *;$$
$$T = 2 \quad *;$$
$$T = 3 \quad \text{IF } (A = 0) \; T \leftarrow 1;$$
$$T = 4 \quad \text{IF } (A = 1) \; T \leftarrow 2 \text{ ELSE } T \leftarrow 1;$$

Assume that *A* is a level supplied as an input; the T_i states are in a binary-ordered sequence starting with $T_1 = 00$; and the symbol * represents an arbitrary production which implies $T \leftarrow T + 1$ but has no other effect on the state generator design or on the input *A*.

(b) Draw the timing diagram for the circuit of part (a); show T_1 through T_4, C_p, and *A*. Starting the circuit in state T_1 before the first clock pulse occurs, present 12 clock cycles for the case where $A = 0$ through the first four clock cycles and $A = 1$ for the remaining cycles.

7.14 Repeat Exercise 7.13(a) for the program shown below, but employ a ripple counter as a basis for the state generator.

$$T = 1 \qquad\qquad *;$$
$$T = 2 \qquad\qquad \text{IF } (A = 0) \; T \leftarrow 2;$$
$$T = 3 \qquad\qquad *;$$
$$T = 4 \qquad\qquad \text{IF } (A = 0) \; T \leftarrow 4;$$
$$T = 5 \text{ through } T = 7 \quad *;$$
$$T = 8 \qquad\qquad T \leftarrow 1;$$

7.15 Write an RTL program to process the following problem. A data source register *D* contains a new number following each clock pulse. If the $S = 1$ level is present, arrange the next three numbers in three registers, R_1, R_2, and R_3, such that $R_1 \geq R_2 \geq R_3$. When the results are ready, generate a level *F*, which lasts one clock period. Then return and look for another $S = 1$ level. Minimize the number of states, T_i employed in the programs; but there is no requirement to recover any data lost during the processing of a group of three numbers.

7.16 Write the RTL program for a unit which produces a variable time delay.
(a) The delay starts whenever the level *S* goes to 1 after having been at 0.
(b) The delay is proportional to the contents of register *D*, but only the outputs of *D*, i.e., its Y_i and $\bar{Y}_i$ values, are available.
(c) When the delay expires, the level *F* is to be set to 1 and stays there for the next 10 clock pulses. After this, the system should be available to start another delay.
(d) Minimize the number of states, T_i, employed.

7.17 Using the array of Figure 7.13(b), list the pairs of 12-bit words that represent the following characters: A, C, E, 3, and 4.

7.18 Using digital logic not part of a computer, write an RTL program, with fewest possible *T* states, for a system which adds the contents of memory locations 10 through 15 and places the sum in location 25. Before the program starts, you may assume that the data are in locations 10–15 and that a register F with the condition $F = 0$ is available. Other needed registers should be labeled A, B, C, etc. The memory employs a nondestructive readout.

7.19 How would the RTL programs of Figure 7.17 and Table 7.6 change if a destructive readout memory were assumed? Be specific concerning what RTL statements would be added or deleted at what state, T_i.

7.20 Using the fewest possible states, T_i, and being consistent with the constraints placed on the pseudo-LINC by Table 3.2, Figure 7.16, and other pertinent sections of the text, write an RTL program for a new instruction: C̲lear on E̲ven location C̲ontents (CEC XXX). If the content of XXX is even, clear that memory location. In either case, the accumulator is unchanged.

8

THE
ARITHMETIC
AND
LOGIC UNIT

The Arithmetic and Logic Unit, ALU, is one of the most important parts of a digital computer because this is where the actual "number crunching" is accomplished. There is a wide range in size and complexity of the ALUs employed in computers. Some only include hardware for the add, complement, and shift operations, with all remaining work being accomplished with software subroutines. On the other hand, there are machines that not only have multiply, divide, and floating-point hardware, but also other special wired features such as square root and fast Fourier transform.

The principles applicable to ALU design will be developed here using the tools of Chapters 4, 5, and 7. Excellent opportunities to produce innovative ideas and to appreciate the original work of others exist in the ALU area. Many examples showing the trade-offs between cost, processing speed, and circuit complexity will be presented.

8.1 BASIC HARDWARE FOR ADDITION AND SUBTRACTION

The adder is the fundamental ALU circuit; it is often employed as a building block for the construction of multiply, divide, and other units. Our goal is to find the combinational circuit, outlined in Figure 8.1, which will realize the $(n + 1)$ sum bits resulting from the addition of two n-bit numbers.

In the Digital Equipment Corporation's PDP-11 family of computers, n is 16; in the IBM System/370, it is 32; and in some other machines, it is even larger. Because of the number of variables involved, we cannot hope to design the adder combinational circuit, even for the PDP-11, by employing the same techniques used in Chapter 4. Thus, a heuristic method would seem to be appropriate—perhaps one starting with an elementary *module* which can then be iterated for as many bits as the adder requires. If we try to *segment* this problem into an individual circuit or module, for each bit of the sum, we will soon conclude that the manual addition process—employing one addend bit and the corresponding augend bit, resulting in a sum and carry output for

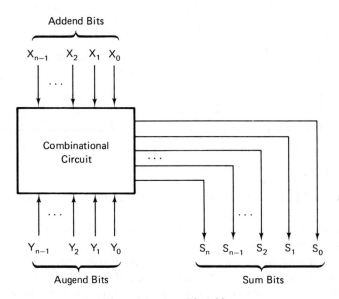

Figure 8.1 An n-Bit Adder.

each stage—can easily be implemented. (The resulting circuit is called a *full adder* in computer terminology.)

The addition process can be represented as:

$$\begin{array}{c} X_i \\ Y_i \\ \underline{C_{i-1}} \\ C_i S_i \end{array}$$

where C_{i-1} represents the carry-in, C_i represents the carry-out, and the other symbols are defined in Figure 8.1. Table 8.1 presents the required logic

Table 8.1   Truth Table for the Full Adder

	Input		Output	
X_i	Y_i	C_{i-1}	S_i	C_i
0	0	0	0	0
0	0	1	1	0
0	1	0	1	0
0	1	1	0	1
1	0	0	1	0
1	0	1	0	1
1	1	0	0	1
1	1	1	1	1

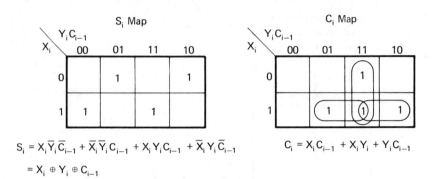

$$S_i = X_i \overline{Y}_i \overline{C}_{i-1} + \overline{X}_i \overline{Y}_i C_{i-1} + X_i Y_i C_{i-1} + \overline{X}_i Y_i \overline{C}_{i-1}$$

$$= X_i \oplus Y_i \oplus C_{i-1}$$

$$C_i = X_i C_{i-1} + X_i Y_i + Y_i C_{i-1}$$

Figure 8.2   Karnaugh Maps and Simplified Equations for the Full Adder.

relationship between the variables, and Figure 8.2 shows the corresponding Karnaugh maps and the resulting simplified functions.

Note that the most concise form for the sum output is the EXOR function. When one of the full-adder circuits is employed for each bit of the adder of Figure 8.1, the resulting circuit is as shown in Figure 8.3. Since there is no carry into the least significant bit, the C_{i-1} input to that circuit is constant at logical 0. Also, the C_i output of the leftmost full adder becomes the S_n bit.

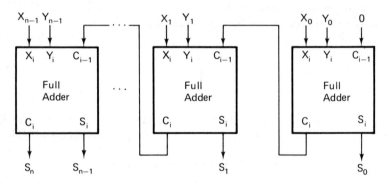

Figure 8.3   The Basic Parallel Adder.

Serial Addition

Is it possible to do n-bit addition without employing n full adders? Again from heuristic considerations, using an analogy with manual addition, we see that by processing the bits serially only one full adder is needed; but an extra flip-flop will be required to store the previous carry. This scheme is shown in Figure 8.4 where the following process occurs: The current LSB of registers X and Y are combined, in the full adder, with the previous carry, yielding a new sum, S_i, and a new carry, C_i, S_i is shifted into the MSB end of

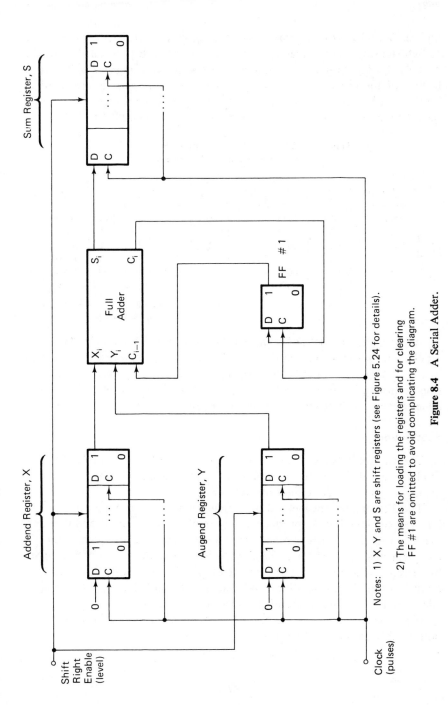

Notes: 1) X, Y and S are shift registers (see Figure 5.24 for details).

2) The means for loading the registers and for clearing
   FF #1 are omitted to avoid complicating the diagram.

Figure 8.4 A Serial Adder.

register S, and C_i is temporarily stored in flip-flop 1; registers X and Y are shifted right by one bit, placing new values in the LSB position; finally, the process is repeated until the complete sum is stored in register S. The time history of the circuit, when adding two four-bit numbers, is presented in Table 8.2. (Assuming the line marked Start is given, the reader should verify the contents of the table by employing the logic of Figure 8.4.)

**Table 8.2   Serial Addition Example (5 + 7 = 12)
Using the Circuit of Figure 8.4**

After Clock Pulse Number	X	Y	$FF1$	C_{i-1}	X_i	Y_i	S_i	C_i	S
Start	0101	0111	0	0	1	1	0	1	00000
1	0010	0011	1	1	0	1	0	1	00000
2	0001	0001	1	1	1	1	1	1	00000
3	0000	0000	1	1	0	0	1	0	10000
4	0000	0000	0	0	0	0	0	0	11000
5	0000	0000	0	0	0	0	0	0	01100

Result = 12_{10}, as expected.

As with the parallel adder, $n + 1$ sum bits are generated by two n-bit inputs, and this is the reason that $n + 1 = 5$ clock pulses were employed in Table 8.2 to get S shifted to the proper position in its register. Further consideration will be given to adders, e.g., gate implementation and the half-adder circuit, in the Exercises at the end of this chapter.

Subtraction

The two main ways to perform subtraction are by the complement method, discussed in Chapter 2, and by mechanizing a special subtractor circuit analogous to the adder circuit in Figure 8.3.

The 2's complement method is particularly simple since the carry-out of bit n is discarded, both in producing the complement and in the addition step that follows. By contrast, an end-around carry must be provided in the 1's complement system. This may be accomplished with the circuit of Figure 8.3 by merely connecting the carry out of the nth bit (signal S_n) into the C_{i-1} terminal of bit 0. (The modification needed in the serial adder, Figure 8.4, to allow for 1's complement numbers is left as an exercise.)

Now consider the alternate approach—a circuit which performs subtraction directly by means of the process

$$\begin{array}{r} X_i \\ -Y_i \\ -B_{i-1} \\ \hline B_i D_i \end{array}$$

where X is the minuend, Y is the subtrahend, B is the borrow, and D is the difference. From heuristic reasoning, similar to what was done in the design of the parallel adder, a *full subtractor* circuit is first designed. The corresponding truth table is shown in Table 8.3, and the resulting Karnaugh maps with their simplified equations are shown in Figure 8.5. A complete parallel subtractor circuit looks exactly the same as the parallel adder in Figure 8.3, except that the sum variables, S, are changed to difference variables, D, and the carry variables, C, are changed to borrow variables, B.

Table 8.3 Truth Table for the Full Subtractor

Input			Output	
X_i	Y_i	B_{i-1}	D_i	B_i
0	0	0	0	0
0	0	1	1	1
0	1	0	1	1
0	1	1	0	1
1	0	0	1	0
1	0	1	0	0
1	1	0	0	0
1	1	1	1	1

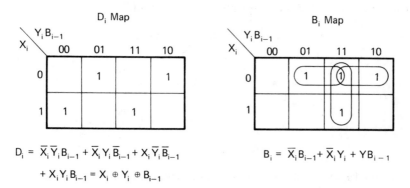

$$D_i = \overline{X}_i \overline{Y}_i B_{i-1} + \overline{X}_i Y_i \overline{B}_{i-1} + X_i \overline{Y}_i \overline{B}_{i-1}$$
$$+ X_i Y_i B_{i-1} = X_i \oplus Y_i \oplus B_{i-1}$$

$$B_i = \overline{X}_i B_{i-1} + \overline{X}_i Y_i + Y B_{i-1}$$

Figure 8.5 Karnaugh Maps and Simplified Equations for the Full Subtractor.

‡8.2 FAST ADDERS

The main disadvantage of the basic parallel adder circuit in Figure 8.3 is that a carry originating at bit 0 may propagate (ripple down) the adder as far as the nth bit. (This occurs, for example, when $+1$ is added to -0 in the 1's complement system.) As Figure 8.2 shows, each carry stage is a two-level circuit; thus the ripple time, T, can be as long as $T = 2n\,\Delta t$ seconds, where

Δt is the propagation time for one level. If $n = 32$ and $\Delta t = 10$ ns, $T = (2)(32)(10) = 640$ ns, which is much too long for most applications. Of course, Δt c n be reduced considerably by the use of ECL or other fast circuitry, but there is a definite physical limit to how small Δt can be made. Moreover, if ECL is employed throughout the computer, there will still be a serious difference between the operating time of the adder and most other units. For these reasons, the basic adder is augmented in most computers by some additional circuitry.

One of the most popular adder improvements is known as the *carry look-ahead principle*. This method reduces the ripple time by employing fewer than $2n$ levels between the first and the nth bits. Since any Boolean function can be represented by a two-level circuit, it is possible to reduce the ripple delay to $T = 2\,\Delta t$; but for reasonable values of n, the cost is prohibitive. Therefore, the current approach seeks a tradeoff between cost and adder speed.

We begin by modifying Table 8.1 to emphasize that any adder stage can produce a carry by two mechanisms—either the carry is *generated* within the stage i, or it was generated previously and stage i is serving to *propagate* it. Both of these conditions are illustrated in Table 8.4. Observe that those cases where the carry-out depends on the carry-in are called *propagation stages*, and the case where $X_i = Y_i = 1$ is called a *generation stage* because a carry-out will be generated independently of the carry-in.

Table 8.4 Carry Conditions

| Input | | Carry Output | |
X_i	Y_i	C_i	Remarks
0	0	0	
0	1	C_{i-1}	Propagation, P_i
1	0	C_{i-1}	Propagation, P_i
1	1	1	Generation, G_i

From Table 8.4, the equation that describes carry generation, G_i, is:

$$G_i = X_i Y_i \tag{8.1}$$

and the equation that describes carry propagation, P_i, is:

$$P_i = \bar{X}_i Y_i + X_i \bar{Y}_i = X_i \oplus Y_i \tag{8.2}$$

Therefore, the carry-out of state i becomes:

$$C_i = G_i + P_i C_{i-1} \tag{8.3}$$

It is also possible to express the sum output in terms of P_i and C_{i-1} by rearranging the equation in Figure 8.2 and then substituting Equation (8.2):

$$S_i = X_i\bar{Y}_i\bar{C}_{i-1} + \bar{X}_i\bar{Y}_iC_{i-1} + X_iY_iC_{i-1} + \bar{X}_iY_i\bar{C}_{i-1}$$
$$= (X_i\bar{Y}_i + \bar{X}_iY_i)\bar{C}_{i-1} + (\bar{X}_i\bar{Y}_i + X_iY_i)C_{i-1}$$
$$= P_i\bar{C}_{i-1} + (\overline{X_i\bar{Y}_i + \bar{X}_iY_i})C_{i-1} \tag{8.4}$$
$$= P_i\bar{C}_{i-1} + \bar{P}_iC_{i-1} = P_i \oplus C_{i-1}$$

Returning to Equation (8.3), we see that this equation defines the carry output recursively. Thus, the equation for C_i may be rewritten in terms of the parameters of previous stages as follows:

$$C_i = G_i + P_i(G_{i-1} + P_{i-1}C_{i-2})$$
$$= G_i + P_iG_{i-1} + P_iP_{i-1}(G_{i-2} + P_{i-2}C_{i-3})$$

Thus, for m bits the result is:

$$C_i = G_i + P_iG_{i-1} + P_iP_{i-1}G_{i-2} + P_iP_{i-1}P_{i-2}G_{i-3} + \ldots$$
$$+ P_iP_{i-1}P_{i-2}\ldots P_{i-m+1}C_{i-m} \tag{8.5}$$

Equation (8.5) can be expressed in the more concise form:

$$C_i = G_i + \left(\prod_{j=i}^{i} P_j\right)G_{i-1} + \left(\prod_{j=i-1}^{i} P_j\right)G_{i-2} + \ldots + \left(\prod_{j=i-m+1}^{i} P_j\right)C_{i-m} \tag{8.6}$$

Finally, by defining the carry-in as

$$G_{i-m} = C_{i-m}$$

we can express Equation (8.6) in the still more concise form:

$$C_i = \sum_{k=i-m}^{i} \left(\prod_{j=k+1}^{i} P_jG_k\right) \tag{8.7}$$

where $0 \leq m \leq i + 1$. (Recall that $\prod_{j=i+1}^{i} P_j = 1$.)

For the case $m = i + 1$, Equation (8.7) produces the carry for any bit i in terms of the previous bits all the way down to bit 0, and in this case $G_{-1} = C_{-1} =$ carry into bit 0. Since P_i is a two-level function, Equation (8.5) reveals that C_i is a four-level expression. Consequently, the carry delay is $T = 4 \Delta t$, which is extremely fast compared to the ripple carry. For example, for $n = 32$ and $\Delta t = 10$ ns, it was previously determined that $T = 640$ ns; but in this case, it would only require $T = (4)(10) = 40$ ns. On the other hand, the logic gates to mechanize Equation (8.7) for all 32 bits would lead to prohibitive costs in terms of numbers of gates and the number of inputs per gate.

A good compromise in a situation like this is to construct carry look-ahead, CLA, logic stages consisting of l bits each, where l is usually a number not more than 6. (To avoid overly complicated diagrams, yet to bring out the basic principles, a stage consisting of four bits will be considered here.) First we will show, in Figure 8.6, how a modified form of the full adder can be produced having the necessary G_i, P_i, and S_i outputs. Now gate 2 generates

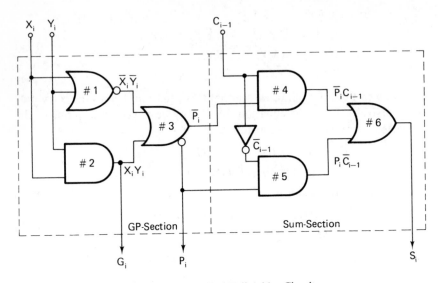

Figure 8.6   Modified Full-Adder Circuit.

the G_i output directly from Equation (8.1); the inverted output of gate 3 yields

$$P_i = \overline{X_i Y_i + \bar{X}_i \bar{Y}_i} = (\bar{X}_i + \bar{Y}_i)(X_i + Y_i) = X_i \bar{Y}_i + \bar{X}_i Y_i$$
$$= X_i \oplus Y_i$$

as expected from Equation (8.2), and the output of gate 6 produces the S_i term directly from Equation (8.4).

Having completed the modified full adder, MFA, it is now necessary to show how the carry signal is generated. Since we have agreed to use CLA stages consisting of four bits each, stage 2 representing bits four through seven is arbitrarily selected as a typical circuit. Here, $i = 7$, $m = 4$, and Equation (8.7) becomes:

$$C_7 = \sum_{k=7-4}^{7} \left(\prod_{j=k+1}^{7} P_j G_k \right) = \prod_{j=4}^{7} P_j G_3 + \prod_{j=5}^{7} P_j G_4 + \prod_{j=6}^{7} P_j G_5$$
$$+ \prod_{j=7}^{7} P_j G_6 + \prod_{j=8}^{7} P_j G_7 \qquad (8.8)$$

$$= P_4 P_5 P_6 P_7 G_3 + P_5 P_6 P_7 G_4 + P_6 P_7 G_5 + P_7 G_6 + G_7$$

Similarly C_6 is found with $i = 6$, $m = 3$:

$$C_6 = \sum_{k=3}^{6} \left(\prod_{j=k+1}^{6} P_j G_k \right) = P_4 P_5 P_6 G_3 + P_5 P_6 G_4 + P_6 G_5 + G_6 \qquad (8.9)$$

C_5 is found with $i = 5$, $m = 2$:

$$C_5 = \sum_{k=3}^{5} \left(\prod_{j=k+1}^{5} P_j G_k \right) = P_4 P_5 G_3 + P_5 G_4 + G_5 \qquad (8.10)$$

Finally, C_4 is found with $i = 4$, $m = 1$:

$$C_4 = \sum_{k=3}^{4} \left(\prod_{j=k+1}^{4} P_j G_k \right) = P_4 G_3 + G_4 \qquad (8.11)$$

(In all of the above equations, it should be noted that $G_3 = C_3 =$ carry into the stage.)

By employing Equations (8.8) through (8.11), and by representing the modified full adder with two blocks, as implied by the dashed sections of Figure 8.6, we can represent the complete CLA circuit for stage 2 as shown in Figure 8.7.* (It should be emphasized that the *GP* section of each MFA generates its outputs independently of the carry into that bit—of course, this is the key to the success of the CLA method.) There are seven other stages which are exactly the same as the one shown in Figure 8.7 except that each subscript has a multiple of 4 either added to or subtracted from it. The complete CLA adder, with all eight stages, is shown in Figure 8.8.

How does the 32-bit CLA adder compare with the ripple adder? It was previously established that each CLA stage is a four-level circuit, and there is ripple carry between stages. Thus, each stage must wait for the previous one to settle before developing the carry output for the next stage. The total propagation time of the adder is therefore

$$T_{\text{CLA}} = (8)(4)\,\Delta t = 32\,\Delta t$$

and this is to be compared with the ripple adder

$$T_{\text{ripple}} = (32)(2)\,\Delta t = 64\,\Delta t$$

which means a factor of 2 in speed has been gained with the CLA adder circuit. However, by employing more bits per stage, a larger improvement is possible at the cost of a more complicated circuit, e.g., with 32 bits and 8 bits per stage (a value which may not be practical) the ratio is:

$$\frac{T_{\text{ripple}}}{T_{\text{CLA}}} = \frac{64\,\Delta t}{(4)(4)\,\Delta t} = 4$$

There are other variations of the basic CLA circuit which can achieve improvements of four or more over the ripple method without costing as much as the 8-bit per stage unit. Some involve grouping CLA stages on two or three

---

*There are several commercial CLA adder units which employ the principles indicated in Figure 8.8, e.g., the 8260 MSI package manufactured by Signetics Corporation.

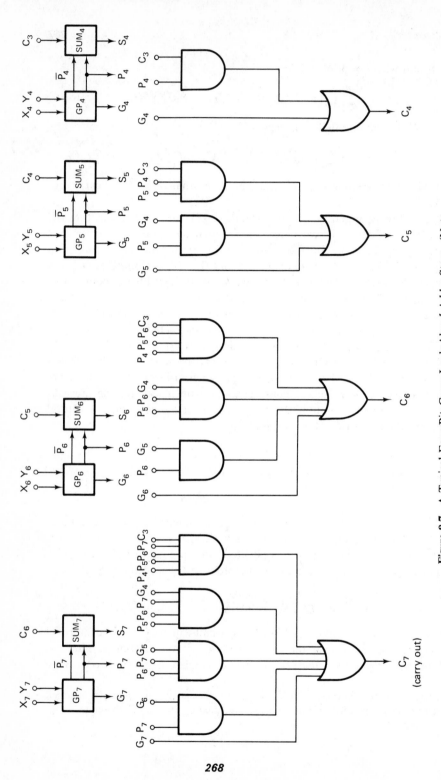

Figure 8.7 A Typical Four Bit Carry Look-Ahead Adder Stage. (Note that GP_i is the generate-propagate half and SUM_i is the sum half of the ith MFA.)

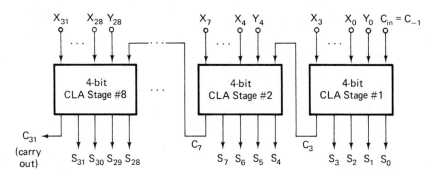

Figure 8.8 A 32-Bit Adder Employing 8 CLA Stages with Ripple Carry Between Stages.

levels instead of the single level employed in Figure 8.7. These and other techniques are extensively discussed in the literature.[3,7,9]

8.3 MULTIPLICATION [5,6,9]

We have previously considered a simple multiplication algorithm using RTL in Figure 7.22*. There are, however, many alternate methods for performing multiplication. It is our purpose here not to catalog the various possible multiplication algorithms, but to present fundamental principles, to develop certain basic circuits, and, most important of all, to convey design techniques so that the reader can learn to arrive at a circuit satisfying a given specification. In the design process, the RTL method of Chapter 7 will assume a very prominent role.

Multiplication of Positive Numbers

Most multiply algorithms are based on the conventional manual method of successive additions and shifting of the multiplicand. Since all of our work will be with binary numbers, some simplification is produced; and if both numbers are positive, a particularly simple case arises. Even here, however, three are alternate possibilities which may lead to better and less expensive circuits. Consider the algorithm presented in Figure 7.22. There, two 12-bit registers, a 6-bit register, and a 12-bit adder were employed in the multiplication of two 6-bit numbers. Unless this equipment is required for another task, significant savings can be achieved by employing the logic diagram shown in Figure 8.9. The major difference between this circuit and the pre-

*If Section 7.5 was previously omitted, it is advisable, at this point, to study Figure 7.22 and the paragraph that discusses it.

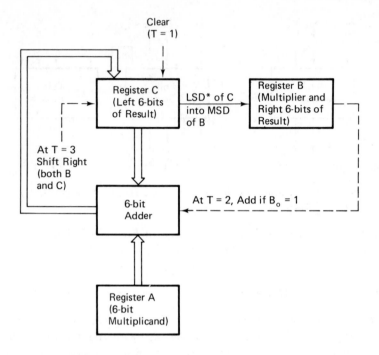

(Note counters for I and states, T, are not shown.)

Figure 8.9   An Alternate Multiply Algorithm.

vious one is that register B serves the dual purpose of storing the multiplier at the beginning of the process and storing the right six bits of the result at the end of the process. The modification is allowable because each time register B is shifted right to obtain a new LSB for testing, there is room at the left end of the register for another bit of the result. Due to this circuit modification, registers A and C, as well as the adder, can each be reduced from 12- to 6-bit units. The RTL program corresponding to Figure 8.9 is shown in Figure 8.10. Note the differences, at $T = 2$ and 3, between the program in Figure 7.22 and the one in Figure 8.10. These differences are due to the structural changes made in going to the new circuit.

$T = 1$ $C \longleftarrow 0; I \longleftarrow 0; A \longleftarrow$ (Multiplicand) ; $B \longleftarrow$ (Multiplier) ;
$T = 2$ IF $(B_0 = 1)$ $C \longleftarrow A + C$; $I \longleftarrow I + 1$;
$T = 3$ $C,B \longleftarrow$ srC,B ; IF $(I \neq 6)$ $T \longleftarrow 2$;
$T = 4$ HALT ;
        (*Note:* Bit 0 is the LSD for all registers.)

Figure 8.10   RTL Program for Multiplication of Positive Numbers.

Sign-and-Magnitude Multiplication

Again by analogy with the conventional manual process, the most obvious method for multiplying signed numbers is to handle the sign and magnitude separately. There are four possible cases for the signs of the multiplicand and the multiplier: (1) $++$, (2) $+-$, (3) $-+$, and (4) $--$. Assuming 1's complement numbers, the four cases can be detected; both numbers can be made positive by complementation; if either case 2 or 3 occurs, a memory bit, M, can be set; positive-number multiplication can be performed; and finally, any correction for the proper sign can be made. Actually, the only modifications needed in Figure 8.9 to accomplish sign-magnitude multiplication are: add the memory bit, M, and provide a means for complementing registers A, B, and C. Figure 8.11 is an RTL program for

$T = 1$ $C \leftarrow 0; I \leftarrow 0; A \leftarrow$ (Multiplicand); $B \leftarrow$ (Multiplier);
$T = 2$ IF ($\bar{A}_5 \wedge \bar{B}_5$) $M \leftarrow 0$;
          IF ($\bar{A}_5 \wedge B_5$) $B \leftarrow \bar{B}, M \leftarrow 1$; IF ($A_5 \wedge \bar{B}_5$) $A \leftarrow \bar{A}, M \leftarrow 1$;
          IF ($A_5 \wedge B_5$) $A \leftarrow \bar{A}, B \leftarrow \bar{B}, M \leftarrow 0$;
$T = 3$ IF ($B_0 = 1$) $C \leftarrow A + C; I \leftarrow I + 1$;
$T = 4$ $C,B \leftarrow$ srC,B; IF ($I \neq 6$) $T \leftarrow 3$;
$T = 5$ IF ($M = 1$) $C,B \leftarrow \bar{C},\bar{B}$;
$T = 6$ HALT;

Figure 8.11   RTL Program for Sign-Magnitude, 1's Complement, Multiplication.

this mode of multiplication, and the following points should be observed when studying the figure:

1. The IF statements, at $T = 2$, detect which of the four sign cases have occurred, and M is changed to remember the sign of the result.

2. The IF statement and the complement operation, at $T = 5$, establish the proper form for the result.

3. Since two n-bit numbers (magnitudes) generate a $2n$ ($2 \times 5 = 10$) bit product and since the MSD of register C, B is employed for the sign, there is one ($12 - 10 - 1 = 1$) unused bit, which is C_4.

4. Registers B and C must be shifted six times to get the LSB of the result into B_0 of the 12-bit C, B register.

‡1's and 2's Complement Multiplication

The multiplication of (3) $\times$ (-5) in the 1's complement system, employing the usual convention, is shown in Figure 8.12. Several points should be observed concerning this example:

$$
\begin{aligned}
-5 &= - \ (101) = \underline{1}\rfloor\,010 \\
3 &= + \ (011) = \underline{0}\rfloor\,011
\end{aligned}
$$

```
11111010
11110100
00000000
00000000
```
① 11101110
⟶ + 1

$\underline{1}\rfloor1101111 = - \ (0010000) = -16 \neq -15$

Figure 8.12 Conventional 1's Complement Multiplication.

1. The symbol $\rfloor$ is used to separate the sign bit from the others.

2. Since the product of two four-bit numbers yields an eight-bit result, the partial products are also shown in eight bits just as they would appear in a register.

3. Since the first two partial products are negative, it is not surprising to find that the leading bits are 1's (For example, the first partial product is 11111010 = −5, as it should be, not 00001010 = +10.)

4. Although it appears that the correct operations were performed, the final result is unexpected.

This counter example is meant to illustrate that one must not make any assumptions, in performing binary arithmetic, that have not been proved.

In order to examine the above problem in detail, a general representation for the multiplication process will be presented. First, consider the conventional multiplication algorithm for two positive integers. The product, c, can be represented as:

$$
c = \sum_{i=0}^{n-1} b_i a 2^i \tag{8.12a}
$$

This may be analyzed as follows:

1. (a) is the multiplicand.

2. $a2^i$ is the multiplicand shifted i bits to the left.

3. b_i is the ith multiplier bit $(b_i = 0, \text{ or } 1)$
   If $b_i = 1$, $a2^i$ is added to the partial result.
   If $b_i = 0$, $a2^i$ is not added.

4. The n partial results are combined by carrying the sum from $i = 0$ to $i = n - 1$.

Equation (8.12a) may be simplified as follows:

$$
c = \sum_i b_i a 2^i = a \sum_{i=0}^{n-1} b_i 2^i
$$

where, by definition:

$$\sum_{i=0}^{n-1} b_i 2^i = b_0 2^0 + b_1 2^1 + \ldots + b_{n-1} 2^{n-1} = b$$

Therefore, $c = ba$ as expected.

Now, examine the following 2's complement product of two, nonzero, n-bit numbers:

$$(b) \times (-a) = \sum_{i=0}^{n-1} b_i (2^{2n} - a2^i) \tag{8.12b}$$

The equation can be analyzed as follows:

1. The partial results contain $2n$ bits; therefore, by definition, the 2's complement of the multiplicand is $2^{2n} - a$.

2. The multiplicand is shifted i bits to the left by the 2^i term to form the 2's complement $2^{2n} - a2^i$.

3. Again, the multiplier for the ith bit, b_i, controls whether or not a particular partial result is included in the sum.

Equation (8.12b) may be written:

$$(b) \times (-a) = 2^{2n} \sum_i b_i - a \sum_i b_i 2^i$$
$$= 2^{2n} \sum_i b_i - ab \tag{8.13}$$

where by definition $b = \sum_i b_i 2^i$. Consider the terms $2^{2n} \sum_i b_i$; since $b \neq 0$, there are two possible cases:

1. $b_i = 1$ for only one i. Here, $2^{2n} \sum_i b_i = 2^{2n}$.

2. $b_i = 1$ for k values of i. Here, addition of the partial products will result in $k - 1$ carry-outs, but these are neglected in the 2's complement system; and again $2^{2n} \sum_i b_i = 2^{2n}$.

Therefore, Equation (8.13) becomes:

$$(b) \times (-a) = 2^{2n} - ba \tag{8.14}*$$

which is the desired product. Now Equation (8.14) proves that direct multiplication is acceptable for $(b) \times (-a)$ in the 2's complement system.

In order to try this proof, we consider the product $(3) \times (-5)$ in Figure 8.13. As expected, the correct result of -15 was obtained.

---

*A more rigorous way to obtain Equation (8.14) follows. By definition, $\sum_t b_i = k$. Thus, Equation (8.13) becomes:

$$b \times (-a) = 2^{2n}k - ba = 2^{2n} + 2^{2n}(k - 1) - ba$$
$$= \underbrace{2^{2n} - ba}_{\substack{\text{2's complement} \\ \text{number}}} + \underbrace{2^{2n}(k - 1)}_{\substack{(k - 1) \text{ carry-outs} \\ \text{which are neglected}}} = 2^{2n} - ba$$

$$-5 = -\ (101) = \underline{1|}\,010 + 1 = \underline{1|}\,011$$
$$3 = +\ (011) \qquad\qquad = \underline{0|}\,011$$

$$
\begin{array}{l}
11111011 \\
11110110 \\
00000000 \\
00000000 \\
\hline
① \ \underline{1|}\,1110001 = -15
\end{array}
$$

Figure 8.13 2's Complement Multiplication Employing the Algorithm of Eq. (8.12b).

Now let us return to the problem which arose in considering Figure 8.12. The product of $(b) \times (-a)$ in the 1's complement system may be expressed in general as:

$$
\begin{aligned}
(b) \times (-a) &= \sum_{i=0}^{n-1} b_i [2^{2n} - (a+1)2^i]^* \\
&= 2^{2n} \sum_i b_i - (a+1) \sum_i b_i 2^i = 2^{2n} \sum_i b_i - ba - b
\end{aligned}
\tag{8.15}
$$

Since in the 1's complement system carry-outs are added to the LSB, the first term in Equation (8.15) becomes

$$2^{2n} \sum_i b_i = 2^{2n} + (k - 1)$$

where $(k - 1)$ is the number of carry-outs. Therefore, Equation (8.15) becomes

$$
\begin{aligned}
(b) \times (-a) &= (2^{2n} - ab - 1) + k - b \\
&= \text{(correct result)} + k - b
\end{aligned}
\tag{8.16\dagger}
$$

In the example of Figure 8.12, $k = 2$ and $b = 3$; therefore, Equation (8.16) yields

$$(b) \times (-a) = (-15) + 2 - 3 = -16$$

which is the exact result obtained.

One way to do $(b) \times (-a)$ 1's complement multiplication is to proceed as we did in the $(3) \times (-5)$ example and to employ a correction of $k - b$. (That process is actually followed in some computers.) The other alternative is to find a modified algorithm which does not require a correction. To

---

*By definition, the 1's complement of $(-a)$ is $2^{2n} - a - 1 = 2^{2n} - (a + 1)$. The quantity 2^i, again provides the required shift of i bits.

†Again $\sum_i b_i = k$; therefore, Equation (8.15) can be expanded as follows:

$$b \times (-a) = 2^{2n}k - ba - b = [\underbrace{2^{2n} - (ba + b - 1) - 1}_{\substack{\text{1's complement} \\ \text{term}}}] + \underbrace{2^{2n}(k - 1)}_{\substack{(k-1)\ \text{carry-outs} \\ \text{to be added to LSB}}}$$

$$= [2^{2n} - (ba + b - 1) - 1] + k - 1$$

which agrees with Equation (8.16).

pursue this latter alternative, consider the result of multiplying the 1's complement number, $-a$, by 2; from the definition, $-a = (2^n - a - 1)$. Now the conventional shift-left procedure will be employed; a correction factor, X, will be added; and the result will be equated to $(2^{n+1} - 2a - 1)$, which is the correct result.

Thus:

$$-2a = 2(2^n - a - 1) + X = 2^{n+1} - 2a - 1$$

Simplifying:

$$2^{n+1} - 2a - 2 + X = 2^{n+1} - 2a - 1$$

Therefore $X = 1$, and we see that to multiply a 1's complement negative number by 2, the usual shift left is accomplished, but a 1, not a 0, must be used to fill the vacated bit.* In a similar way, the reader should verify the following multiplication rules:

1. To multiply a 1's complement number by 2, shift left and replicate the sign bit in the vacant position.

2. To multiply a 2's complement number by 2, shift left and fill the vacant position with a 0.

Returning to the general case of $(b) \times (-a)$ in the 1's complement system, in place of Equation (8.15) we write the new algorithm:

$$(b) \times (-a) \equiv \sum_{i=0}^{n-1} b_i(2^{2n} - a2^i - 1)$$
$$= 2^{2n} \sum b_i - a \sum b_i 2^i - \sum b_i \tag{8.17}$$

where the term within the parentheses should be recognized as the multiplicand $(-a)$ properly shifted i bits to the left. Again

$$\sum_i b_i = k \quad \text{and} \quad 2^{2n} \sum b_i = 2^{2n} + (k - 1)$$

Therefore, Equation (8.17) becomes:

$$(b) \times (-a) = 2^{2n} + (k - 1) - ba - k = 2^{2n} - ba - 1$$

which indicates that the correct product is obtained.

All of this means that Figure 8.12 can be corrected merely by shifting a 1 into the vacancy left in the second partial product, which would be 11110101. (Showing that the correct product now result in Figure 8.12 is left as an exercise.)

We must be careful in finding not only the product $(b) \times (-a)$, but also $(-b) \times (a)$ and $(-b) \times (-a)$. For the 2's complement, consider $(-b) \times (a)$:

$$(-b) \times (a) = (\bar{b} + 1)a = \left(\sum_{i=0}^{n-1} \bar{b}_i a2^i \right) + (1)a \tag{8.18}$$

*In general, when a 1's complement number is shifted left i places, multiplied by 2^i, the result is $-2^i a = 2^{n+i} - 2^i a - 1$; since leading 1's may be eliminated in a 1's complemented number, this reduces to $-2^i a = 2^n - 2^i a - 1$.

In Equation (8.18), the 2's complement of (*b*) is formed by complementing the individual bits and adding 1. Thus, the first term is the partial sum created by shifting $\bar{b}$ the proper number of times; the second term accounts for the 1 which must be added to the 1's complement to get the 2's complement. Now,

$$\bar{b}_i = 1 - b_i \tag{8.19}$$

This may be seen as follows:

$$\text{If } b_i = 1, \qquad \bar{b}_i = 1 - b_i = 1 - 1 = 0$$
$$\text{If } b_i = 0, \qquad \bar{b}_i = 1 - 0 = 1$$

Thus, the relation is true in general since it holds for all possible cases.

Now we substitute Equation (8.19) into Equation (8.18) and simplify:

$$(-b) \times (a) = \left(\sum_i (1 - b_i) a 2^i \right) + a$$
$$= a \sum_{i=0}^{n-1} 2^i - a \sum_i b_i 2^i + a = a(2^n - 1) - ba + a \tag{8.20}$$
$$= a2^n - ba \neq 2^{2n} - ba$$

Since the desired result, $2^{2n} - ba$, is not obtained, *the straightforward method of Equation (8.18) is not a correct procedure for 2's complement multiplication.*

A fairly simple alternate procedure which does produce the correct result is shown in Equation (8.21):

$$(-b) \times (a) = [-(-b)] \times [-(a)] = (b) \times (-a) \tag{8.21}$$

Equation (8.21) indicates that by multiplying both numbers by (-1), which does not upset the equality, a form is obtained matching an algorithm already developed. Changing the sign of the two numbers, of course, requires the 2's complement operation. At first it appears that this operation would require additional time. However, the multiplicand, (*a*), can be 2's complemented by adding 1 to the 1's complement. The actual process is accomplished by gating *a* into the adder forming the partial sum and at the same time employing a carry-in of 1 to the LSB. In the multiplication process, the multiplier bits are tested in sequence to determine whether or not the multiplicand is added to the partial product. Thus, the 2's complement of the multiplier can be determined sequentially, and no additional time is required to change the sign of either number.

The following algorithm for the sequential conversion of a number (*a*), with bits (a_i), to its 2's complement (a^*) can be established:

1. Examine the (a_i) bits from right to left, and place a 0 in each (a_i^*) bit until the first $a_i = 1$ bit is found.

2. Corresponding to this first $a_i = 1$ bit, make $a_i^* = 1$.

3. Continue to examine the a_i bits, and in each case produce the following transfer: $a_i^* \leftarrow \bar{a}_i$.

(As an exercise, show that the above algorithm produces 10100 as the correct 2's complement of 01100.)

The only multiplication possibility that has not been covered is $(-b) \times (-a)$; but this case follows the same pattern as $(-b) \times (a)$. Similar to Equation (8.21), we have:

$$(-b) \times (-a) = [-(-b)] \times [-(-a)] = (b) \times (a) \qquad (8.22)$$

Again, both numbers must be complemented, but the same process as employed for $(-b) \times (a)$ can be used.

Figure 8.14 is the RTL program that forms the product of 2's complement

$T = 1$	$C \leftarrow 0$; $I \leftarrow 0$; $A \leftarrow$ (Multiplicand); $B \leftarrow$ (Multiplier);
$T = 2$	IF $(B = 0)$ $T \leftarrow 9$;: Test for multiplier equal to zero;
	IF $(B_{n-1} = 1)$ $T \leftarrow 5$;: Test for multiplier negative;
$T = 3$	IF $(B_0 = 1)$ $C \leftarrow A + C$; $I \leftarrow I + 1$;: Positive multiplier employed;
$T = 4$	$C,B \leftarrow$ scC,B; IF $(I \neq n)$ $T \leftarrow 3$ ELSE $T \leftarrow 9$;
$T = 5$	IF $(B_0 = 0)$ $C,B \leftarrow$ srC,B, $T \leftarrow 5$; $I \leftarrow I + 1$;: Continue executing until $B_0 \neq 0$;
$T = 6$	$C \leftarrow \overline{A} + C + 1$; IF $(I = n)$ $T \leftarrow 9$;: First $B_0 = 1$;
$T = 7$	$C,B \leftarrow$ scC,B; IF $(I = n)$ $T \leftarrow 9$;
$T = 8$	IF $(B_0 = 0)$ $C \leftarrow \overline{A} + C + 1$; $I \leftarrow I + 1$; $T \leftarrow 7$;
$T = 9$	HALT;

Figure 8.14 Program For Finding the Product $(C = B \times A)$ of 2's Complement Numbers.

numbers. Examination of the figure reveals that it treats both positive and negative numbers by means of the algorithms just discussed. At T_4 and T_7 note how the scale, instead of the shift, operation preserves the sign bit. The block diagram for this system is similar to Figure 8.9, but to avoid overflow* of the $n - 1$ magnitude bit into the sign bit during intermediate stages of multiplication for large numbers, registers A and C are made $n + 1$ bits wide. The multiplicand is transferred to A with the sign in both bits $n + 1$ and n; thus, as usual, bit $n - 1$ is actually the first magnitude bit.

‡8.4 FAST MULTIPLICATION

Although much has been accomplished in the area of fast multiplication since 1950, the number of recent journal articles indicate that advancements in theory as well as in practice have continued to make this a rich field for research and development. (The progress in MSI and LSI have certainly had an important influence.)

*Overflow is considered in detail in Section 8.6.

Methods for obtaining multiplication speedup can be classified into the following three main areas:

1. *Combinational Logic*—Methods employing combination circuits to directly yield the product of two numbers.

2. *Adder Speedup*—Methods for decreasing multiply time by reducing the time required to add the partial products.

3. *Multiple Bit Processing*—Techniques in which multiplication time is decreased by simultaneous processing of more than one multiplier bit.

This section will present representative members of each of the three areas. (For simplicity, most techniques will consider only positive numbers.)

Combinational Logic

It is certainly possible to find the logic function for each bit of a product by constructing a truth table for all combinations of the multiplier and the multiplicand. This brute-force approach requires a truth table with 2^{2n} rows, where n is the number of bits per data word, and it is not a reasonable design technique for most practical values of n. On the other hand, another combinational approach called the *array multiplier* does permit a design description which can easily be specified for any desired value of n.

Employing the requirements segmentation heuristic, examine the manual multiplication process, for binary numbers, shown in Figure 8.15. It is obvious that each partial product, P_{ij}, can be formed from the individual multiplicand, A_j, and multiplier, B_i, bits simply by means of an AND gate.

				A_3	A_2	A_1	A_0
				B_3	B_2	B_1	B_0
				P_{03}	P_{02}	P_{01}	P_{00}
			P_{13}	P_{12}	P_{11}	P_{10}	
		P_{23}	P_{22}	P_{21}	P_{20}		
	P_{33}	P_{32}	P_{31}	P_{30}			
R_7	R_6	R_5	R_4	R_3	R_2	R_1	R_0

Figure 8.15 Manual Multiplication—Partial Products and Results.

These gates and the array of full adders required to sum the partial products into the result bits, R_i, may be systematically arranged as shown in Figure 8.16. The first row of full adders in the figure basically combines the top two rows of partial products. As expected, the sum outputs from one row of full adders provide one input to the full adder one row below, but the carry outputs are shifted one bit to the left. The last row of adders provides a ripple carry and picks up the last, p_{nn}, partial product.

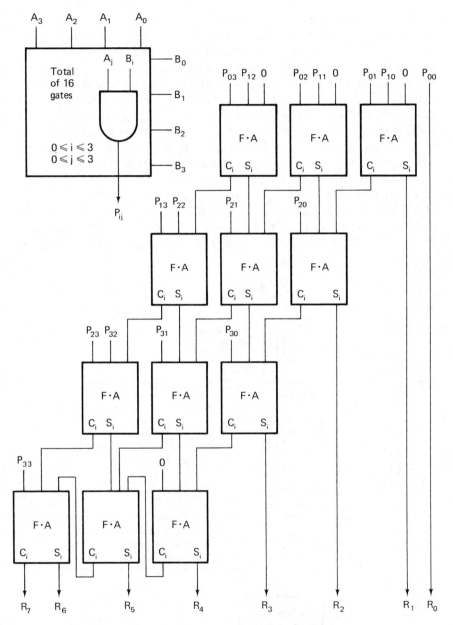

Figure 8.16 A Four-Bit Array Multiplier

Adder Speedup

Since the time required for the carry bits from the addition of each partial product to ripple through the adder is a major function of the total multiply time, anything that can be done to make addition more efficient is beneficial. Obvious improvements are to employ fast circuitry and to implement the techniques mentioned earlier for fast addition, e.g., carry look-ahead and similar methods. On the other hand, there are aspects of the multiply process that, in themselves, provide an opportunity for adder speedup, and one of the best of these is the *carry-save method*. Here, the carries from the addition of two partial products are saved in a set of flip-flops and used as one of the adder inputs when the next partial product is determined. By employing this scheme, it is not necessary to wait for the carries to ripple through the adder.

Figure 8.17 shows the block diagram for a carry-save multiplier. Except for the adder itself and a few new control signals, the diagram is the same as that for the conventional multiplier shown in Figure 8.9. It is in Figure 8.18, however, that the details of the carry-save principle are evident. In this figure,

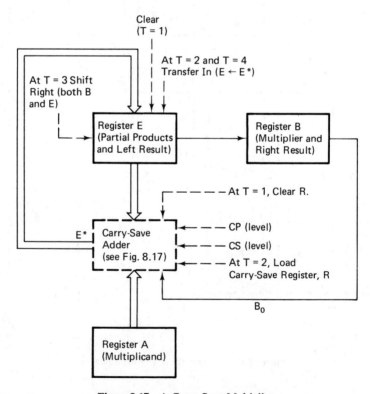

Figure 8.17 A Carry-Save Multiplier

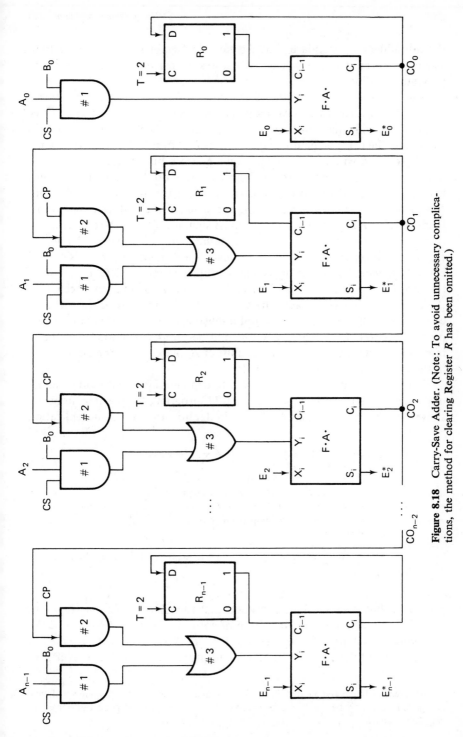

Figure 8.18 Carry-Save Adder. (Note: To avoid unnecessary complications, the method for clearing Register R has been omitted.)

the full adders are standard; but the carry-in terminals, C_{i-1}, receive their signals from register R, which actually stores the carry information from the previous addition.

The multiply cycle consists of two modes: carry-save, CS, and carry-propagate, CP. The carry-save mode employs steps similar to those in our previous multiply algorithms in that a shift-and-add scheme is used; but in this mode, the carry is temporarily stored in R. After the last partial product is completed, R will usually contain some nonzero bits. These must be combined with the final partial product before the true product is obtained. Thus. it is the purpose of the carry-propagation mode to perform the operation: $E \leftarrow E + R$. This time, however, the carry-out of each adder is not transferred to R, but is passed to the next most significant bit in the conventional ripple-and procedure. In Figure 8.18, note how the CS level controlling gate 1 of each bit places the system in the carry-save mode, while the CP level controlling gate 2 places it in the carry-propagate mode. (Recall that since the two modes are mutually exclusive, $CS \neq CP$.) It might at first seem that the carry-out signal for each bit, CO_i, should be transferred to the R_{i+1} carry-save flip flop ($R_{i+1} \leftarrow CO_i$); but it must be remembered that the partial product register, E, is shifted one bit right before the next addition. Therefore, the transfer $R_i \leftarrow CO_i$, with no shift of R, places the carry information in the correct position for the next step.

The detailed operations of Figure 8.18 can be understood more completely after a careful examination of the RTL program, for positive numbers, shown in Figure 8.19. This figure is similar to other RTL multiplication

$T = 1$ $E \leftarrow 0$; $CS \leftarrow 1$; $CP \leftarrow 0$; $A \leftarrow$ (Multiplicand);
           $B \leftarrow$ (Multiplier); $I \leftarrow 1$; $R \leftarrow 0$;

$T = 2$ IF ($B_0 = 1$) $E \leftarrow E \oplus R \oplus A$ ELSE $E \leftarrow E \oplus R$;
           $R \leftarrow CO$; $I \leftarrow I + 1$;: Carry-save phase;*

$T = 3$ $E,B \leftarrow$ srE,B; IF ($I \neq n + 1$) $T \leftarrow 2$ ELSE $CS \leftarrow 0$, $CP \leftarrow 1$;

$T = 4$ $E \leftarrow E + R$;: Carry propagation phase;

$T = 5$ HALT;

Figure 8.19 RTL Program For the Carry-Save Multiplier (Positive Numbers).*

programs; but it should be understood that the EXOR function employed at $T = 2$ produces modulo-2 addition of each bit. [In $E \leftarrow E \oplus R \oplus A$ the individual bits follow the expression $E_i \leftarrow (E_i \bar{R}_i \bar{A}_i) \lor (\bar{E}_i R_i \bar{A}_i) \lor (\bar{E}_i \bar{R}_i A_i) \lor (E_i R_i A_i)$ and the carry is not allowed to propagate.] For example, if $n =$

---

*It is reasonable to assume that B_0 will have been available and the full adders will have completed operation before $T = 2$ occurs, so that the new data can be transferred to registers E and R.

2, $A = 10$, and $E = R = 01$, the process arranged in the form for conventional addition is

$$A = 10$$
$$E = 01$$
$$\underline{R = 01}$$
$$E = 10$$
$$R = 01$$

where the numbers shown below the horizontal line represent the new values of E and R. By contrast, the equation at $T = 4$, $E \leftarrow E + R$, indicates conventional carry-propagation addition, which the algorithm demands and which the circuit in Figure 8.18 realizes when $CP = 1$. It should also be understood that the CS and CP signals are actually levels originating in flip-flops, so that once they are set at state $T = 1$, they remain constant until externally changed at state $T = 3$.

Table 8.5 shows the individual steps in the carry-save multiplication of two positive numbers. It is instructive to follow the program in Figure 8.19 while verifying the table. Note that the table lists the contents of the registers as well as the T, I, CP, and CS signals for all steps of the multiplication process. Examine each line of the table and check the calculations shown in the rightmost column. (The values of T and I are key references to the program in Figure 8.19.) When performing the calculations, particular attention should be given to the fact that, through the quantity $A \wedge B_0$, multiplier bit B_0 controls whether the contents of register A or a zero is added to the partial product. Also, study the way $(A \wedge B_0)$, E, and R are combined to produce the noncarry sum E and the carry information R.

Multibit Processing

There are actually several methods that fall under the classification of multibit processing; three of the most popular will be considered here. Instead of the usual test, add, and shift cycle for each multiplier bit, considerable time can be saved if advantage is taken of certain simplifying multiplier bit patterns.

Shifting Over Zeros. Probably the most obvious technique is to keep checking the remaining portion of the multiplier for zero as the multiplication process continues. When this condition is detected, no further additions to the partial product will be required, and the final result can be quickly established. (In practice, it often happens that the multiplier is a small positive number so that the left half, or more, of its bits are zeros and a considerable savings is achieved.)

For the elementary multiplication process in Figure 7.15, a particularly simple case occurs, and once the multiplier zero is detected, the final result is

Table 8.5 Carry-Save Multiplier Example (1011 × 1111 = 11 × 15 = 10100101 = 165$_{10}$)

At End of State, T	I	A	E	B	R	CS	CP
1	1	1111	0000	1011	0000	1	0
2	2	1111	[1111]	1011	[0000]	1	0
3	2	1111	0111	1101	0000	1	0
2	3	1111	[1000]	1101	[0111]	1	0
3	3	1111	0100	0110	0111	1	0
2	4	1111	[0011]	0110	[0100]	1	0
3	4	1111	0001	1011	0100	1	0
2	5	1111	[1010]	1011	[0101]	1	0
3	5	1111	0101	0101	0101	0	0
4	5	1111	[1010]	0101	[0101]	0	1
5 (HALT)	5	1111	1010	0101	0101	0	1

Calculations* (for boxed values of E and R)

$A \wedge B_0$ = 1111
E = 0000
R = 0000
E = 1111
R = 0000

$A \wedge B_0$ = 1111
E = 0111
R = 0000
E = 1000
R = 0111

$A \wedge B_0$ = 0000
E = 0100
R = 0111
E = 0011
R = 0100

$A \wedge B_0$ = 1111
E = 0001
R = 0100
E = 1010
R = 0101

E = 0101
R = 0101
E = 1010
(carries propagate)

Final result = 165$_{10}$, as expected

*The numbers above the horizontal line are input values from the previous row and those below the line are results.

284

immediately available in register C. Figure 8.20 shows the minor modifications needed in the RTL program of Figure 7.15 to convert it to an elementary multibit processor. (Note that here we are using our standard notation with B_0 as the LSB.)

$T = 1$ $C \leftarrow 0$; $I \leftarrow 1$; $A \leftarrow$ (Multiplicand); $B \leftarrow$ (Multiplier);

$T = 2$ IF $(B = 0)$ HALT; IF $(B_0 = 1)$ $C \leftarrow A + C$; IF $(I = 6)$ HALT;

$T = 3$ $A \leftarrow slA$; $B \leftarrow srB$; $I \leftarrow I + 1$; $T \leftarrow 2$;

Figure 8.20   An Elementary Multibit Processor.

On the other hand, the multiplication process in Figure 8.9 is a somewhat more complicated case because registers B and C must still be shifted as many times as there are bits in the multiplier in order to get the result into the proper position. Also, since register B contains bits from the result as well as from the multiplier, it is somewhat more difficult to test the remaining multiplier bits. (A multibit processing version of Figure 8.9 is left as an exercise.)

Of course, it is not necessary that all the remaining multiplier bits be zero before the above technique can be employed; as long as a group of zeros occurs in a string, a savings can be achieved. The method of simultaneously examining two or more bits and, if they are zero, performing a multiple shift over them is known as *shifting over zeros*. (A similar method for simultaneously processing groups of ones is called *shifting over ones*.)

Consider a multiplication unit similar to one in Figure 8.9 but with n-bit registers, where n is of the magnitude employed in large-scale computers— say between 30 and 50. We will evaluate a shifting-over-zeros scheme in which the next four multiplier bits, B_0 through B_3, are examined, and if they are all zero, the registers are shifted by four bits.*

On the other hand, if the four bits are not zero, or if there are less than four multiplier bits remaining, the usual single-bit processing is initiated. (A practical multiplication system would probably employ a more extensive version of this method by including a shifting-over-ones scheme and perhaps also simultaneous examination of other groups of multiplier bits, with the largest group satisfied being the one that determines the actual shift.) Since one cannot afford the circuitry to mechanize shifts of all possible magnitudes. a shift of four is chosen in this example as being sufficiently probable while saving a considerable amount of multiply time. Figure 8.21 illustrates the RTL program for the shift-of-four algorithm. To account for the adder operating time, an extra state, $T = 5$, is introduced to provide a delay when the multiplicand is added to the partial product, but not when shifting over one or more zeros. Note that C^* is the raw output of the adder, which is developed during the delay and is transferred to register C at $T = 5$.

---

*Multibit shift registers can be constructed in a manner similar to the one-bit versions discussed in Chapter 5.

$T = 1$ $C \leftarrow 0$; $I \leftarrow 1$; $A \leftarrow$ (Multiplicand); $B \leftarrow$ (Multiplier);

$T = 2$ IF $(B_{0-3} = 0$ AND $I \leq n - 3)$ $C,B \leftarrow$ sr3C,B, $I \leftarrow I + 3$;

　　　　IF $(B_0 = 1)$ $C^* \leftarrow A + C$, $T \leftarrow 5$;

$T = 3$ $C,B \leftarrow$ srC,B; $I \leftarrow I + 1$; IF $(I \neq n)$ $T \leftarrow 2$;

$T = 4$ HALT;

$T = 5$ $C \leftarrow C^*$; $T \leftarrow 3$;: Delay for adder to reach completion;

Figure 8.21   A Shift-of-Four Algorithm for Positive Numbers.

It might at first appear that there is a contradiction at $T = 3$, since the variable I is simultaneously incremented and checked with an IF statement. However, if I is implemented by means of master-slave flip-flops, both actions can take place without mutual interference. Furthermore, the close location of these operations is dictated by the requirement for a fast and efficient algorithm, When a string of four zeros is found by the first IF statement in $T = 2$, only three shifts are produced because the fourth shift is accomplished at $T = 3$, where I is checked for multiplication complete. If the number 2 is the multiplier in a 12-bit machine, the arrows and brackets in the following diagram indicate how the shifting-over-zeros scheme of Figure 8.21 groups the individual bits of the multiplier and what the processing sequence is:

0   0   0   0   0   0   0   0   0   0   1   0

Start

Multiplicand Multiples. A slightly more complicated scheme called *multiplicand multiples* employs precalculated values of the multiplicand which are gated into the adder when the proper pattern of multiplier bits occurs. Since it is expensive to store many precalculated values, advantage can be taken of the fact that a multiple that is a power of 2 greater than one already stored can be produced simply by shifting the original number to the left the proper number of places as it enters the adder. This method can be employed when two, three, or more bits are processed, but for purposes of our discussion the three-bit version, shown in Table 8.6, will be considered. For three bits, there are eight cases, and in all but two of them the full three bits are processed simultaneously. Since 101 and 111 would require storage of multiples not used in any other case, only two of the bits are processed, B_0 and B_1, while B_2 is left for the next group. (The fact that only two bits are processed is indicated by a double rather than a triple shift in column 4.) With the above simplification, column 2 shows that only one special multiple need be calculated; this is designated by the symbol T, since it is three times the contents of the register A.

Consider the bit pattern 100. The table indicates that it can be satisfied by an input from register A, which has been shifted left two bits before entering the adder, in order to multiply by 4, and that once the partial product has reached register C, registers B and C are shifted three places to the right. In the same way, verify the other lines in the table.

Table 8.6 Multiplication By Triple Bit Processing*

Multiplier Bit Pattern $(B_2B_1B_0)$	Adder Input	Shift into Adder (Bits Left)	Shift of B and C (Bits Right)
000	0	0	3
001	A	0	3
010	A	1	3
011	T	0	3
100	A	2	3
101	A	0	2
110	T	1	3
111	T	0	2

*Where register A contains the multiplicand, register T contains 3 times A, and 0 represents a zero input.

Observe that the shift-left operations of column 3 require versions of the adder and register C that are two-bits wider than normal.

The Booth Algorithm.   A very efficient multiplication algorithm for all 2's complement numbers was conceived by A.D. Booth.[2] His scheme simultaneously tests two bits of the multiplier but shifts the partial produce only one place at a time; it requires both an add and a subtract capability. The algorithm is based on the fact that a multiplier consisting of n consecutive 1's can be implemented by means of a sequence of n additions and shifts; or since n 1's equals $(2^n - 1)$,* first a subtraction, then n shifts, and finally an addition can be performed. This means that if the multiplier, B, is a string of 1's, the following sequence will be produced:

1. The right end of the string is detected—even a single 1 satisfies the criterion for a string.

2. The multiplicand, A, is subtracted from the partial product, E. Initially assume $E = 0$; then $E \leftarrow -A$.

3. Registers B and E are shifted right and B_0 is examined each time. (Since the string contains n 1's, B and E are shifted n times.)

4. The left end of the string is detected; the contents of register A are added to the shifted E register, generating the term $2^n A$; and the final result is $E \leftarrow 2^n A - A = A(2^n - 1)$. (It is interesting to note that if the multiplier had been all 1's, which is (-1) in 2's complement, register A would *not* have been added, and the final result would be $E = -A$. However, this is exactly what is expected for the product $-1 \times A$.)

Through superposition, the above approach also applies to multipliers consisting of groups of 1's separated by groups of 0's, e.g., 11100 would be represented by one subtraction and one addition as follows: $-2^2 + 2^5$.

*For example: 1111 has $n = 4$; therefore, $(2^4 - 1) = 16 - 1 = 15$ as expected.

In order to facilitate implementation of this algorithm, an extra bit, B_{-1}, is added to the right end of register B. Thus, $B_0B_{-1} = 10$ indicates that the right end of a string of 1's has been detected; $B_0B_{-1} = 01$ indicates that the left end of a string has been detected; $B_0B_{-1} = 11$ indicates the continuation of a string; and $B_0B_{-1} = 00$ indicates the absence of a string, in this region. From the above concepts, a general proof of the Booth algorithm may be derived;[2] a concise summary of the algorithm, however, appears in Table 8.7.

Table 8.7 Statement of the Booth Algorithm

Case	Condition	Required Action
1	$B_0B_{-1} = 00$ or $B_0B_{-1} = 11$	Shift B and E right
2	$B_0B_{-1} = 10$	Subtract A from E; then shift B and E right
3	$B_0B_{-1} = 01$	Add A to E; then shift B and E right

The basic block diagram, Figure 8.22, for this multiplier is quite similar to previous ones except that circuitry is required to implement the logic of Table 8.7 and a dual add/subtract unit is needed to operate between the partial product and multiplicand registers.

Table 8.8 demonstrates the step-by-step operation of the algorithm for the product $9 \times 5 = 45$. The reader should examine the table in detail while

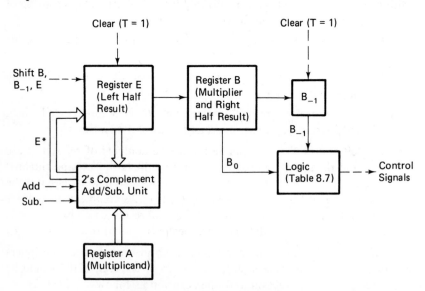

Figure 8.22 The Booth Multiplier.

Table 8.8 Booth Multiplication Example: 9 × 5 = 45

T	I	Case	Register A Multiplicand	Register E	Register B	B₋₁	Calculations
1	0	—	0|0 1 0 1	0|0 0 0 0	0|1 0 0 1	0	0|0 0 0 0
2	0	2	0|0 1 0 1	1|1 0 1 1	0|1 0 0 1	0	1|1 0 1 1
5	0	—	0|0 1 0 1	1|1 0 1 1	0|1 0 0 1	0	1|1 0 1 1
3	1	—	0|0 1 0 1	1|1 1 0 1	1|0 1 0 0	1	1|1 1 0 1
2	1	3	0|0 1 0 1	0|0 0 1 0	1|0 1 0 0	1	0|0 1 0 1
5	1	—	0|0 1 0 1	0|0 0 1 0	1|0 1 0 0	1	0|0 0 1 0
3	2	—	0|0 1 0 1	0|0 0 0 1	0|1 0 1 0	0	
2	2	1	0|0 1 0 1	0|0 0 0 1	0|1 0 1 0	0	
3	3	—	0|0 1 0 1	0|0 0 0 0	1|0 1 0 1	0	0|0 0 0 0
2	3	2	0|0 1 0 1	1|1 0 1 1	1|0 1 0 1	0	1|1 0 1 1
5	3	—	0|0 1 0 1	1|1 0 1 1	1|0 1 0 1	0	1|1 0 1 1
3	4	—	0|0 1 0 1	1|1 1 0 1	1|1 0 1 0	1	1|1 1 0 1
2	4	3	0|0 1 0 1	0|0 0 1 0	1|1 0 1 0	1	0|0 1 0 1
5	4	—	0|0 1 0 1	0|0 0 1 0	1|1 0 1 0	1	0|0 0 1 0
4	4	HALT	0|0 1 0 1	0|0 0 1 0	1|1 0 1 0	1	

Result = 55₈ = 45₁₀ as expected.

Notes: 1) − A = − (0\|0101) = 1\|1011

2) Once the algorithm starts, the sign of Register B is meaningless. Thus, the ⌐ symbol is dropped; but an underline is employed to mark the location of the original MSB as it is shifted.

3) In accomplishing the right shift operation, at each T = 3, again observe that the scale, instead of the shift, operation is employed to preserve the sign bit.

4) The column labeled "Case" refers to the information in Table 8.7.

289

following the RTL program of Figure 8.23. (The first three columns of the table provide a convenient set of references to the program and the algorithm statement.) Again, it should be noted that the RTL inserts a delay between bit testing and register shifting, *only* in those situations (cases 2 and 3) where an arithmetic operation requires time for the bits to ripple down. Thus, considerable time will be saved for multipliers containing large strings of consecutive 1's or 0's. Each line of the table represents conditions after completion of the action indicated by the RTL state in column 1.

$T = 1$ $E \leftarrow 0$; $B_{-1} \leftarrow 0$; $I \leftarrow 0$; $B \leftarrow$ Multiplier; $A \leftarrow$ Multiplicand;
$T = 2$ IF $((B_0, B_{-1} = 0,0$ OR $1,1)$ AND $I = n)$ $T \leftarrow 4$;
         IF $(B_0, B_{-1} = 1,0)$ $E^* \leftarrow E - A$, $T \leftarrow 5$;
         IF $(B_0, B_{-1} = 0,1)$ $E^* \leftarrow E + A$, $T \leftarrow 5$;
$T = 3$ $E,B,B_{-1} \leftarrow$ scE,B,B_{-1}; $I \leftarrow I + 1$; $T \leftarrow 2$;
$T = 4$ HALT;
$T = 5$ IF $(I = n)$ $T \leftarrow 4$ ELSE $T \leftarrow 3$; $E \leftarrow E^*$;:   Delay for add/subtract circuit to operate;

Figure 8.23   RTL Program for the Booth Algorithm.

The final result contains eight bits, not counting sign, since two 4-bit numbers were multiplied. It should be noted from column 1 that five $B_0 B_{-1}$ test and add/subtract operations were performed ($T = 2$), but only four right-shifts ($T = 3$). The fact that the number of shifts equals the number of bits in each number is not surprising since this is the only way for the LSB of the first partial product to reach the LSB of the $2n$-bit result. On the other hand, the multiplier contains five bits, including sign, and if only four ($B_0 B_{-1}$ test and add/subtract) operations were performed (none after the last shift), the sign bit of the multiplier would not have a role in determining the result.† Thus it is apparent that, in general, two n-bit numbers required $n + 1$ test operations, but only n right-shifts.

‡8.5   DIVISION

Our discussion of addition and multiplication has already provided an adequate introduction to hardware design from algorithms. Moreover, many of the multiplication techniques are also applicable to division. Thus, this section will be of moderate length and will only consider positive numbers; however, after completing the material, the reader should have a good basic understanding of division methods.

†If only four operations were performed for the table's example, the last (case 3) addition would not appear. This results in $111011101 = -43_8 = -35_{10}$, which, of course, is incorrect.

The three approaches to division to be considered differ mainly in the way it is decided whether or not the divisir goes into the partial remainder:

1. *Comparison*—A method in which the divisor is compared with the partial remainder one bit at a time, starting with the MSB, much as is done in manual arithmetic.

2. *Restoration*—The divisor is subtracted from the partial remainder to determine go/no-go, and if the result is negative the old remainder is restored before continuing.

3. *Nonrestoration*—When divisor subtraction produces a negative result, the old remainder is not restored; but an alternate quicker process is followed.

Comparison

The manual division below should be studied and note taken that it is basically a subtraction and shifting process; but a comparison is always made between the partial remainder and the divisor to insure that the new remainder will always be positive. For example, in the first step the comparison

$$
\begin{array}{r}
01011. \\
110\overline{)1000010.} \\
110 \\
\hline
1001 \\
110 \\
\hline
110 \\
110 \\
\hline
000
\end{array}
$$

100/110 is made; since the first number is smaller than the second, a 0 is added to the quotient and the divisor is shifted. The new compairson is 1000/110; since the divisor goes into the partial remainder, a 1 is added to the right end of the quotient. Subtraction is performed to get a new partial remainder $(N \leftarrow N - D)$, where N is the partial remainder and D is the divisor. Then the process repeats.

It is not difficult to implement the comparison method as the block diagram in Figure 8.24 shows. Contrary to the procedure for the manual method, here it is more convenient to shift the dividend to the left while the divisor remains fixed. In many other respects, the block diagram in this figure is similar to the previous block diagrams for multiplication except that subtraction is achieved by adding the 2's complement of D, a comparison between D and N is required, and as already mentioned, the partial result (along with the remainder) is shifted left.

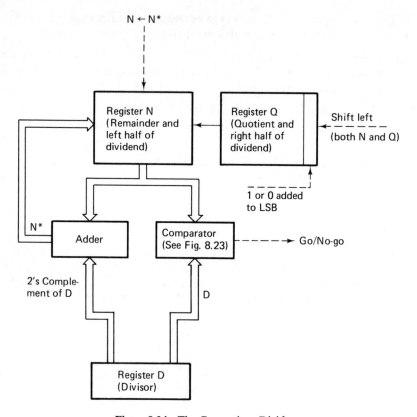

Figure 8.24 The Comparison Divider.

Of course, the comparison circuit is at the heart of this type of divider. Let us demonstrate the design of a comparator by employing the heuristic method for a four-bit circuit. The requirements segmentation and information analysis heuristics can be employed to achieve a module for each bit, and then combine the modules with logic which may be iterated to produce a circuit for as many bits as desired.

Consider the manual comparison process during division; the leftmost bits for the dividend and its nonzero divisor are examined. If $N_{n-1} = 1$ and $D_{n-1} = 0$, we have a match, M, and immediately know that the divisor will go into the dividend, completing the comparison process. On the other hand, if $N_{n-1} = 1$ and $D_{n-1} = 1$, or if $N_{n-1} = 0$ and $D_{n-1} = 0$, the situation is undecided, U, indicating that the next place on the right must be compared. These basic comparator module conditions are summarized in Table 8.9 for an arbitrary bit i.

Once the M_i and U_i outputs are available for each of the n bits, all that remains is to provide the logic between them for developing the final go/no-

Table 8.9   Logic Conditions for a Comparator Module

N_i	D_i	M_i	U_i
0	0	0	1
0	1	0	0
1	0	1	0
1	1	0	1

go signal. The go-condition, as bits further to the right are considered, depends on all previous positions having developed an undecided signal and the current position indicating a match. (Of course, when the above go condition exists, it is immaterial what the values for M_i and U_i are for all positions further to the right.) The logic just described, which is actually for the relation $N \geq D$, is realized by the circuit shown in Figure 8.25. The following points concerning the figure should be studied:

1. Iteration of the circuit for as many bits as desired is a straightforward process. For each additional bit, another comparator module and another AND gate must be added.

2. In completing the heuristic design process (the step of checking for circuit simplification), it will be observed that the AND gate producing M_2 can be combined with AND gate 1, resulting in the saving of one gate and one input. Moreover, the same savings can be achieved for M_1 and M_0. (The construction of a simplified version of Figure 8.25 is left as an exercise.)

3. The output of gate 4 indicates whether the contents of the two registers are equal.

4. Some typical register contents and the resulting M_i, U_i, and go/no-go signals are:

$N = 1011$	$N = 1000$	$N = 0101$
$D = 1000$	$D = 1000$	$D = 1000$
$M = 0011$	$M = 0000$	$M = 0101$
$U = 1100$	$U = 1111$	$U = 0010$
$G = 1$, from	$G = 1$, from	$G = 0, N < D$
gate 2	gate 4	
	(equality)	

Restoring Division

The basic system for *restoring division* is presented in Figure 8.26. The process is similar to comparison division except that the go/no-go decision is based on the sign of register N after the divisor is subtracted. Thus, a negative sign indicates that D did not go into N, and the register N must be

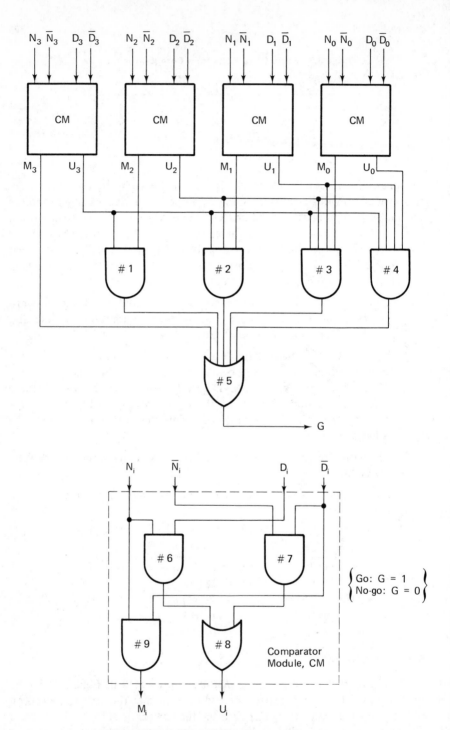

Figure 8.25 A Four-Bit Comparator ($N \geq D$).

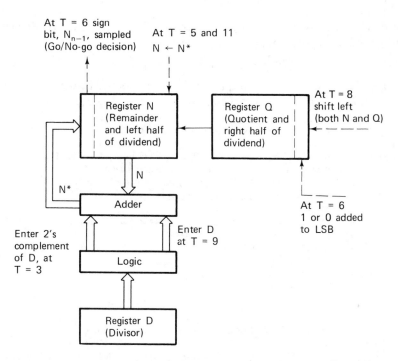

Figure 8.26 The Restoring Divider. (The T values shown are discussed later.)

restored to its original value, through the step $N \leftarrow N + D$, before the zero can be entered in the quotient and the left-shift produced. The figure reveals that addition is accomplished by entering D directly into the adder, and subtraction is accomplished by entering the 2's complement of D. The 2's complement is produced by the same method employed previously: $\bar{D}$ is entered into one set of adder inputs (N is entered into the other set), and a 1 is entered into the carry input. Note how the sign bit of N is used to make the go/no-go decision and how each new bit of the quotient is entered into the LSB of Q. The hardware shown is for integer numbers as is the basic division instruction for such machines as the IBM System 360/370 and the PDP-11. Thus, it is the programmer's responsibility to keep track of the binary point if the numbers contain a fractional part, and to perform any scaling which is desirable. It can be shown that the relationship between the bits in the various registers in the division process is as follows:

$$(Q) = (NQ) - (D) + 1 \tag{8.23}$$

where (Q) is the number of bits in the quotient register, $(NQ)^*$ is the bits in

---

*NQ is employed to represent the dividend because the bits are divided between the N and Q registers.

the dividend, and (D) is the bits in the divisor. Thus, with a $2m$-bit dividend and an m-bit divisor, the quotient requires:

$$(Q) = 2m - m + 1 = m + 1 \text{ bits} \tag{8.24}$$

The details of the restoration algorithm are shown in the program of Figure 8.27. It is assumed that the registers N, D, and Q each contain n bits

$T = 1$	$I \leftarrow 0; E \leftarrow 0; N \leftarrow$ (Left half of dividend); $Q \leftarrow$ (Right half of dividend);
$T = 2$	IF $(N_{n-1}, N_{n-2} \neq 0,0$ OR $D_{n-1} \neq 0)$ $E \leftarrow 1$, $T \leftarrow 12$ ELSE $N,Q \leftarrow$ slN,Q;: Test for legal dividend and divisor;
$T = 3$	$N^* \leftarrow N + \bar{D} + 1$;: Subtract using 2's complement addition;
$T = 4$	: Delay for adder bits to ripple down;
$T = 5$	$N \leftarrow N^*$;
$T = 6$	IF $(N_{n-1} = 1)$ $T \leftarrow 9$, $Q_0 \leftarrow 0$ ELSE $Q_0 \leftarrow 1$; $I \leftarrow I + 1$;
$T = 7$	IF $(I = n)$ $T \leftarrow 12$;
$T = 8$	$N,Q \leftarrow$ slN,Q; $T \leftarrow 3$;
$T = 9$	$N^* \leftarrow N + D$; IF $(I = n)$ $T \leftarrow 12$;
$T = 10$	: Delay for adder bits to ripple down;
$T = 11$	$N \leftarrow N^*$; $T \leftarrow 8$;
$T = 12$	HALT;

Figure 8.27 RTL Program for Restoring Division.

(bits 0 to $n - 1$); but the fact that N can go negative in the subtraction process requires that the MSB bit be reserved for sign, leaving $n - 1$ bits for the magnitude. When $m = n - 1$ is substituted into Equation (8.24),

$$(Q) = (n - 1) + 1 = n$$

which means that the magnitude bits occupy the entire Q registers. Now from Equation (8.23),

$$(NQ) = (Q) + (D) - 1 = n + (n - 1) - 1 = 2n - 2$$

and since we assume that the dividend is right-justified (the number is as far right in the register as possible), the number of dividend bits (B) originally in the N register is:

$$(B) = (NQ) - (Q) = (2n - 2) - n = n - 2$$

Therefore, there is an extra bit between the MSB and sign bit of the dividend, which means that the divisor and dividend are misaligned by one bit. For these reasons, at $T = 2$ in the program, N and Q are shifted left, which also conveniently makes room for the first bit of the quotient.

Observe that the IF statement at $T = 2$ checks that the dividend and divisor are both positive—sign bit equals 0. In addition, it checks that the extra bit which will become the sign bit, after shifting, is also 0. When these conditions are violated, the error bit, E, is set to 1. (The error bit is not shown in Figure 8.26, but in most computers there are several types of error flags (bits) which are automatically set by the hardware and can be read with a special instruction.)

In the sequence of instructions, $T = 3$, 4, and 5, the transfer to the adder output, N^*, at $T = 3$ is merely a way of indicating that the 2's complement of D is switched to the adder input and the sum will start to be formed. After the bits ripple down at $T = 4$, the actual transfer into N occurs at $T = 5$. (A similar sequence occurs at $T = 9$, 10, and 11.) Carefully study the individual steps of Figure 8.27—the comments within the program should be helpful in following the algorithm.

As a further guide to understanding the restoration algorithm, Table 8.10 has been prepared. Again, the first two columns provide a convenient reference to the RTL program, and these two should be studied together. Note that:

1. Underscores are used to show the progress of the MSB of the quotient as it is shifted along.

2. An X is used as a temporary filler in bit Q_0 after shifting and before the next bit of the quotient is ready. (The hardware would actually place a zero there after shifting, but the X is a useful device for explanation purposes.)

3. The E bit is not shown explicitly; but it would be zero in this example since the given input data would not generate an error.

4. In order to keep the table as concise as possible, states 3–5 (also 9–11) are shown on a single line with the entries representing results of the action initiated by $T = 5$ ($T = 11$).

5. The remainder always appears in register N, while the quotient always appears in register Q.

Nonrestoring Division

The method of *nonrestoring division* is very similar to restoring division. Except for the states, T, it has the same block diagram as Figure 8.26; but since negative partial remainders are not explicitly restored, the system runs somewhat faster.

The following is a summary of basic steps in the nonrestoring process:

1. Subtract the divisor from the partial remainder.

2. If the result is positive, make $Q_0 = 1$, shift N and Q left, and return to step 1.

3. If the result is negative, instead of restoring, make $Q_0 = 0$, shift N and Q left, and add D.

4. If the partial remainder continues to be negative, repeat step 3; otherwise, make $Q_0 = 1$, shift N and Q left, and return to step 1.

Table 8.10   Restoring Division Example: $\dfrac{111100}{110} = \dfrac{60}{6} = 10_{10}$, $n = 4$

(2's complement of D = $-(0110) = 1001 + 1 = 1010$)

T	I	Register D (divisor)	Register N (Remainder and left half of dividend)	Register Q (Quotient and right half of dividend)	Calculations
1	0	0|1 1 0	0|0 1 1	1 1 0 0	
2	0	0|1 1 0	0|1 1 1	1 0 0 X	0|1 1 1 / 1|0 1 0 / 0|0 0 1
3-5	0	0|1 1 0	[0|0 0 1]	1 0 0 X	
6	1	0|1 1 0	0|0 0 1	1 0 0 1	
7	1	0|1 1 0	0|0 0 1	1 0 0 1	
8	1	0|1 1 0	0|0 1 1	0 0 1 X	0|0 1 1 / 1|0 1 0 / 1|1 0 1
3-5	1	0|1 1 0	[1|1 0 1]	0 0 1 X	
6	2	0|1 1 0	1|1 0 1	0 0 1 0	1|1 0 1 / 1|1 1 0 / 0|0 1 1
9-11	2	0|1 1 0	[0|0 1 1]	0 0 1 0	
8	2	0|1 1 0	0|1 1 0	0 1 0 X	0|1 1 0 / 1|0 1 0 / 0|0 0 0
3-5	2	0|1 1 0	[0|0 0 0]	0 1 0 X	
6	3	0|1 1 0	0|0 0 0	0 1 0 1	
7	3	0|1 1 0	0|0 0 0	0 1 0 1	
8	3	0|1 1 0	0|0 0 0	1 0 1 X	0|0 0 0 / 1|0 1 0 / 1|0 1 0
3-5	3	0|1 1 0	[1|0 1 0]	1 0 1 X	
6	4	0|1 1 0	1|0 1 0	1 0 1 0	1|0 1 0 / 0|1 1 0 / 0|0 0 0
9	4	0|1 1 0	[0|0 0 0]	1 0 1 0	
12	4	0|1 1 0	[0|0 0 0]	[1 0 1 0]	

(HALT)

Remainder = 0, as expected.   Result = $12_8 = 10_{10}$, as expected.

Consider the mathematical rationale for nonrestoring division. Let NQ represent the $2m$-magnitude bits of the dividend, i.e., the m bits of register N, not including the sign bit, and the m bits of register Q. (Recall that since there are $m + 1$ bits in the quotient, Q_0 is not included in the NQ bits.) When we subtract (add the 2's complement), the divisor from the remainder the mathematical operation actually is $NQ - 2^m D$, where the 2^m factor accounts for the fact that the LSB of the D register is really shifted m bits to the left of the LSB of NQ.

Assume that the subtraction leads to a negative result. Then, according to the above steps for nonrestoring division, the new remainder, R, is formed by shifting the old remainder left and adding D:

$$R = 2(NQ - 2^m D) + 2^m D$$
$$= 2NQ - 2^m \cdot 2D + 2^m D \qquad (8.25)$$
$$= 2NQ - 2^m(2D - D) = 2NQ - 2^m D$$

Therefore, shifting and adding D has had the effect of shifting NQ one bit to the left and subtracting D, which is equivalent to restoring NQ, shifting, and subtracting D. Of course, if the remainder of Equation (8.25) is still negative, the process may be repeated by shifting and adding D as many times as necessary. It can easily be seen that if k shifts and additions are employed, the result is:

$$R = 2^k NQ - 2^m D \qquad (8.26)$$

which is exactly the desired result.

Now, when the divisor is subtracted from the remainder and a positive result occurs, NQ is shifted to the left and the effect is the same as starting a new problem with a dividend of $NQ - 2^m D$.

Since nonrestoring division is so similar to restoring division, the RTL program, examples, and other details are included as part of the exercises at the end of this chapter.

‡8.6 OVERFLOW

One potential problem with which the programmer must be concerned is that of register overflow during arithmetic operations. Since the possibilities are so varied, the correction of an overflow condition is usually the responsibility of the software. The detection of overflow conditions can, however, be very efficiently made by the ALU hardware.

The term *overflow* can be somewhat misleading; but in the computer field it is commonly employed to represent the condition where the result of some arithmetic operation is too large for its register. This most frequently occurs when large numbers with like signs are added and when large numbers of opposite sign are subtracted.

Table 8.11 shows examples for each of the eight possible cases that exist for the three sign bits involved in the addition process. Overflow is produced only in cases 2 and 7, which are large-magnitude numbers of the same sign. In these two cases, note that the result is too large to be contained in the three magnitude bits, and this error causes the sign bit to be changed. Overflow, of course, is not correlated with the carry-out of the sign bit alone in that cases 3, 5, 7, and 8 have carry-outs, but all except case 7 yield the correct result. On the other hand, the overflow condition may be detected either from the values of the three sign bit involved or from a function of the carry-in and carry-out of the sum sign bit. (The adder carry-out bits, C_i, are listed under the sum bits in Table 8.11.)

Table 8.11   The Eight Possible Combinations of Adder Data Types
(2's complement numbers)

Case 1	Case 2	Case 3	Case 4
$\begin{bmatrix}\text{Input: Small}\\ (+)\text{ numbers.}\end{bmatrix}$	$\begin{bmatrix}\text{Input: large}\\ (+)\text{ numbers.}\end{bmatrix}$	$\begin{bmatrix}\text{Input: large}\\ (+)\text{ and large}\\ (-)\text{ numbers.}\end{bmatrix}$	$\begin{bmatrix}\text{Input: small}\\ (+)\text{ and large}\\ (-)\text{ numbers.}\end{bmatrix}$
$x_i = \underline{0}|011 = 3$	$\underline{0}|110 = 6$	$0|110 = 6$	$\underline{0}|011 = 3$
$y_i = \underline{0}|011 = +3$	$\underline{0}|110 = +6$	$\underline{1}|010 = +(-6)$	$\underline{1}|010 = +(-6)$
$S_i = \underline{0}|110 = 6$	$1|100 \neq 12$	$①0|000 = 0$	$1|101 = -3$
$C_i = 0011$	0110	1110	0010
	Overflow occurs		

Case 5	Case 6	Case 7	Case 8
$\begin{bmatrix}\text{Input: small}\\ (-)\text{ and large}\\ (+)\text{ numbers.}\end{bmatrix}$	$\begin{bmatrix}\text{Input: large}\\ (-)\text{ and small}\\ (+)\text{ numbers.}\end{bmatrix}$	$\begin{bmatrix}\text{Input: large}\\ (-)\text{ numbers.}\end{bmatrix}$	$\begin{bmatrix}\text{Input: small}\\ (-)\text{ numbers.}\end{bmatrix}$
$x_i = 1|101 = -3$	$1|010 = -6$	$1|010 = -6$	$1|101 = -3$
$y_i = 0|110 = +6$	$0|011 = +3$	$\underline{1}|010 = +(-6)$	$\underline{1}|101 = +(-3)$
$S_i = ①0|011 = 3$	$1|101 = -3$	$①0|100 \neq -12$	$①1|010 = -6$
$C_i = 1100$	0010	1010	1101
		Overflow occurs	

Table 8.12 is the truth table which shows all possible combination of the three sign bits correlated with the eight cases presented in Table 8.11. Thus, the Boolean equation to detect the overflow condition, Ov, is given by:

$$Ov = \bar{x}_3\bar{y}_3 s_3 + x_3 y_3 \bar{s}_3 \qquad (8.27a)$$

Table 8.12   Overflow Conditions Based on Value of Sign Bits

Case	x_3	y_3	s_3	Overflow
1	0	0	0	0
2	0	0	1	1
3	0	1	0	0
4	0	1	1	0
5	1	0	0	0
6	1	0	1	0
7	1	1	0	1
8	1	1	1	0

Detection of overflow from the carry-in and carry-out states of the sum sign bit is represented in the truth table of Table 8.13 and the resulting equation is:

$$Ov = \bar{c}_3 c_2 + c_3 \bar{c}_2 \qquad (8.27b)$$

The selection of Equation (8.27a) or Equation (8.27b) as the equation to be mechanized depends on the detailed nature of the other circuits employed in the ALU. In either case, however, the application is straightforward.

Table 8.13 Overflow Conditions Based on Carry-In and Out of the Sum Sign Bit

Case	c_3	c_2	Overflow
1, 4, 6	0	0	0
2	0	1	1
7	1	0	1
3, 5, 8	1	1	0

8.7 BIBLIOGRAPHY

1 ATKINS, D. E., and H. L. GARNER (Guest eds.), "Special Section on Computer Arithmetic." *IEEE Trans. Comput.*, Vol. C-22, 1973, pp. 549–610.

2 BOOTH, A. D., "A Signed Binary Multiplication Technique." *Quart. Journ. Mech. and Applied Math.*, Vol. 4, Pt. 2, 1951, pp. 236–40.

3 FLORES, I., *The Logic of Computer Arithmetic.* Englewood Cliffs, N. J.: Prentice-Hall, Inc., 1963.

4 HABIBI, A., and P. A. WINTZ, "Fast Multipliers." *IEEE Trans. Comput.*, Vol. C–19, 1970, pp. 153–57.

5 HILL, FREDRICK J., and GERALD R. PETERSON, *Digital Systems: Hardware Organization and Design.* New York: John Wiley & Sons, Inc., 1973.

6 LEWIN, DOUGLAS, *Theory and Design of Digital Computers.* New York: John Wiley & Sons, Inc., 1972.

7 MacSORLEY, O. L., "High-Speed Arithmetic in Binary Computers." *Proc. IRE*, Vol. 49, 1961, pp. 67–91.

8 RHYNE, V. THOMAS, *Fundamentals of Digital Systems Design.* Englewood Cliffs, N.J.: Prentice-Hall, Inc., 1973.

9 SHIVELY, R. R. (Guest ed.), "Special Issue on Computer Arithmetic." *IEEE Trans. Comput.*, Vol. C-19, 1970, pp. 679–757.

8.8 EXERCISES

8.1 (a) Draw the circuit for a basic full adder with AND, OR, and NOT gates; then repeat the process with NOR gates.

(b) Design a modified full-adder stage similar to that in Figure 8.6 using the fewest possible NAND gates.

8.2 An arithmetic module called a *half adder* has sum and carry outputs but only addend and augend inputs; carry-in is not present.

    (a) Employing NORs, design a logic circuit for the module with a minimum number of gate inputs.

    (b) Employing half-adder modules and any needed gates, draw the minimum block diagram for a three-bit parallel adder.

8.3 What must be done to Figure 8.4 so that 1's complement numbers may be processed?

8.4 Write an RTL program which employs the serial algorithm to convert a number in register X to its 2's complement in register Y.

8.5 Design a two-level combinational circuit which performs the addition of two-bit numbers. Discuss the advantages and disadvantages of the circuit along with its possible extension to more bits.

8.6 Employ the usual add and shift process indicated by Equation (8.18) and a *correction term* to develop an algorithm for 2's complement multiplication of $(-b) \times (a)$. Show that this gives the proper results for $(-5) \times (3)$.

8.7 Prepare a table similar to Table 8.5 for the carry-save multiplication of 3×7.

8.8 *Prove* the rule, given above Equation (8.17), for multiplication of a 2's complement number by 2.

8.9 Employing the Booth algorithm, complete a table showing the processing steps for the example $(-3) \times (5)$.

8.10 Repeat Exercise 8.9 for the example $(3) \times (5)$.

8.11 Design a comparator circuit that detects the condition $A < B$.

8.12 Write an RTL program for comparator division.

8.13 Construct a table showing the processing steps of comparator division for the example $Q = \frac{10000100}{0110}$. (The table should correlate with the RTL program of Exercise 8.12.)

8.14 Register A contains $\underline{1}|1101$.

    (a) What will the result be after the operation sc$A$?

    (b) What will the result be after sc2A?

    (c) What will the result be after rl$A$?

    (d) Are the expected results obtained for division and multiplication? Explain.

    (Assume 1's complement numbers.)

8.15 Using restoring division, construct a table similar to Table 8.9 for $Q = \frac{110111}{101}$.

8.16 Write the RTL program for nonrestoring division. Include a check for improper divisor and dividend.

8.17 Construct a table similar to Table 8.10 for nonrestoring division using $Q = \frac{111100}{110}$.

8.18 Verify (prove) the sequential 2's complement method given above Equation (8.22).

9 | SEMI-CONDUCTOR AND CORE MEMORY SYSTEMS

There has been a truism in the computer field that as memory developments go, so goes computer technology. This is still valid today in that a significant part of a general-purpose computer's cost is memory. Also, the processing speed of most present computers is strongly limited by memory speed.

What portions of the computer are classified as memory? Certainly the *main store*, which holds the current instructions and data, is part of it, with typical members being magnetic core and semiconductor types. The *mass memory* (or *bulk storage*), which contains additional routines and data that must be available at short notice, is also part of it, with disk and drum being representative types. Finally, *archival storage*, for vast quantities of data files and other information, typically on magnetic tape or photographic film, is another form of memory. Moreover, the categories are not as clearly defined as is implied above because, for example, many minicomputer users employ magnetic tape as a form of mass memory while many large-scale computer users employ slow, relatively inexpensive core for the same purpose.

Many different physical phenomena have been successfully employed as a memory medium. These range from simple schemes of punching holes in paper or cards to sophisticated cryogenics.[9, 10] Moreover, there is reason to believe that many engineers and scientists will continue to be employed for a long time in research and development on new types of memory systems.

Even though the phenomenon used may be very different, there are a few basic characteristics by which all memories can be compared. One of these is *volatility*. To say that a memory is *nonvolatile* means that the storage principle employed does not require electrical power to sustain it. Magnetic cores have this property since, once the core is magnetized in a particular direction, it permanently retains that state until changed by an external agent. The catch here is that, in turning off the power to a computer, one must take precautions that electrical transients do not become that external agent. Such precautions, however, usually are not hard to take.

Another basic memory characteristic is whether *nondestructive* readout is possible. It happens that some memory types, which are very good in other respects, destroy the memory contents of a word as it is read. This is not as serious as it at first seems since it is possible to immediately rewrite the word in question from the memory buffer register. The main disadvantage of the rewrite process is that it costs a certain amount of time and money.

The various storage methods are often classified on the basis of *memory access time*, which is the interval between the occurrence of a request for a certain data word and the time that it is available to the CPU. This should not be confused with *memory cycle time*, which is the total time it takes to read and, if necessary, to restore a memory having a destructive readout. The memory access time, in turn, depends on the storage access method. *Random-access memory*, abbreviated RAM, is usually the fastest; and the term, random, itself implies that each word requires the same time for retrieval. Core and semiconductor memories fall into the random-access category, while disk and drum storage are *cyclic access*, since the word of interest comes under the read head in a cyclic manner and the access time depends on where the system is on the cycle when the request occurs. If the word is just about to come under the read head, the access time will be very short, but on the average it is one-half the cycle time. Because information on tape is stored in a sequential fashion, it and similar methods are called *serial access*. Another parameter to consider here is the memory write capability. A memory section for storing permanent data, e.g., a trigonometric sine table, does not require a write capability once it is loaded; thus, cost and time may be saved by omitting the write feature. Such a storage device is called a *read-only memory*, ROM. Moreover, for data only occasionally changed, there are even cases between full write capability and the ROM. One such type is called a *programmable read-only memory*, PROM.

Because of the frequency with which computer engineers are involved in the area and because of the opportunity to present many basic principles, this chapter will be devoted to the design of the main store and related topics.* We begin with core storage and follow with semiconductor storage. Next, the ROM and several of its applications are presented. Finally, we consider a special unit, called a *control memory*, which has become a powerful way to implement various types of sequential circuits including the CPU of several well-known computers.

9.1 MAGNETIC CORE STORAGE[1,11]

Core storage has a long history of being the best trade-off between memory cost and speed. For many applications, it is currently being replaced by semiconductor memories; but it does appear that core storage will remain

---

*The reader who is interested in details concerning mass memories and archival storage, which are more specialized topics, should consult Bibliography entries 9, 10, and 12.

important for a long time. Moreover, once the principles of core memory design are understood, it is easy to apply them to semiconductor memories.

The principle of core memory operation is that very small (typically 22 mils, outside diameter), donut-shaped ceramic ferrites permanently retain magnetic flux in either a clockwise or counterclockwise direction, i.e., they store either a logical 1 or 0. Moreover, by means of wires passing through the core, it is fairly easy to electrically switch it to a desired state, to write, or to read the contents. Most memories employ one core for each bit stored; thus an 8K (actually $2^{13} = 8192$) word 16-bit memory requires $16 \times 8192 = 131,072$ cores. (Such a single-layer memory array of cores and their wiring, without electronics, will easily fit within an area of 8 inches square.)

In order to understand the principles of ferrite core operation, consider the experiment shown in Figure 9.1. An input winding is employed to

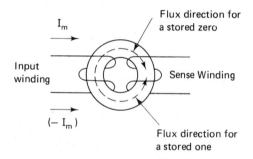

Figure 9.1 Ferrite Core Experiment.

establish a magnetic field in the core having either direction 1 or 0; these numbers represent the logical value stored. The sense winding develops an induced voltage (in the millivolt range) wherever there is a flux change within the core, and it is this signal which indicates the logical state of the core during the read cycle.

The key to successful operation of a core memory system is the fact that there are ferrite materials available with very square hysteresis loops, such as the one shown in Figure 9.2. Assume that the core is in the 0 state with no current flowing in the input winding. Then the component will be stable with remanent flux $+B_r$ shown in the hysteresis diagram. Now, if a current $-I_m$ is placed on the input winding, a magnetic field intensity $-H_m$ will be produced, resulting in the flux density $-B_m$. After the input current returns to 0, the core will go to point $-B_r$ on the hysteresis loop, and a logical 1 will have been written. Because the core acts as a permanent magnet, it will remain at the point $-B_r$ until the next electrical input occurs. Thus, we see that ferrite core storage is nonvolatile.

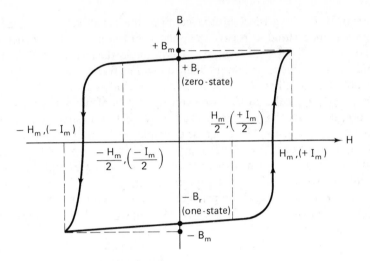

Figure 9.2   Ferrite Core Hysteresis Loop.

To determine what state a core is in, the current $+I_m$ is placed on the input winding. From the hysteresis diagram, we can see that if the core is at $+B_r$, the current $+I_m$ will cause only a small flux change in establishing $+B_m$, and only a small noise voltage, δ, shown in Figure 9.3, will be developed. On the other hand, if the core is at $-B_r$, the input current $+I_m$ will cause a large flux change, resulting in the relatively large signal voltage, S, shown in the diagram. Not only is the noise voltage, δ, smaller than S, but the maximum values occur at a different time. Thus, by sampling the sense winding after a proper delay, a good signal-to-noise ratio can be established between the signal received for a stored 1 as opposed to the signal for a stored 0. It should be observed that the above-mentioned read process leaves all cores in the 0 state; therefore, core memories are said to exhibit a destructive readout.

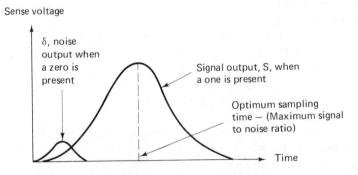

Figure 9.3   Sense Voltage Waveforms.

Coincident-Current (3-D) Memory Systems

Another important fact observable from Figure 9.2 is that a current of $\pm I_m/2$ has no effect on the state of the core, e.g., with the core in $-B_r$, a current of $+I_m/2$ leading to a flux of $H_m/2$ will not cause the core to switch. Thus, when the current goes to zero, the core will remain in the 1 state. The fact that $\pm I_m/2$ will not cause the core to switch is the basis of the coincident-current (or 3-dimensional) memory. This memory consists of rectangular arrays, or planes, of cores. Each plane represents one bit of the memory, and within a plane, each core represents a different word. Thus, for a 10^4-word, 16-bit memory, the rectangular array of cores would be 100 cores on a side, and there would be 16 such planes. If the planes are visualized as being arranged in a stack with bits from corresponding words aligned over one another, all the bits from one word will lie on a straight line which pierces each plane. Now, by placing an X- and a Y-input select winding through an array of cores, as shown in Figure 9.4, we can select a particular word,

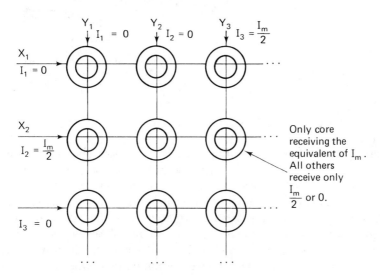

(Note: the sense winding is omitted for simplicity.)

Figure 9.4 A Small Section of One Memory Plane.

W_{ij}, in the array by placing current $I_m/2$ through winding X_i and $I_m/2$ through winding Y_j. Then words $W_{i1}, W_{i2}, \ldots, W_{ij}, \ldots, W_{ip}$ (where the array has dimensions p by p) receive $I_m/2$ and $W_{1j}, W_{2j}, \ldots, W_{ij}, \ldots,$ W_{pj} also receive $I_m/2$; but only word, W_{ij}, receives the full current, I_m, needed to switch the core. In Figure 9.4 it is W_{23} that receives the equivalent of current I_m. If we examine Figure 9.5, the reason for calling the coincident-current scheme a 3-D memory becomes obvious, and we can more easily

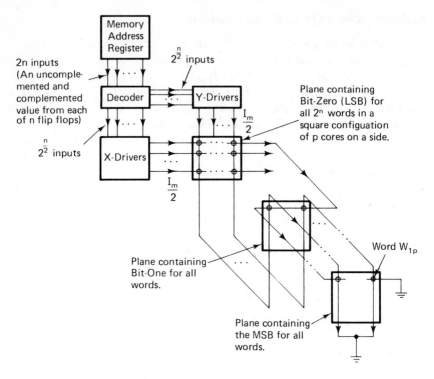

Figure 9.5   Organization of a Coincident Current (3-D) Memory.

visualize the wiring organization required to select one word from the $N = p^2$ words available in the memory. Note how the same two selection wires required to select word W_{1_p} intersect on each plane.

Now that the method of selecting the individual words is understood, we consider how to write any desired bit pattern into that word. If a current $-I_m/2$ were placed on windings X_i and Y_j, all bits of the selected word would have a 1 stored. To prevent this, the bits that are to remain at zero must be inhibited so that they do not receive the full current. One way to provide inhibition is to employ a separate circuit for each plane which is connected to a wire through the cores in such a way that only the plane to be inhibited (stay at zero) receives a current $+I_m/2$. Then the resulting current, I, on the inhibited core is

$$I = \frac{-I_m}{2} - \frac{I_m}{2} + \frac{I_m}{2} = \frac{-I_m}{2}$$

Thus, the core for the word being considered does not switch. (All of the other cores in that plane receive either a resultant current of zero or $+I_m/2$ and do not switch.)

Each plane must also have a sense circuit for the read operation. There-

fore, a typical memory requires four wires through each core—two wires for
the word selection, one for the sense circuit, and one for the inhibit circuit.
One of the common wiring schemes for these four circuits is shown in Figure
9.6. Observe that each memory plane has its own sense amplifier and its own

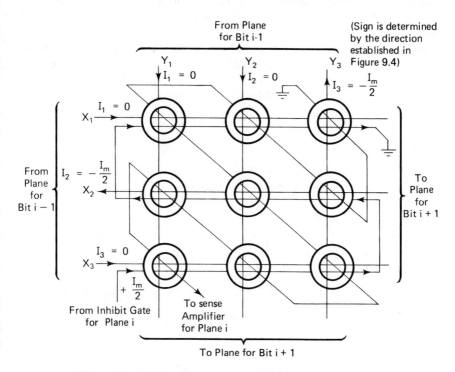

Figure 9.6 Bit Plane i for a Very Small (9 word) 3-D Memory.

inhibit gate; but as previously indicated by Figure 9.5, one set of X and Y
selection wires threads through all planes. As was true in Figure 9.4, word
W_{23} has again been selected. In this case, however, the inhibit current is
also on and it opposes the selection signal so that the resultant current,
$-I_m/2$, is not sufficient to write a 1 into the memory plane. The pattern shown
for the inhibit and sense windings in Figure 9.6 tends to minimize interference between the various circuits—particularly the delta noise which occurs
on the sense windings. Note that, if only the sense amplifiers for bit planes
0, 3, 4, 9, and 11 of a 12-bit memory detect signals during sampling, the
octal number 5031 will appear in the memory buffer register. Moreover,
after the read operation, the selected memory location will have all bits at
0. During the restore operation, the inhibit gates for bits 1, 2, 5, 6, 7, 8, and
10 must be on to preserve the 0's, while the inhibit gates for bits 0, 3, 4, 9,
and 11 are off so that these bits are set to 1 by the X and Y select currents.

2-D Memory Systems

Most core memories are of the 3-D design just described, but for practical as well as academic reasons some other schemes are also worthy of consideration. Core switching speed depends on core size, and one way to make a core smaller is to reduce the number of wires that must pass through it. The so-called 2-D system is an improvement in this regard over our previous system since it requires two wires instead of four. The 3-D method uses the cores themselves in the coincident selection process, while the 2-D method uses a separate circuit and driver for each memory word. Thus, for an N-word memory, N selection drivers are required instead of $\sqrt{N}$ drivers for X and $\sqrt{N}$ for Y, resulting in $2\sqrt{N}$ total drivers required by the 3-D system. This means that the 3-D system requires much less electronics.

The basic configuration for 2-D storage is shown in Figure 9.7. Note

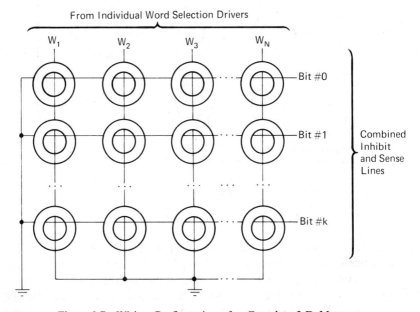

Figure 9.7   Wiring Configuration of a Complete 2-D Memory.

that in contrast to the 3-D system this is a planar arrangement. The bit (horizontal) lines serve the dual purpose of carrying the sense signal during read and supplying the inhibit signal during write. Thus, much like the 3-D memory, only as many sense amplifiers are required as there are bits in a word, and the same number of inhibit drivers are required. The two memory types, of course, differ in their geometric arrangement of the sense-inhibit lines in that 2-D storage employs a linear configuration and 3-D storage employs a planar configuration. A more serious electronic difference exists, however, in that the shared sense-inhibit lines in the 2-D memory make it more difficult to provide electrical isolation between each sense amplifier and the corresponding inhibit driver. The heart of the problem is that the sense amplifier must be very carefully designed so that the current pulse (which may approach 1 ampere) occurring during inhibition does not saturate the sense amplifier and prevent it from responding to a signal during the next read cycle.

$2\frac{1}{2}$-D Memory Systems

The term $2\frac{1}{2}$-D is not a paradox but merely a way of indicating that this storage method is a kind of hybrid between the 2-D and 3-D methods. The $2\frac{1}{2}$-D system is a compromise in many respects:

1. It employs fewer lines through each core than the 3-D system; thus, it can use smaller cores.
2. It employs a plane for each bit and a coincident-current technique; but it requires a different X-driver for each plane, which makes the electronics more expensive than for 3-D storage.

Figure 9.8 shows how the $2\frac{1}{2}$-D memory is organized. The Y-select and

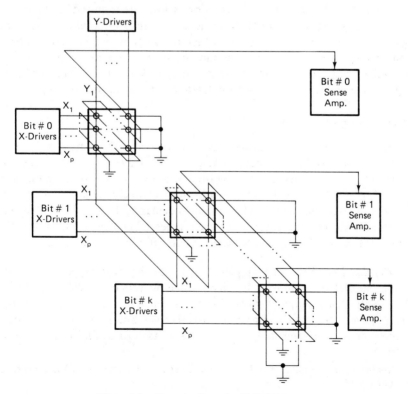

Figure 9.8 Organization of a $2\frac{1}{2}$-D Memory.

the sense circuits are essentially unchanged from the 3-D method, and during the read cycle, the coincidence current between X_i and Y_j signals will still select word W_{ij}. Here, however, the X_i line is not threaded continuously from plane to plane, but it originates in a separate set of drivers for each plane. At first this seems wasteful, but during the write cycle an inhibit signal is not needed because only those X_i lines for planes that are to contain 1's in the word W_{ij} receive the current $-I_m/2$. With the advent of fairly inexpensive MSI circuits, the $2\frac{1}{2}$-D system has been employed by some memory manufacturers.

9.2 MEMORY LOGIC CIRCUITS AND ELECTRONICS

Although space is conserved through extensive use of integrated circuits in the memory electronics, a typical core memory has approximately 70% of its printed circuit board area devoted to electronics and only 30% devoted to the cores themselves. Registers are required to contain the memory address and to buffer the data as they are received from the cores. Extensive logic is required to decode the memory address into the required select signals. Electronics are required for the various drivers, inhibit circuits, sense amplifiers, power supplies, and several other tasks. It is beyond our scope to consider specialized electronic design problems, such as arise with sense amplifiers, but the Bibliography does contain some basic references in this regard. We will, however, place moderate emphasis on memory logic.

Decoder Circuits

It has already been indicated that an N-word 2-D memory requires N selection lines, and a 3-D memory requires $2\sqrt{N}$ lines. These are generated from the n-bit memory address register, where $2^n = N$. For example, with $n = 12$, $2^n = 2^{12} = 4096$. Even with this relatively small memory, the number of selection signals to be generated demands that an efficient decoder be employed. There are three common types of decoders currently being employed: *parallel, tree,* and *balanced*; these will now be considered.

Parallel Decoder. The simplest, but by far the most expensive, decoder is the parallel type, and for brevity a three-bit version is shown in Figure 9.9.* Note that it employs an AND gate for each of the selection lines that are required. A particular gate in turn receives one input, either complemented or uncomplemented, from all bits in its address register. It is easily

*Since only one of eight ($2^n = 2^3 = 8$) lines is selected at a time, this is frequently called a 1-of-8 decoder.

X-Address Register

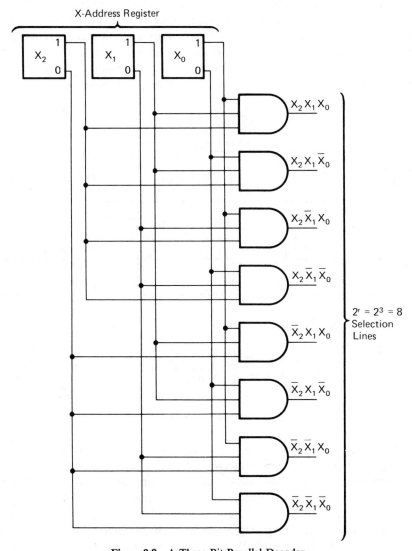

Figure 9.9   A Three-Bit Parallel Decoder.

seen that the total gates required by the parallel decoder, G_p, is the same as the number outputs:

$$G_p = 2^r \tag{9.1}$$

where r represents the bits in the address register. Also, the total gate inputs, I_p, which is one measure of decode cost, is given by

$$I_p = r2^r \tag{9.2}$$

In the case of a 2-D memory, there is only one decoder circuit, and all n bits of the memory address register supply signals to the decoder. In the case of a $2\frac{1}{2}$-D or 3-D memory, however, the memory address register splits into X-address and Y-address sections, each with its own decoder, as shown in Figure 9.10. Thus, each decoder receives inputs from only $r = n/2$ bits of the memory address register instead of receiving $r = n$ bits.

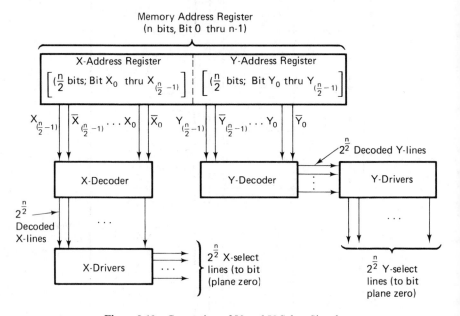

Figure 9.10   Generation of X and Y Select Signals.

‡*Tree Decoder.*   Several other methods are more economical than the parallel decoder because they employ multilevel logic to compose the output function from subfunctions, which in turn contribute to more than one output. The *tree decoder* is a representative member of this approach. Here the output is produced, through several levels of AND gates, by adding one new variable (bit) to the previous subfunction, at each level. Figure 9.11 shows how individual bits of the address register contribute to a typical output. Note the $(n - 1)$ levels of AND gates and how the output function is formed by adding the contribution from one variable at a time. (The resulting tree-like structure is the source of the method's name.)

Figure 9.12 represents the complete gating structure for decoding a four-bit address register. The correspondence between the full logic structure and the abbreviated structure in Figure 9.11 should be apparent. It can easily be seen why the number of gates increases by a factor of 2 in each level of gating as the output is approached.

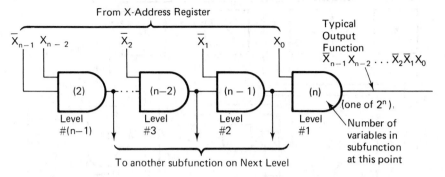

Figure 9.11 Multilevel Logic for a Typical Tree Decoder Output.

Let us employ the number of gate inputs required by the tree decoder, I_T, as a measure of system cost. Now any level whose subfunction contains k variables possesses 2^k output AND gates, and since the gates have two inputs, each level requires a total of $(2)(2^k)$ inputs. Thus,

$$I_T = \underbrace{(2)(2^2)}_{\substack{\text{From} \\ \text{level} \\ \text{(n-1)}}} + \underbrace{(2)(2^3)}_{\substack{\text{From} \\ \text{level} \\ \text{(n-2)}}} + \ldots + \underbrace{(2)(2^{n-1})}_{\substack{\text{From} \\ \text{level} \\ 2}} + \underbrace{(2)(2^n)}_{\substack{\text{From} \\ \text{level} \\ 1}} \qquad (9.3)$$

(Note that the terms in Equation (9.3) are in the same sequence as the levels in Figure 9.11.)

By the following reasoning, it is easy to obtain a closed-form expression for the equation. Factoring 2^3 from the expression:

$$I_T = 2^3(1 + 2 + 2^2 + \ldots + 2^{n-3} + 2^{n-2})$$

But the terms within the parentheses form a geometric progression whose sum, S, is given by:

$$S = \frac{rl - a}{r - a}$$

where the first term is $a = 1$, the last term is $l = 2^{n-2}$, and the common ratio is $r = 2$. Thus,

$$I_T = 8\left[\frac{(2)2^{n-2} - 1}{2 - 1}\right] = 8[2^{n-1} - 1] \qquad (9.4)$$

For the case $n = 4$ shown in Figure 9.12, Equation (9.4) yields $I_T = 56$; this is the exact number of gate inputs appearing in the figure.

‡*Balanced Decoder.* One of the most economical circuits has been given the name *balanced decoder* because of the way it partitions the variables between groups of gates on each level. Again the decoder is composed entirely of AND gates, with the general partition scheme shown in Figure 9.13 for a typical output function. In contrast to the tree decoder, all variables

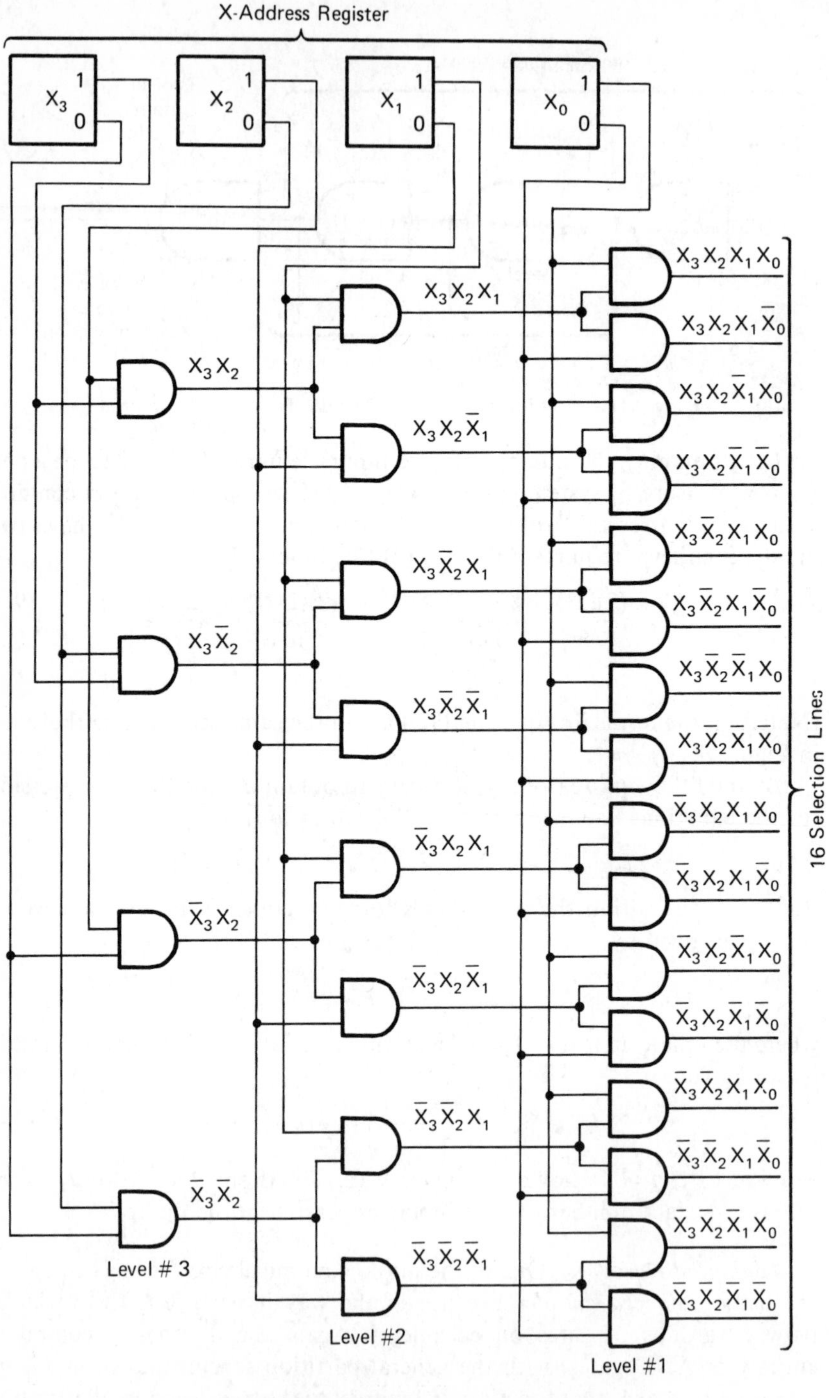

Figure 9.12 A Four-Bit Tree Decoder.

316

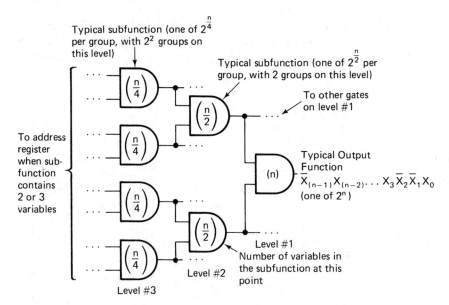

Figure 9.13 General Partition Diagram for the Balanced Decoder.

exist on each level, but they are in groups of subfunctions with the number of variables per group being indicated within the AND gate symbol. From the figure, it can be seen that as the level number increases, the variables in each subfunction decrease, and to compensate, the number of groups of subfunctions increases. Moreover, within each group, a complete set of minterms exists for the variables. Note that, except for level 1, all AND gates supply more than one input for the next level.

A partition diagram for the case of $n = 4$ is shown in Figure 9.14. This is a specific version of the general diagram in Figure 9.13, except that some of the details, such as the AND symbols, have been eliminated for conciseness. The numbers at each level, however, still indicate how many variables are in the particular subfunction group at that point. The complete logic circuit corresponding to the $n = 4$ partition diagram is shown in

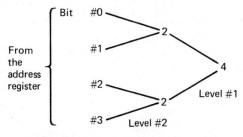

Figure 9.14 Partition Diagram for $n = 4$.

Figure 9.15. The two groups of subfunctions at level 2 are easily identified; each contains $2^{4/2} = 4$ outputs. Level 1 contains $2^4 = 16$ outputs. The three shaded gates demonstrate how a typical output, $\bar{X}_3 X_2 X_1 \bar{X}_0$, is generated,

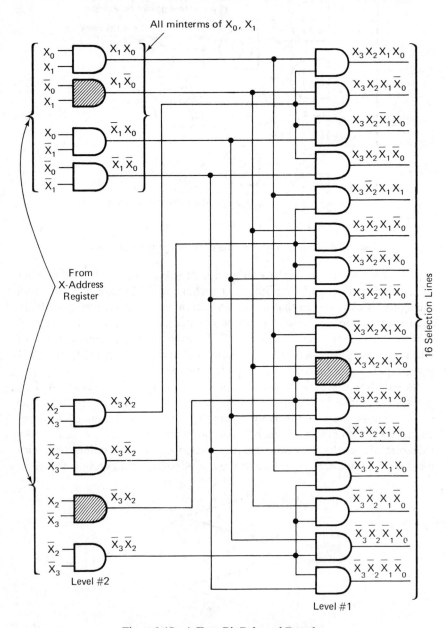

Figure 9.15   A Four-Bit Balanced Decoder.

and they clarify the way Figure 9.14 concisely represents the complete decoder.

By using the partition diagram of Figure 9.14 alone, let us determine the number of inputs, I_B, required by a four-bit balanced decoder:

$$I_B = \underbrace{(2)(2^2) + (2)(2^2)}_{\text{Level 2}} + \underbrace{(2)(2^4)}_{\text{Level 1}} = 8 + 8 + 32 = 48$$

We see that the 2^4 factor at level 1 is merely the number of outputs, and this is doubled to yield the inputs. Similar reasoning applies to the level-2 calculation, except that there are two groups and each contains only two variables. Of course, the calculation can easily be verified by inspection of Figure 9.15.

A more challenging case is that of the seven-bit decoder, i.e., $n = 7$; its partition diagram is shown in Figure 9.16. Because n is an odd number,

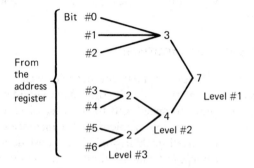

Figure 9.16 Partition Diagram for $n = 7$.

it is necessary to use a four-three partition at level 1. Moreover, the best way to generate the three-variable group on level 2 is to employ three input gates. Thus, as the figure shows, the three-variable group receives its inputs directly from the address register at level 2, but the four-variable group at level 2 must be partitioned another time. The calculation for the required gate inputs becomes:

$$I_B = \underbrace{(2)(2)(2^2)}_{\text{Level 3}} + \underbrace{(3)(2^3) + (2)(2^4)}_{\text{Level 2}} + \underbrace{(2)(2^7)}_{\text{Level 1}} = 328$$

An interesting comparison can be made between the gate input requirements for the three types of decoders as a function of the number of bits in the address register. Table 9.1 shows such a comparison for three values of n, and it also includes, in parentheses, the percentage of improvement available by going to the balanced decoder. Note that for $n = 4$ there is not a significant difference between the circuits; but for a small increase in n, to $n = 7$, there is a major savings in going from the parallel decoder to the three decoder and good savings in going from the tree decoder to the balanced decoder. In addition, with an increase in n, to $n = 10$, the differences between the decoders are accentuated. Also, remember that $n = 10$ is a rather modest decoder for a 2-D memory, e.g., $2^n = 2^{10} = 1024$. On the other hand, it will serve a rather large 3-D memory, e.g., $\sqrt{N} = 2^{10}$; therefore, $N = 2^{20} \approx 10^6$ (but, of course, two 10-bit decoders are required). The above calculation gives an impression of the additional electronics required by the 2-D system.

Table 9.1   Comparison of Decoders

Decoder Type	Required Gate Inputs (Percentage Savings of Balanced Decoder)		
	$n = 4$	$n = 7$	$n = 10$
Parallel	64 (33.)	896 (173.)	10,240 (357.)
Tree	56 (17.)	504 (54.)	4,088 (83.)
Balanced	48 (0)	328 (0)	2,240 (0)

9.3 SEMICONDUCTOR MEMORIES[4, 5, 7, 14]

Semiconductor memories are now competitive with magnetic core in most applications. As a matter of fact, they have significant advantages, particularly when it comes to speed and to the cost of small memories. It must be understood, however, that semiconductor memories are not homogeneous; they are quite varied in their objectives, construction, and performance. They range in purpose from the random-access memory, RAM, to the read-only memory, ROM. (The latter type will be discussed in Section 9.4.) Bipolar memories are available which are more than an order of magnitude faster than core, while some less-expensive MOS memories are only comparable in speed to core.

There is considerable similarity between the way core memories and RAMs are structured. Both require a memory cell to store each bit; in magnetic memories the cell is a core, while in RAMs it is one of many possible types of bistable devices. Both have two main ways for addressing individual bits: either a linear technique (employed in 2-D core memories) or a coincident technique (employed in 3-D core memories).

A photomicrograph of a typical 256-word bipolar RAM (the 74S201) is shown in Figure 9.17. The regular pattern in the figure is produced mainly by the transistors and their interconnections forming the memory cells; but a portion of the pattern is due to decoders and other supporting circuitry.

First, consider how the idealized *linear selection* technique shown in Figure 9.18(a) is employed to develop the output data. Note that, by means of the decoder, one of the 16-word lines (w_0 through w_{15}) selects the outputs of its four flip-flops to be placed on the four common data lines, and that a wired OR connection allows the AND gates servicing each line to be tied

Figure 9.17 Photomicrograph of a 256 Word Bipolar RAM (Photo courtesy of Texas Instruments, Inc.).

together. Here all $16 \times 4 = 64$ bits of the memory are represented on a single plane.

The other addressing scheme, *coincident selection*, is shown in Figure 9.18(b). By contrast with linear selection, here four planes are required to represent the same 64-bit memory. Moreover, the reason for the name *coincidence selection* is obvious from the diagram since an AND gate can place information on the data output line only when a logical 1 exists on both its X and Y inputs from the decoder. Since each 1-of-4 decoder has a single logical-1 output, only one memory cell can be selected on each plane, and the result at the data output terminal is the state of the selected memory cell in that plane.

Thus far, only the read process has been discussed; we now consider the complete system. The semiconductor memory industry has invested a great amount of effort in developing efficient memory cells since the savings achieved is multiplied by the number of cells in the memory, and this can be a large number indeed—several thousand or more. Various practical memories will be discussed shortly; but to establish the basic principles, the

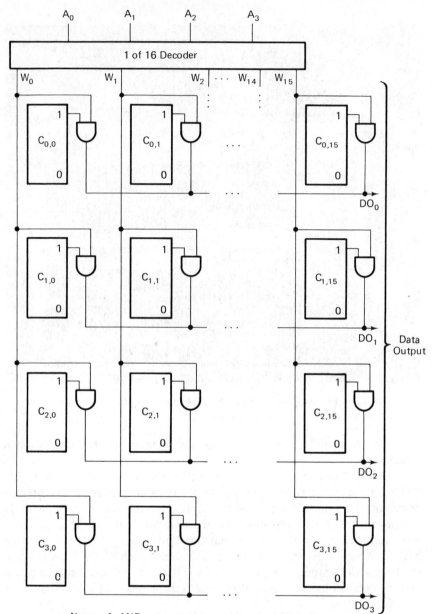

Notes: 1. AND gates are connected by wired OR logic.
2. A_0 through A_3 are from a four bit address register.

(a) Linear Selection System (all output bits)
for 16 Word x 4 Bit Memory

Figure 9.18 Memory Selection Techniques

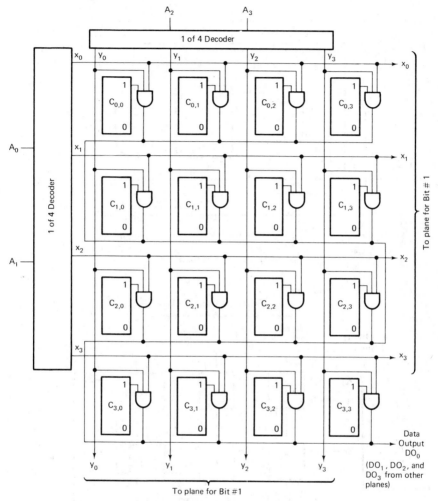

Notes: 1. The other three planes are similar.
2. AND gates are connected by wired OR logic.
3. A_0 through A_3 are from four bit address register.

(b) Coincident Selection System
for Bit Zero of a 16 Word x 4 Bit Memory

Figure 9.18—*Cont.*

idealized coincident cell and its plane, shown in Figure 9.19, are considered. The cell in part (a) of the figure accomplishes data read with an AND gate exactly as was done in Figure 9.18. Moreover, information is stored with a similar coincidence technique in that the x_i and y_j decoder signals combined with the write pulse to control whether or not the clock input of the flip-flop is actuated. Finally, the input information is present on a data input

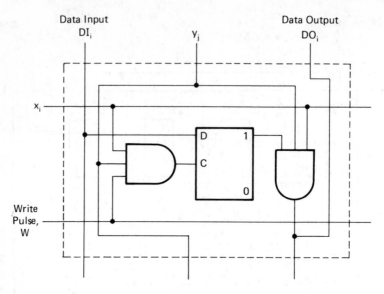

(a) Idealized Memory Cell

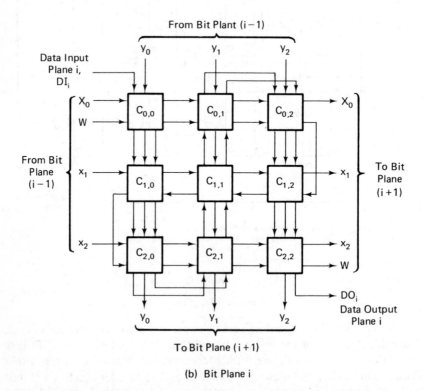

(b) Bit Plane i

Figure 9.19 A Nine-Word Semiconductor Memory Employing Coincident Addressing.

line; but of course it is not stored unless a pulse reaches the C terminal. Part (b) of the figure shows how cell wiring is organized to produce a memory plane. (In order to really understand memory operation, the reader should carefully trace all wiring paths.) Each plane has separate data input and output lines, which thread through each cell in a way analogous to the method employed by the sense lines in a core memory. The x, y, and write lines, however, are common to all planes. Note that, as a means of simplifying the memory plane circuits, a "feed-through" wiring convention is employed. This means that all lines supplying a particular cell pass directly through the cell and emerge in the same position on the opposite side. (Refer to part (a) of the figure for details.) As a consequence of this technique, there is no preferred direction for signal flow; it is left to right or right to left, depending only on convenience (up and down are similar). The only time a line does not emerge from a cell is when there is no further requirement for it. The termination of the data input signal within $C_{2,2}$ is typical. (A similar situation occurs with the origination of the data output signal in $C_{0,0}$.)

Consider the situation where a 1 is to be stored in $C_{2,1}$. Here the decoder outputs would be $x_2 = y_1 = 1$ and $x_i = x_j = 0$ for all other i and j. $DI_i = 1$ and the write pulse would be present. Since only $C_{2,1}$ is selected, it would be the only cell which would store the data. The output from $C_{2,1}$ would appear at DO_i first, indicating the original bit contents and then, after the write pulse, the new bit contents.

‡Practical Memory Cells

We intend to describe only a few basic types of memory cells here since this is currently an area of active development. The reader is referred to the literature and particularly to various manufacturers' data sheets for information on specialized high-performance memories.

A typical bipolar memory cell, employing linear selection, is shown in Figure 9.20. In overall structure, it should be recognized as an elementary flip-flop (similar to that in Figure 6.30) in which one device (gate or transistor) is on and the other is off. Normally the word selection line is held near 0 volts and the data lines are held near 1.5 volts by their amplifiers; therefore, only the emitters connected to the word line are forward biased. The result of this is that the full current from the conducting transistor is carried by the word line and not by either of the data lines. In order to read the contents of the cell, it is selected by bringing the word line up to, say, 3 volts. This places the emitters connected to the word line in the reverse-biased state, while the data line emitters remain at 1.5 volts, and one of them becomes forward biased. The result is that the current from the conducting transistor is now switched from the word line to one of the data lines, resulting in either the *Data* or $\overline{Data}$ line carrying current. With a current sensing amplifier in

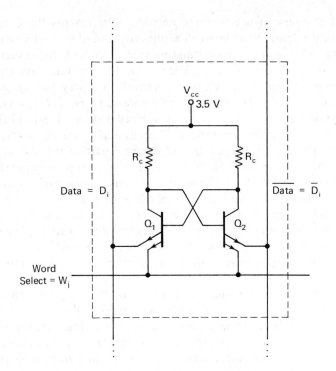

Figure 9.20 Bipolar Memory Cell Employing Linear Selection.

each of the two data lines, it is easy to determine whether the sensed cell is in the 1 or 0 state.

Writing data into a cell is accomplished in a similar way. Again the cell is selected by raising the word line to 3 volts; but now the data lines are unbalanced by raising one of them to, say, 3 volts and lowering the other to, say, 0.7 volt. This forces one transistor into cutoff and the other into saturation. Observe, however, that as long as the word select line is low, it sinks all the current from the conducting transistor, and unselected cells are effectively disconnected from the data lines, from the standpoint of reading or writing information.

If the cell is storing a logical 1, Q_1 will be conducting; after selection, the D_i line will be carrying the current, resulting in a logical 1 being developed by the sense amplifier. (If a 0 is stored, the $\bar{D}_i$ line will be carrying current after selection.) In contrast, if a logical 1 is to be written, the following conditions are required to turn Q_1 on:

$$D_i = 0.7 \text{ volt} \quad \text{and} \quad \bar{D}_i = 3 \text{ volts}$$

Thus, assuming positive logic, the write amplifiers must invert the input signal; but the sense amplifiers must not invert the output. This fact, as well as the general structure of the bipolar memory, is revealed by Figure 9.21. Note how the sense and write amplifiers terminate both ends of the feed-through circuitry. Also note that the overall memory configuration is very economical in the number of transistors it requires—only two per memory cell, plus a few in the read and write amplifiers.

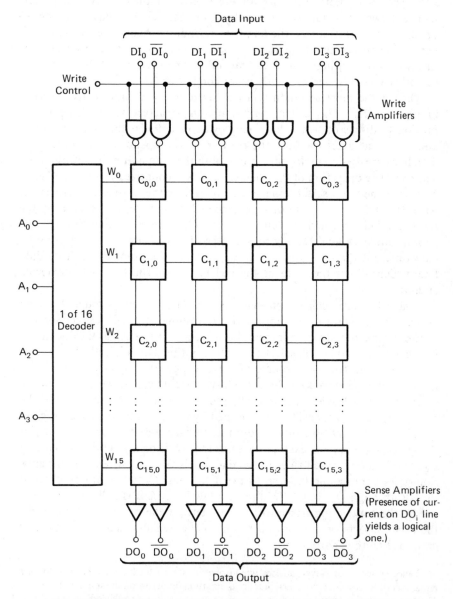

Figure 9.21   Bipolar Random Access Memory Employing Linear Selection.

Since MOS transistors occupy much less area on a silicon chip, it is not surprising that MOS memories have become very popular. It was mentioned in Chapter 6 that information can be stored in MOS circuits either by means of a dynamic charge storage or by means of a static technique. As a matter of fact, the master-slave flip-flop circuits of Figures 6.43 and 6.44 could be employed as memory elements. In the interest of bit density and performance, however, a number of variations of those circuits have been developed. For large memories, the number of support circuits, e.g., sense amplifiers, is small compared to the number of memory cells. The designer is therefore often willing to accept a tradeoff of small cell size at the expense of the other circuitry being slightly more complicated.

One of the most common types of static MOS memory cells is shown in Figure 9.22.* This cell is very similar in structure to the bipolar cell previously discussed—they both employ cross-coupled flip-flops. Here Q_1 and Q_6 act as gates for both the read and write operations. Also, the two data lines are again employed to carry both the read and write information. The complete operation of this memory cell is summarized in Figure 9.22(b). Note that the word select line must go high for either a read or write cycle. Also, the data lines are both high except during write, at which time the transistor being turned on (either Q_2 or Q_4) receives a low through the appropriate data line. During sensing, the off transistor has no effect on its data line, but the on transistor draws part of its current through its data line, which allows the state of the cell to be transmitted to the external circuitry.

Since the static cell is continuously supplied current through the load devices Q_3 and Q_5, it does not require refreshing. But it is less economical in power consumption than comparable dynamic cells—such as the one shown in Figure 9.23. When the select line is low in this dynamic cell, the flip-flop itself is isolated from any source of current, and its state is maintained by means of charge stored on the gate stray capacitance, shown dashed in the diagram. Since leakage paths exist, the capacitors must occasionally be refreshed; but this is easily accomplished by opening the transmission gates Q_1 and Q_4. At that time these two transistors also serve as load resistors for the Q_2-Q_3 flip-flop. For example, assume that Q_2 is on and Q_3 is off, but that C_2 is no longer charged to $+V$, the data line's voltage. When the transmission gates connect the flip-flop to the data lines, C_2 will again be charged to $+V$, through Q_4; but C_3 will remain low because it is still being shorted out by the conducting transistor, Q_2. Except for the periodic refresh cycle, the other aspects of memory operation of the dynamic cell follow the diagram of Figure 9.22(b).

---

*Since either n- or p-type devices would operate satisfactorily in the circuits for the remainder of this chapter, the usual practice of omitting the arrow on the source is followed. The text description, power supplies, etc., are presented as if n-type devices are employed.

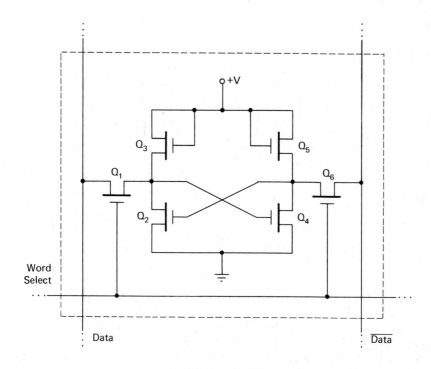

(a) Memory Cell

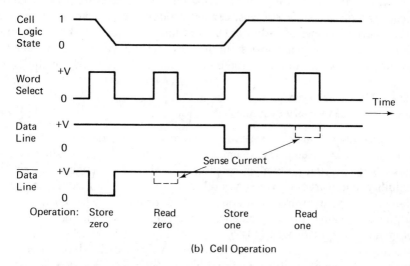

(b) Cell Operation

Figure 9.22 A Static Memory Cell Employing Six Transistors.

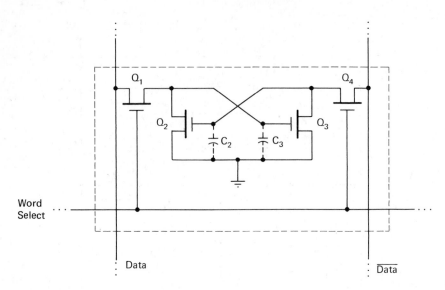

Figure 9.23   Dynamic Memory Cell Employing Four Transistors.

There are many ways to still further reduce the number of transistors required by a dynamic memory cell. A three-transistor cell is fairly common in practice. It employs a single capacitor instead of a flip-flop to store charge. Moreover, it has even been possible to construct a cell with a single transistor. But this technique has a number of disadvantages—the readout is destructive, and the writing speed is slow due to the need for a fairly large capacitor to deliver sufficient sense current. Undoubtedly, electronic engineers will achieve a number of additional major improvements in both bipolar and MOS random-access memories in the next several years. The reader is encouraged to keep abreast of these developments by regularly reading the literature.

Semiconductor Memory Systems

Semiconductor memory systems are commonly constructed from groups of moderate-size, elementary memory chips, frequently dual-in-line packages. By employing a number of these packages, the designer has great flexibility in constructing memories of various word and bit configurations. Manufacturers' literature should be consulted for the many practical devices currently available. For convenience, we will employ a hypothetical eight-bit memory, the TUtorial 1 (TU1), shown in Figure 9.24, which is organized into an array consisting of four words of two bits each. (It is similar in structure to several small commercial bipolar memories, e.g., the Signetics. 8225, which is a 16-word by 4-bit array.)

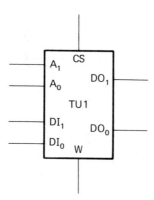

Figure 9.24 Logic Diagram of the TU1 Memory Package.

Examination of the diagram reveals two address bits (that actuate a 1-of-4 on-chip, built-in, decoder) allowing any one of the four words to be selected Since the memory handles two bits in parallel form, at both the input and output, for each addressed word, these are available as DI_0, DI_1, DO_0, and DO_1, respectively. A chip select, CS, and a write, W, control are also provided. The condition $CS = 1$ effectively actuates the chip and allows its data lines to receive and transmit information. (With $CS = 0$, $DO_0 = DO_1 = 0$, independent of the chip's contents as well as the values of A_0 and A_1.) Even though $CS = 1$, a pulse must exist on the W line, with a $1 \rightarrow 0$ transition, before the input data are written into the specified memory address. Finally, the chip has the wired-OR capability, which allows the outputs to be tied to corresponding terminals on other chips so that quite complicated memory configurations can be produced.

In wired-OR situations like this, where a chip select signal is employed to actuate the output, an electronic configuration called *tristate logic* has come into use. Here, a chip which is not selected presents essentially an open circuit at its output terminals so that it has no effect on any other circuits to which it is connected. On the other hand, the selected chip controls the logic state of its output terminals just like an ordinary gate does.

As an example of the way memory systems can be constructed from elementary chips, consider Figure 9.25, where an 8-word × 4-bit memory has been constructed from four TU1 chips. Here two TU1's are employed to increase the number of bits per word to four by paralleling both the input and the output lines. In addition, the number of words is increased to eight by having a third address line, A_2, control another pair of parallel chips through an inverter, while the first pair of chips receive the direct A_2 line. (Note that corresponding bits between the pairs of TU1's are OR-tied.) Further consideration will be given to memory systems in the Exercises.

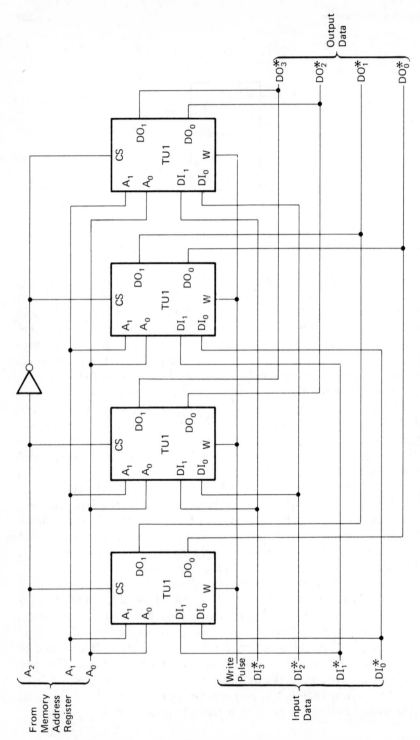

Figure 9.25 An 8-Word by 4-Bit Memory System Constructed From 4-Word By 2-Bit Packages.

9.4 READ-ONLY MEMORIES

Its low cost, versatility, and high operating speed have elevated the read-only memory, ROM, to the status of a basic building block for digital circuit design. Typical applications include: control sequence generator for a large central processor through a microprogram; storage of frequently used subroutines; character generators; table lookup schemes; large combinational circuits; and translators from one code to another, e.g., from Gray code to BCD.

A ROM is a memory with many of the same properties of a RAM, except that the data are either permanently or semipermanently stored; therefore, it is not subject to destructive readout, volatility, and similar problems. It can be constructed from diodes, bipolar transistors, MOSFETs, and many other types of devices; also, both linear as well as coincident word selection are employed. ROMs constructed for high-volume applications having permanently stored information are made from photographic masks in large production runs much like ordinary ICs. Moreover, versions with a capacity of thousands of bits are commonly produced.

In order to gain physical understanding of how a ROM may be constructed, consider the elementary diode memory shown in Figure 9.26. Observe that the ROM transmits a logical 1 to the output if there is a diode connecting the active word select line with the output line; otherwise, it transmits a 0. (The disconnected diodes are shown with one of their leads cut to represent the omission of part of conductor such as would be done with integrated-circuit fabrication.) The diode ROM operates according to the same principles as diode gates, and the reader should have no trouble verifying that word W_1 in the figure does indeed produce voltages at the DO_i output terminals corresponding to 1101. It should also be observed that the diodes connected to the unselected word have no effect on the output since these diodes do not carry current.

Programmable Read-Only Memories. For small production runs and for developmental systems, it is necessary to place information into a read-only memory in the field quickly and conveniently. For these reasons the *programmable read-only memory* or PROM was developed.

There are two main classes of PROMs: the one-time programmable type and the reprogrammable type. The former involves cutting a special link in the memory array for each 0 created. Typically, this is accomplished by manufacturing an array with all available positions filled with diodes or other components. Then the word of interest is selected by the usual addressing method, and a large-amplitude current pulse is placed on each bit terminal where a 0 is to be created. The pulse causes the fuse link to be broken

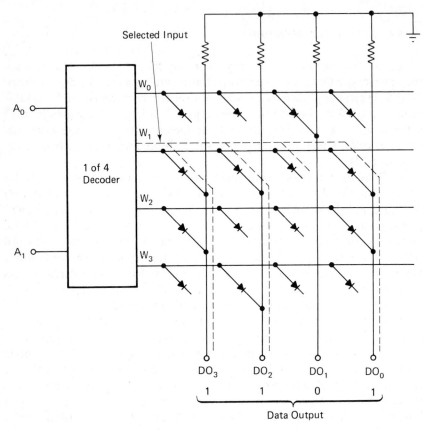

Selected Input

A_0

A_1

1 of 4
Decoder

W_0

W_1

W_2

W_3

DO_3 DO_2 DO_1 DO_0

1   1   0   1

Data Output

Note: For both word select and data output lines, W_i and DO_i:
       "0" = 0 volts and "1" = 5 volts. (Neglect diode drop)

Figure 9.26 A 16-Bit Diode ROM Employing the Linear Selection
Technique.

at the location of each desired 0; but the other diodes are undisturbed. After
programming, the memory can be operated in the same way as a factory-
fabricated version, and it will produce a 1 output where a diode is intact and
a 0 output where the fuse has been broken.

Typically, the reprogrammable PROM employs MOSFETs as memory
elements, in place of diodes. The gate of each device, however, is electrically
insulated from all other parts of the transistor. Programming is accomplished
in a similar way to the fuse type—the desired word is selected as usual, but
this time instead of breaking a link to create a 0, an avalanche effect is
produced in the selected FET causing a charge to be deposited on the gate,
which stores a 1. (Since the gate is electrically well-insulated, the charge will
remain for years.) The gates that did not receive a pulse remain uncharged

and thus store 0's. After programming, the PROM can be read in the usual way by selecting one word at a time. The FETs storing a 1 have a low-impedance path on the transmission gate between the select line and the bit line, while those storing a 0 have an open circuit.

The charge on the FET's gate requires an energy of about 5 electron-volts to escape; but sunlight and thermal excitation associated with normal operating temperatures provide for less than the required amount. On the other hand, ultraviolet light has sufficient energy to liberate the charge from the gate, thereby clearing the memory. This process is facilitated by means of a small window in the top of the PROM. With commonly available ultraviolet lamps, it requires about 10 minutes to complete the process.

Applications of Read-Only Memories

In applying ROMs, there are two alternate paths that can be taken: either a *word approach* or a *Boolean function approach*. Both methods can be thought of as generating a truth table with the input word forming the left half of the table. In the word approach, the designer takes a horizontal view of the truth table and specifies an output word in correspondence to each input word. This view is illustrated in Table 9.2(a), where the ROM design information for converting from BCD to excess-3 code is given. The boxed line shows that 0110 in BCD converts to 1001 in excess-3. On the other hand, the function approach leads the designer to take a vertical view of the

Table 9.2 Comparison of ROM Design Methods

(a)
Word Approach:
BCD to Excess-3 Converter

Input Word (BCD) $A_3A_2A_1A_0$	Decoded Word	Output (Excess-3 Word) $DO_3DO_2DO_1DO_0$
0 0 0 0	W_0	0 0 1 1
0 0 0 1	W_1	0 1 0 0
0 0 1 0	W_2	0 1 0 1
0 0 1 1	W_3	0 1 1 0
0 1 0 0	W_4	0 1 1 1
0 1 0 1	W_5	1 0 0 0
0 1 1 0	W_6	1 0 0 1
0 1 1 1	W_7	1 0 1 0
1 0 0 0	W_8	1 0 1 1
1 0 0 1	W_9	1 1 0 0
Other cases not allowed	W_{10} through W_{15}	0 0 0 0

(b)
Function Approach:
Multifunction Generator

Word $A_2A_1A_0$	Decoded Word	F_0	F_1	F_2	F_3
0 0 0	W_0	0	0	0	0
0 0 1	W_1	0	1	1	1
0 1 0	W_2	0	1	1	1
0 1 1	W_3	0	0	1	0
1 0 0	W_4	0	1	1	1
1 0 1	W_5	0	0	1	1
1 1 0	W_6	0	0	1	0
1 1 1	W_7	1	1	1	0

truth table, where the truth values of a single output bit for all combinations of the input variables are specified. This view is illustrated in Table 9.2(b), where the ROM information for generating several functions is specified. The boxed column presents the EXCLUSIVE-OR function

$$F_1 = \bar{A}_2\bar{A}_1A_0 + \bar{A}_2A_1\bar{A}_0 + A_2\bar{A}_1\bar{A}_0 + A_2A_1A_0,$$

which is the sum output equation for a full adder. It only requires eight ROM bits ($2^3 \times 1 = 8$); but if mechanized by conventional AND-OR logic, the function requires five gates. In this example, an extremely favorable ratio of $\frac{8}{5}$ bits per gate is obtained. The function F_0, however, has a ratio of $\frac{8}{1}$; and on the average a ratio of about 12 bits per gate is typical. Moreover, in unfavorable cases the ratio can be more than 50 bits per gate. Nevertheless, it is often advantageous to mechanize combinatorial circuits by means of ROMs.

By the proper combination of ROM packages, it is possible to extend both the number of functions available and the number of input variables. Such an expanded circuit composed of three ROMs is shown in Figure 9.27.

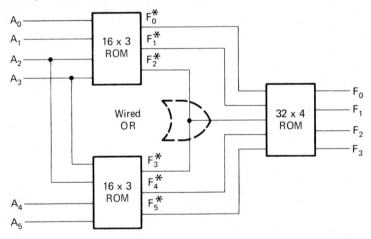

Figure 9.27 A ROM Combinatorial Circuit with Extension of Both Variables and Functions.

How are truth tables converted to actual ROM circuitry? The process is really quite straightforward, as shown in Figure 9.28, which is the mechanization of the BCD-to-excess-3 converter from Table 9.2(a). Basically, the realization process consists of placing a circuit element (in this case a diode) between the word select line and the output bit line wherever a 1 appears in the truth table and omitting a component in a position corresponding to a truth table 0. The resulting hardware is essentially a direct mapping of the original truth table onto a ROM circuit similar to that in Figure 9.26, except

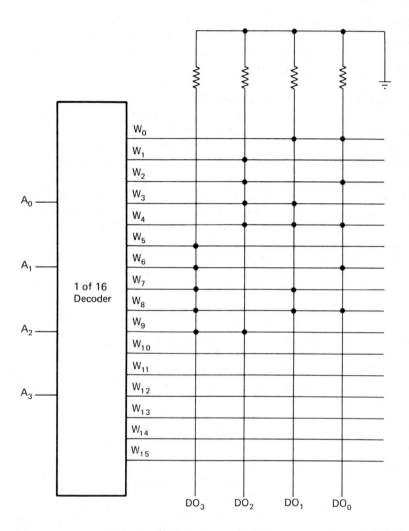

Note: Each dot represents a diode connected in
the manner of Figure 9.26

Figure 9.28   Diode ROM Converter for BCD to Excess-3.

that a more concise nota ion is employed—connected diodes are represented
as dots whereas open diodes merely produce line crossings. We will *always*
assume that the decoder output line is selected from a straight binary representation of the input bits independent of the actual nature of the input
code. For example, W_5 is always selected by $A_3A_2A_1A_0 = 0101$, and the
input bits happen to represent a 2 in excess-3, but they represent a 5 in BCD.
The reader should verify that Figure 9.28 is an exact reproduction of Table

9.2(a) by using the decoded word column in the table as a guide. Observe that the input numbers 1010 through 1111, which are not allowed BCD codes, correctly yield 0000 outputs since word select lines W_{10} through W_{15} are not connected to the data lines.

Implementation of Arithmetic with ROMs.[2]   It is frequently very economical to implement arithmetic operations with ROMs, and IC manufacturers currently have many modules for this purpose.

Let us consider a module for multiplying two 4-bit numbers, which results in an 8-bit product. The resulting truth table will be large ($2^{4+4} = 256$ decoded words), but its implementation will be straightforward. The resulting ROM will be 256 words by 8 bits and yields $256 \times 8 = 2048$ bits, which is well within the capabilities of current technology. Single chips such as this are currently available from Fairchild Camera, Texas Instruments, National, and others. (The National MM 523 is a typical example.) On the other hand, since the number of memory bits increases with n as $n2^{2n+1}$, single chips for large values of n are difficult to obtain, e.g., $n = 8$ requires 1,048,576 bits, which is considerably beyond current capability for a single chip.

The above result does not prevent ROMs from being employed for multiplication of large numbers; it just means that more sophisticated techniques must be used. Consider the multiplication of two numbers, N_1 and N_2, each consisting of a lower four bits L_i and an upper four bits U_i. (For example, N_1 can be expressed as $N_1 = 2^4(U_1) + L_1$, where U_1 is shifted right four places before summing because it is the upper four bits.) Then the resulting product is:

$$N_1 \times N_2 = [2^4(U_1) + L_1] \times [2^4(U_2) + L_2]$$
$$= 2^8(U_1 \times U_2) + 2^4(U_1 \times L_2) + 2^4(U_2 \times L_1) + L_1 \times L_2 \quad (9.5)$$

This equation reveals that the desired $N_1 \times N_2$ product can be achieved with four multipliers (256×8-bit ROMs) each accepting two 4-bit numbers and developing an 8-bit result. In addition, as Figure 9.29 shows, five 4-bit adders are also required. Now, each multiplier yields an 8-bit result, but the terms in Equation (9.5) are separated by shifts ranging from zero to eight bits. Thus, care must be taken in summing the correct groups of bits in the adders. (The bit sequence labeling at the output of each multiplier and other places within the diagram should be helpful in checking the circuit.) Note that bits 4–7 emerge from the multipliers in three sections, which is why they require two adders. (A similar situation occurs with bits 8–11.) Finally, observe how the various carry outputs (C_{out}) are properly passed to adders summing the next higher sequence of bits; but since two 8-bit numbers never give a result with more than 16 places, the leftmost adder always has $C_{out} = 0$.

Eight Bit Multiplier and Multiplicand (Divided into 4 Bit Sections, L_i and U_i.)

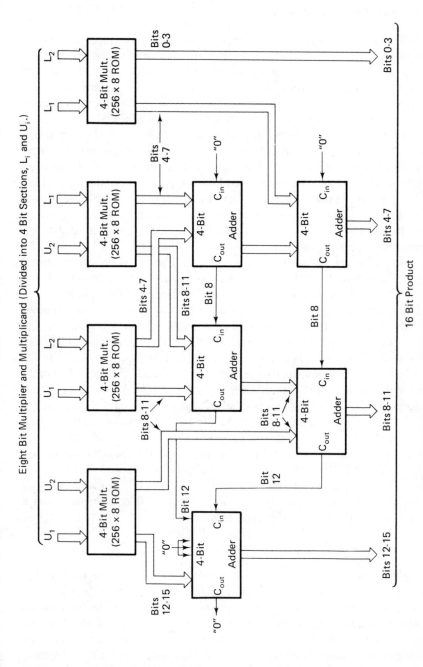

Figure 9.29 An Eight-Bit Multiplier Constructed From Four-Bit Multiplier ROMs.

9.5 CONTROL MEMORIES

Although the memories in the previous section were employed only in combinational design, they are certainly also applicable to sequential problems. This latter application is known as a *Control Memory*, CM, since the memory is used to control a sequence of digital circuit actions. The state generator, the logic and timing module, and some of the other circuitry of Chapter 7 could conveniently be replaced by means of a CM unit.

The CM idea was conceived in 1951 by M.V. Wilkes,[15] of the Cambridge University Mathematics Laboratory, as a means of adding systematic order to methods for designing digital computers. His original concept involved a ROM in which each word was divided into two parts. The first part specified the location in memory where the next instruction word could be found; the second part specified the states of certain signals which were employed to control individual logic circuits in the ALU and other parts of the computer. Because the control existed over very basic, low-level, ALU actions, the instructions became known as *microinstructions*. The instruction words stored in our CM will henceforth be called *microinstructions*. In addition to being employed here, they will also be considered in great detail in Chapter 10.

These ideas were soon further developed by Wilkes and other people; and one of the IBM series 7000 computers was microprogrammed in the early 1960s. Moreover, the IBM Systems 360 and 370, the Honeywell H4200, and several machines since have employed CM techniques. The current availability of inexpensive memories, however, has made this technique more popular than ever before. In addition to its use purely as a design method, it also possesses some very practical advantages. For example, the presence of a CM allows the configuration of the digital system to be easily changed simply by changing the contents of the memory instead of completely rewiring the control unit. In fact, a writeable store such as a RAM can be employed as the CM so that the configuration can be changed about as fast as a program is changed in a standard computer. (The term *Control Read-Only Memory*, CROM, is frequently used when the CM is constructed from a ROM rather than from another type of memory.)

An Elementary CM System

Figure 9.30 shows that the major components of a CM system are the memory itself, the memory address register, and the set of input/output signals. (In more complex systems, particularly those discussed in Chapter 10, another component called the *next-address logic* is added.) The response to each current address is the contents of that memory location which consists of an address output, $A_m \ldots A_1 A_0$, and a group of control signals, $\mathcal{C}_k \ldots \mathcal{C}_1 \mathcal{C}_0$. The address output is combined with some external signals

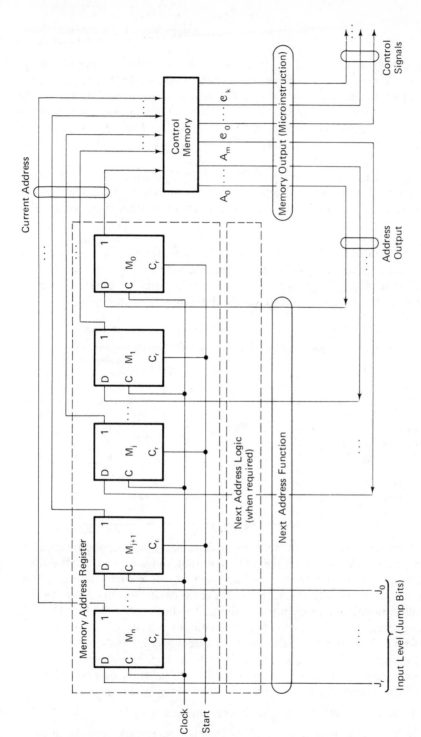

Figure 9.30 An Elementary CM System.

called *jump bits*, $J_R \ldots J_1 J_0$, to produce the next-address function, which for our simple system becomes the current address after the occurrence of the clock pulse. It is easy to visualize how a single jump bit, J_0, may be employed to produce conditional branching, much as was done with the machine language instruction APO in Chapter 3. When $J_0 = 0$, the next-address function becomes $0A_m \ldots A_1 A_0$; and when $J_0 = 1$, the function becomes $1A_m \ldots A_1 A_0$.

Note that the start signal in Figure 9.30 (operating through the clear, C_r, input of each flip-flop) sets the memory address to zero. Thus, in this system, the first microinstruction is always taken from memory location zero.

CM Implementation of an Up-Down Counter and Decoder. In order to demonstrate the CM method of design in a very simple way, consider the implementation of a two-bit up-down counter and its binary-to-decimal decoder. The CM required to satisfy this problem is shown in Table 9.3.

Table 9.3   CM Design of an Up-Down Counter and Decimal Decoder

Current Memory Address, MA			Memory Contents (Microinstruction)					
Input M_2 (J_0)	Counter M_1	M_0	Address Output A_1	A_0	Control Signals (Decimal Decoder Output) e_3	e_2	e_1	e_0
0	0	0	0	1	0	0	0	1
0	0	1	1	0	0	0	1	0
0	1	0	1	1	0	1	0	0
0	1	1	0	0	1	0	0	0
1	0	0	1	1	0	0	0	1
1	0	1	0	0	0	0	1	0
1	1	0	0	1	0	1	0	0
1	1	1	1	0	1	0	0	0

The left side rows are labeled: first four rows "Up-Counter", last four rows "Down-Counter".

As expected, each current memory address has a corresponding memory contents, which consists of the address output and a set of control signals. While only a two-bit counter is implemented, the memory address, MA, requires an extra bit, M_2, to provide the up-down branching for the two counter modes. As an example, if the unit is in the state $MA = 001$, it is acting as an up-counter; and the address output is $A_1 A_0 = 10$, which means that following the clock pulse the next state would be $MA = 010$. Observe that the action taken by the counter at each clock pulse is determined by the current memory output (more specifically, the address output). The unit may be changed to a down-counter by the following sequence:

1. Between clock pulses, the J_0 level changes to $J_0 = 1$.

2. The clock pulse occurs and the counter increments since the memory

address register is still set for the up-counter mode. At the same time, the M_2 bit is being loaded from J_0.

3. After a short delay, the current memory address becomes that of a down-counter, $M_2 = J_0 = 1$, and the address output shows a decrement.

4. The next clock pulse occurs and the counter performs the first decrement.

(In practice, the counter could be made to decrement immediately; but it would require a somewhat more complicated circuit than that shown in Figure 9.30) The control signals are straightforward, and as required, they are a decimal translation of the counter's binary state, $M_1 M_0$.

A CM, such as in Table 9.3 may be designed through the following steps:

1. From the statement of the problem, construct a CM table with the required number of memory addresses, including the jump bits.

2. Add the required address output columns and the control signal columns to the table.

3. Starting at a convenient memory address, fill in the corresponding address output as dictated by the problem statement. Using this output as the current memory address, fill in its corresponding address output. Continue in the natural sequence of the problem itself until the address output entries are complete for the entire table.

4. Review the entries generated in step 3 and be sure they correctly satisfy the requirements of the problem.

5. Complete the control signal columns of the table by determining what values are required by the problem statement for each memory address.

In order to get a clear and concise picture of the original problem requirements, it may be desirable to construct a flow diagram or write an RTL program before starting step 1 in the above procedure. The next example will demonstrate the value of an RTL program as a starting point for CM design.

CM Implementation of a Multiply Algorithm. The multiply algorithm of Figure 7.22 is an elementary but interesting possibility for CM design. Since that algorithm involves six add/shift steps and each step consists of two parts, 12 microinstructions are required for the basic program. To these must be added the $T = 1$ initiation step and the $T = 4$ halt step, for a total of 14 microinstructions. Table 9.4 shows these microinstructions and the other aspects of the CM program. The reader should now go through the entire program and be sure it does indeed correctly implement the multiply algorithm. We see that four control signals are employed; these are in a one-

Table 9.4   CM Implementation for the Multiply Algorithm of Figure 7.22

		Current Memory Address				Memory Contents (Microinstruction)						
						Address Output					Control Signals			
	M_3	M_2	M_1	M_0	A_3	A_2	A_1	A_0	$\mathfrak{C}_4$ (Halt)	$\mathfrak{C}_3$ (Shift)	$\mathfrak{C}_2$ (Add)	$\mathfrak{C}_1$ (Clear)

	M_3	M_2	M_1	M_0	A_3	A_2	A_1	A_0	$\mathfrak{C}_4$	$\mathfrak{C}_3$	$\mathfrak{C}_2$	$\mathfrak{C}_1$
Start →	0	0	0	0	0	0	1	0	0	0	0	1
$I = 1$	0	0	1	0	0	0	1	1	0	0	1	0
	0	0	1	1	0	1	0	0	0	1	0	0
$I = 2$	0	1	0	0	0	1	0	1	0	0	1	0
	0	1	0	1	0	1	1	0	0	1	0	0
$I = 3$	0	1	1	0	0	1	1	1	0	0	1	0
	0	1	1	1	1	0	0	0	0	1	0	0
$I = 4$	1	0	0	0	1	0	0	1	0	0	1	0
	1	0	0	1	1	0	1	0	0	1	0	0
$I = 5$	1	0	1	0	1	0	1	1	0	0	1	0
	1	0	1	1	1	1	0	0	0	1	0	0
$I = 6$	1	1	0	0	1	1	0	1	0	0	1	0
Halt →	1	1	0	1	—	—	—	—	1	0	0	0

to-one correspondence, subscriptwise, with the T signals shown in Figure 7.22(a), e.g., $T = 2$ is the same as $\mathfrak{C}_2$.

As shown in Figure 7.22(a), the $T = 2$ statement is a conditional one. This may be satisfied in two possible ways. One method is to use the condition, i.e., the B_6 signal, as a jump bit in the current address. The second method is to use *auxiliary logic* to realize the conditional statement. Since the first method would almost double the size of our CM program, we have elected to employ the second method with the conditional logic shown in Figure 9.31(a). There, the $\mathfrak{C}_2$ state is ANDed with the B_6 signal to control

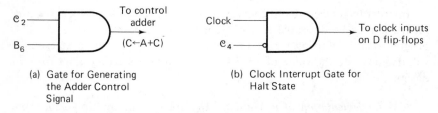

(a) Gate for Generating the Adder Control Signal

(b) Clock Interrupt Gate for Halt State

Figure 9.31   Auxiliary Logic for Use with CM of Table 9.4.

the adder exactly as it is done in Figure 7.22(b). (The halt state is also accomplished by auxiliary logic—see Figure 9.31(b); but another method will be considered in the Exercises.)

Although the two simple applications described above were of a tutorial nature and may not be sufficiently complex by themselves to justify the use of a CM, the current low price of semiconductor memories, particularly

ROMs, have made them competitive with other types of circuitry for many relatively small digital systems. Certainly, the RTL programs for the CPU of a small computer, such as those shown in Table 7.6, could be directly implemented by means of a CM and several MSI packages composing the ALU, the various registers, and any needed auxiliary logic. An alternate to this approach will be considered in Chapter 10, where the CM and microinstruction techniques will be further developed to produce a so-called microprocessor, which often allows the entire CPU to be placed on a single LSI chip.

9.6 BIBLIOGRAPHY

1 ABRAMS, MARSHALL D., and P. G. STEIN, *Computer Hardware and Software.* Reading, Mass.: Addison-Wesley Publishing Company, Inc., 1973.

2 BARNA, ARPAD, and DAN I. PORAT, *Integrated Circuits in Digital Electronics.* New York: John Wiley & Sons, Inc., 1973.

3 CARR, WILLIAM N., and JACK P. MIZE, *MOS/LSI Design and Applications.* New York: McGraw-Hill Book Company, 1972.

4 EIMBINDER, JERRY (ed.), *Semiconductor Memories.* New York: Wiley-Interscience, 1971.

5 HODGES, DAVID A. (ed.), *Semiconductor Memories.* New York: IEEE Press, 1972.

6 KOHONEN, TEUVO, *Digital Circuits and Devices.* Englewood Cliffs, N.J.: Prentice-Hall, Inc., 1972.

7 LUECKE, GERALD, J. P. MIZE, and W. N. CARR, *Semiconductor Memory Design and Application.* New York: McGraw-Hill Book Company, 1973.

8 PARHAMI, BEHROOZ, "Associative Memories and Processors: An Overview and Selected Bibliography." *Proc. IEEE,* Vol. 61, 1973, pp. 722–30.

9 RENWICK, W., and A. J. COLE, *Digital Storage Systems* (2nd ed.), London: Chapman and Hall, 1971.

10 RILEY, WALLACE B. (ed.), *Electronic Computer Memory Technology.* New York: McGraw-Hill Book Company, 1971.

11 "Session on Memory Technology." *Proc. 1968 Fall Joint Computer Conference.*

12 "Special Issue on Large Capacity Digital Storage Systems." *Proc. of IEEE,* Vol. 63, August, 1975.

13 "Special Issue on Memories." *IEEE Trans. Comput.,* Vol. EC-15, August, 1966.

14 "Special Issue on Semiconductor Memories and Digital Circuits." *IEEE Journal of Solid-State Circuits* (Annual October issue starting with 1971).

15 WILKES, M. V. "The Best Way to Design an Automatic Calculating Machine." *Manchester University Computer Inaugural Conference,* 1951, p. 16.

9.7 EXERCISES

9.1 (a) In a 3-D, 2048-word 32-bit core memory, how many sense amplifiers are required?
(b) How many inhibit line drivers are needed?
(c) How many memory planes, how many cores per plane, and how many X-selection drivers and Y-selection drivers are needed?

9.2 Repeat Exercise 9.1 for a 2-D memory.

9.3 Repeat Exercise 9.1 for a $2\frac{1}{2}$-D memory.

9.4 Find the number of gate inputs required by each of the three types of decoding schemes for a 3-D memory having 2^{18} different addresses.

9.5 If we construct a 1024-word, 32-bit memory, how many flip-flops will be required for the memory address register?

9.6 Find the number of gates needed by a nine-bit balanced decoder.

9.7 Consider the possibility of constructing a semiconductor memory from shift registers. Present your ideas in terms of a block diagram and a word description. Can this memory truly be called a RAM? Explain. Evaluate your system, e.g., give the advantages and disadvantages of it.

9.8 Using SR flip-flops, draw the basic memory cell and the array for a 4-word × 2-bit coincident-addressing semiconductor memory.

9.9 How can a chip-select signal be implemented for the memory in Figure 9.18?

9.10 Sketch a memory cell similar to that in Figure 9.20 which allows coincident selection.

9.11 Sketch a 16 × 4 memory array to employ the memory cell of Exercise 9.10; be sure to indicate all of the data-in and data-out signals, as well as other pertinent signals.

9.12 Sketch a static MOSFET memory cell employing the coincidence-selection principle.

9.13 Draw the logic symbol similar to our TU1 symbol but for a 16-word × 3-bit memory.

9.14 Using TU1 chips, draw the circuit for a 16-word × 2-bit memory.

9.15 Draw a 4-word × 4-bit linear-selection ROM using MOS transistors.

9.16 If a single ROM is employed, how many bits are required to multiply an n_1-bit number by an n_2-bit number?

9.17 Design a 2-bit binary multiplier showing both the equivalent logic diagram and ROM truth table.

9.18 How can Figure 9.28 be redrawn so that the bits from the middle two multipliers are not immediately combined?

9.19 Similar to Figure 9.28, draw the ROM circuit for a Gray-code to binary-code converter. (Employ the Gray-code version shown in Table 5.18.)

9.20 Design a CM to implement the modulo-3 counter of Figure 5.3, with reset being achieved by meas of the jump bit, i.e., $X_c = J_0$. (Present your results in a form similar to Table 9.3.)

9.21 Find and describe, in detail, another method for producing the halt state of Table 9.4 without employing auxiliary logic.

9.22 For the up-down counter of Table 9.3, find two ways to modify Figure 9.30 so that the counter decrements on the first pulse input, rather than the second, following the level change from $J_0 = 0$ to $J_0 = 1$. What are the advantages and disadvantages of each method?

9.23 Design a CM to implement the RTL program of Figure 8.19. Use $n = 2$ and *do not employ any auxiliary control logic* other than that explicitly shown in Figure 8.18. (Present your results in a form similar to Table 9.3.)

9.24 If the auxiliary logic of Figure 9.31(a) is not employed, i.e., a jump bit is used instead, list the new entries for the CM of Table 9.4 for the $I = 2$ step. Assume $M_3 M_2 M_1 M_0 = 0101$ at the end of the $I = 1$ step. *Hint:* What is the advantage in connecting B_6 itself as the MSB in the memory address register rather than going through the intermediate operation of having the clock transfer $J_0 = B_6$ to a new M_4 flip-flop?

COMPUTER
10 ARCHITECTURE

Part II
Organization,
Microprocessors,
and Large-Scale
Systems

Some of the architectural tools, RTL and simulation, as well as basic processor design itself, were considered in Chapter 7. This chapter presents a further development of architectural principles beginning with a consideration of instruction organization, then two sections on microprocessor implementation, and concluding with a brief discussion of large-scale computer systems.

10.1 ORGANIZATION OF INSTRUCTION WORDS[2,5,6]

The organization of the instruction words has a tremendous effect on the power and the overall characteristics of a computer. Not only does the instruction determine the type of operation being performed, but it also determines the way the operands are found, the number of data words available for addressing, and special types of operations which can be performed, e.g., convenient ways of establishing loops. It is the purpose of this section to develop the principles involved in designing instructions and to convey some of the flavor of the area, but not to provide an exhaustive treatment. The various manufacturers' manuals, the current literature, and specialized texts can be consulted for additional details.

Thus far we have been dealing with single-address instructions. These are the simplest to understand and they are very popular—particularly in small machines. In a general way, consider what is required when a binary operation* like ADD, requiring two operands, is to be executed? The locations of two operands, as well as the location where the result will be returned, must be specified, which yields three addresses. In addition, the address of the next instruction is also needed, resulting in a total of four required addresses. But this was not done in the programs written in Chapter 3, or was it? First, in single-address instructions, one address is contained in the instruction itself; the other is in the accumulator; and the results are left in the accumulator. Thus, the presence of an accumulator has implicitly pro-

---

*This means an operation with two operands; a uniary like COM requires only one operand.

vided two locations. Moreover, all the machines previously considered contained a program counter which automatically advanced to the next memory location unless a branch instruction was specified.

If the above scheme worked so well and is employed on many minicomputers, one might wonder why we consider other possibilities. However, as occurs frequently in practice, external constraints often have an overwhelming influence on what is best. For example, one reason for allowing the location of the next instruction to be an explicit address is that in certain memories, such as the drum type, true random access is not possible; by specifying the proper address next, it is possible to increase the speed at which instructions are fetched.

In order to provide more information concerning the possibilities in the number of addresses employed, Figure 10.1 has been constructed. The zero-address machine shown in part (a) of the figure may seem to be an impos-

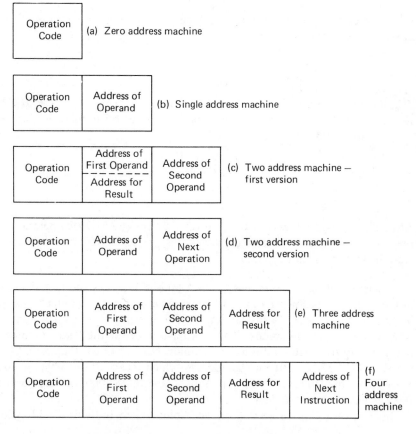

Figure 10.1 Several Possible Instruction Formats.

sibility, but it differs from a single address machine only in that both operands are contained in some kind of register configuration, e.g., the presence of two accumulators. Parts (c) and (d) of Figure 10.1 indicate some variations in the way two addresses may be employed. In part (d) an accumulator is employed for the second operand and to receive the results, but a program counter is not required. In part (c) an accumulator is not needed since the memory provides for both operands and for the result, but a program counter is required. The IBM Systems 360 and 370 are typical computer families using the method of part (c).

We should question the fact that the structure in Figure 10.1(c) indicates that the first operand is destroyed when the result is stored. This is of course true; but the programmer can always move the first operand into a free location so that the original operand is preserved. Suppose it is desired to determine the following sum:

$$Z = W + X + Y$$

In a two-address machine this could be accomplished by means of the instruction sequence:

> MOVE Z,W
> ADD Z,X
> ADD Z,Y

(The mnemonic MOVE A,B means transfer the contents of register B into A and leave B unchanged.) This may be compared to the following single-address instructions for the same task:

> LOAD W
> ADD X
> ADD Y
> STC Z

Although the two-address instruction set requires the MOVE operation to preserve the operand W, it actually requires one less instruction because a store instruction is not needed.

Finite Word Size Limitations. What limitations does the fact that the instruction word is limited to, say, 18 bits impose on a minicomputer? Assume it is desired to employ 64 instructions, which is really a very small number when compared to the hundreds that are available for the large computers made by Control Data and other companies. Now $2^6 = 64$; therefore, there are $18 - 6 = 12$ bits left for the address, which means $2^{12} = 4096^*$ words. Since many minicomputers require more than 32K (32,000) of fast storage, there is obviously a problem.

Instead of trying to furnish ad hoc solutions to individual addressing

---

*The reasoning here is somewhat simplified, as we shall soon see. But this example does expose many of the basic principles.

problems, it is no doubt better to expose the whole family of desired improvements and then provide an integrated approach. One improvement would be a hardware method to accomplish repetitive operations rather than going through software, as was done in Table 3.5, to sum a group of numbers. Another desired improvement would be a way to carry out a set of operations on several sets of data located in different parts of memory without explicitly changing all of the addresses that refer to the location of the data each time. There are certainly many other desired improvements; but if it is possible to accomplish the above, they may at least provide some insight to ways of accomplishing additional ones.

The key to the problem of obtaining additional memory addressing capability is the use of indirect addressing. Suppose that the instruction word is now broken into three parts, as shown in Figure 10.2(a), with the new part

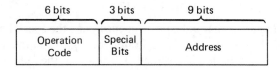

(a) Instruction Format

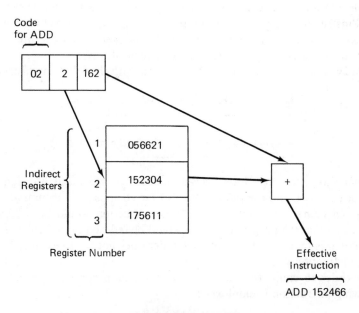

(b) Indirect Addressing Example
(All numbers shown are in octal)

Figure 10.2 Instruction Word Organization.

providing special bits which refer to a group of indirect registers. Figure 10.2(b) shows how the 9-bit instruction address is added to the 18 bits in the indirect register to form an effective instruction with the capability of addressing over 260K memory locations. Note how the three special bits point to one of three indirect registers, it is that register which is combined with the instruction address.

Since the three special bits would permit an instruction to point to one of eight indirect registers, it might appear that one bit was wasted. However, this is not the case because the most significant bit in the special group (i.e., numbers 100 through 111) provides an index feature, which is employed to solve the problem of a hardware method for creating a loop. For example, suppose it is desired to sum the numbers in the memory locations starting at 100200. With the indirect register 2 set at 100000, this may be accomplished with the instruction

$$ADD \ i \ 2, \ 177$$

where the symbol i means that indexing is in effect, i.e., the special bits would be set to 110 and the comma merely separates the register number from the address. Before this instruction is executed, 1 is added to the indirect register, making the effective instruction ADD 100200. When the instruction is executed a second time, the register is again indexed, resulting in ADD 100201 being the effective instruction. The process continues until the sum has been completed for the entire list. Of course, a practical system requres some additional instructions, e.g., an instruction to initialize the indirect registers and another to count the number of times the loop has been executed. These details may be easily accomplished in a number of ways.

One way to initialize an indirect register is with the following instruction:

Memory Address	Instruction Mnemonic
n	Set y
$n + 1$	(Octal number)

where n is any memory address, y is a number between 1 and 3 to signify the register being set, and the value to be transferred appears in memory location $n + 1$. Of course, the hardware for this instruction would be designed in such a way that the program counter would be indexed twice before the next instruction is fetched, so that the computer does not try to execute the data as an instruction.

In order to count the number of times a loop has been executed, the following instruction is employed:

$$DSK \ XXXX$$

This instruction says decrement the contents of memory location XXXX by 1 and skip the next instruction if the contents equal 0. (Since XXXX is a conventional memory location, it may be filled in the same way as any other location.) The program developed in Table 3.5 to sum the numbers in memory locations 200 through 225 is repeated in Table 10.1 using the new instructions. Note that as the loop is being executed the 26th time, the instruction at address 24 decrements $C(10)$* from 1 to 0. Thus, the instruction in 25 is skipped and the contents of the accumulator, which by now contains the sum of $C(200)$ through $C(225)$, is stored at address 300. Comparing the two methods, we observe that Table 10.1 requires only seven instructions, not counting data locations, as opposed to Table 3.5 which requires 12. In addition, the Table 10.1 version appears to be much clearer and, once the notation is mastered, easier to understand.

Let us return to the notation employed for indirect addressing and indexing. Using the ADD operation with an address of 5, the instructions below demonstrate four variations:†

ADD 5 $\equiv$ Direct addressing with the first operand contained in the accumulator and the second contained in memory location 5. (The corresponding machine code is 020005,‡ where 0_8 for the special bits indicates that indexing and indirect registers are not employed.)

ADD 3,5 $\equiv$ Indirect addressing with the number 5 being added to C(register 3) to form the address of the second operand. (Machine code: 023005.)

ADD i 3,5 $\equiv$ Register 3 is indexed and the number 5 is added to it to form the address of the second operand. (Machine code: 027005.)

ADD i 5 $\equiv$ The content of location 5 is indexed and then it becomes the address of the second operand. (Machine code: 024005.)

These examples reveal two general rules:

1. If indexing is employed, an i will follow the mnemonic, otherwise. a blank is inserted.

2. If indirect addressing is employed, the register number will be followed by a comma and that will be followed by the address. Otherwise, no register number or comma will appear.

---

*Recall that the notation $C(X)$ was defined in Chapter 3 to mean the contents of memory location X.

†Other variations are also possible, e.g., ADD # could be implemented to add the contents of the next memory location to the accumulator, which is immediate addressing.

‡Note that the same basic operation codes are employed here as appeared in Table 3.2, i.e., ADD = 2 = 02, JMP = 6 = 06, except that the use of 18-bit instead of 12-bit words allows the three special bits to be inserted, and the addition of three bits to the operation code permits eight times as many memory reference instructions to be defined.

Table 10.1 An Index Register Program to Sum a Group of Numbers

(a) Program Itself

Memory Address	Mnemonic
10	26
Start 20	SET 1
21	177
22	CLR
23	ADD i 1,0
24	DSK 10
25	JMP 23
26	STC 300
27	HLT
200–225	Numbers to be summed
300	Storage for results

(b) Machine Contents While Running

Number of Loop Completions	C(10)	C(Register 1)	C(Accumulator)	C(300)
Start	26	—	—	—
1	25	200	$C(200)$	—
2	24	201	$C(200) + C(201)$	—
...	...	...	...	...
25	1	224	$\sum_{i=200}^{224} c(i)$	...
26*	0	225	0	$\sum_{i=200}^{225} C(i)$

*After machine halts

354

With the above notation, it is not difficult to rewrite the program of Table 10.1(a) with still fewer instructions. The new program is listed in Table 10.2. Note that instructions that do not refer to the memory are in the same form as they were defined in Chapter 3.

Table 10.2 Program of Table 10.1 Rewritten

Memory Address	Mnemonic
10	26
11	177
Start 20	CLR
21	ADD *i* 11
22	DSK 10
23	JMP 21
24	STC 300
25	HLT

10.2 BUS STRUCTURES

Instead of following the philosophy shown in Figure 7.11 of running 24 lines* for each transfer between the MB and the many other subsystems, including the input/output peripherals, with which it communicates, a more elegant scheme called *busing* is frequently employed. The concept of a bus structure is to have a single set of lines, one for each bit, which runs to each subsystem much like a party line employed by the telephone company. Control lines then select which register will be the transmitter of information and which will be the receiver. These principles are expressed in Figure 10.3, which shows the bus structure for four 2-bit registers using D flip-flops. The following points should be observed concerning Figure 10.3:

1. To transmit data from register W to register Y, for example, a control pulse would be placed into the Transmit W terminal and into the Receive Y terminal.

2. Wired-OR logic is employed at the point where each AND gate joins the bus. If this is undesirable in a practical application, it is quite simple to use an OR gate at the junctions.

3. By allowing more than one receive terminal to become 1 simultaneously, the same information can be received by more than one register. In contrast, no more than one register can simultaneously transmit information to the bus.

*The figure 24 comes from transmitting the signal for 12 bits in double-rail form, i.e., both the 1 and 0 outputs of each flip-flop. Even if single-rail transmission is suitable, and accounting for the fact that in a few cases not all 12 bits need be sent, there are still a tremendous number of lines involved.

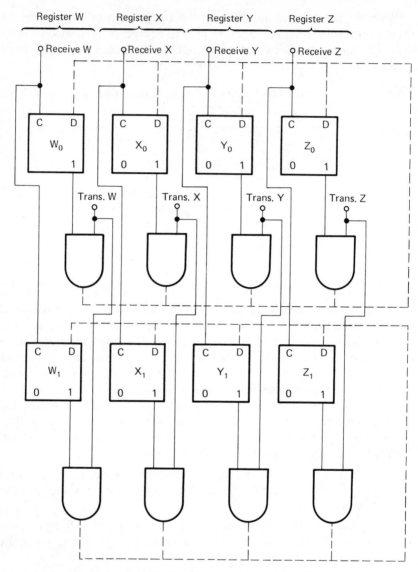

Figure 10.3 Bus Structure For Four Two-Bit Registers (Bus ≡ ------;
Logic and Control ≡ ———).

4. Because of the nature of D flip-flops, the signal at the D terminal
   must be settled a few nanoseconds before the occurrence of the signal
   at the C terminal. Consequently, the timing must be such that the
   transmit signal is generated several nanoseconds before the receive
   signal(s).

5. Only one data word may be transmitted at a time. Thus, there is a trade-off between the simple bus structure and the restriction on the rate at which words may be transferred between registers.

It is instructive to consider Figure 10.4, which shows the basic organization of a computer employing a single-bus structure. This figure should be compared with Figure 7.11, which does not make use of a bus. Again, the reader will find it valuable to review Figure 7.17 and Table 7.6 in light of the structure of Figure 10.4. Note that most subsystems both transmit and receive information, as indicated by the two arrows at each box, but that certain units, e.g., the input peripherals, only handle data in one direction.

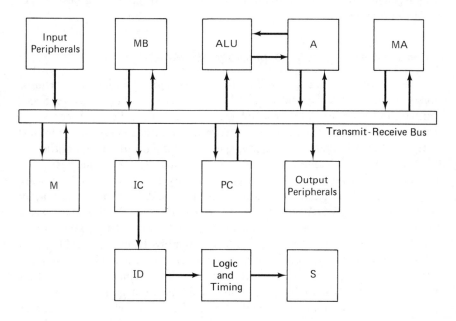

(Note that for simplification the control signals have been omitted)

Figure 10.4 Basic Computer Organization Employing a Single Bus Structure.

The way the ALU accesses data during, say, an add instruction is interesting in that one operand must enter from the memory buffer, the other must enter from the accumulator, and the result must return to the accumulator. Consequently, three simultaneous data paths are required, but the single bus is capable of providing only one path. Thus, the transmit-receive lines directly between the ALU and *A* are required to furnish the extra data paths.

We are now aware of one of the advantages of the bus—the fact that there is a considerable savings in equipment possible. But what is the disadvantage?

A major drawback is associated with the above-mentioned fact that only one data path is available at a time; yet frequently, particularly in large machines, it is desirable to carry out certain operations in parallel. For example, one might want to bring data to the memory from an input peripheral at the same time an arithmetic instruction is being completed by the ALU. One solution to this problem is to employ a multiple-bus system with certain devices on each bus and with the possibility of communicating between the buses. Another alternative is to provide a dual set of buses, with each device having access to both, and with the proper controls. Of course, the decision concerning the best bus structure depends on the particular requirements; but it certainly does open up a lot of interesting possibilities to the computer architect.

10.3 INTRODUCTION TO MICROPROCESSORS[11,14,19]

Digital computers are classified on the basis of scale into five groups: micro, mini, midi, large, and super.* These classes are not rigorously defined; but computing power and cost are usually the main criteria. A good correlation frequently exists, however, between computing power and physical size, word length, memory capacity, speed, and input/output capabilities.

Minicomputers came into wide use in the middle 1960s and have been one of the most popular computer types for the last several years. They are employed for data acquisition, process control, and for problem solving in moderately complicated situations. A typical machine in this class has a 16-bit word size, a 16,000 word memory, a cost well under $10,000, and a good assortment of software including assembly language, BASIC, and FORTRAN. A representative member of this group is the Digital Equipment Corp. PDP 11/20. Computers in the midi category typically have a 24-bit word size, a cost in the neighborhood of $300,000, a good assortment of peripherals, more abundant software, extensive hardware features such as double-precision instructions, floating-point hardware, and automatic error detection features. A representative member of this group is the IBM 370/145. Large-scale computers typically have a 36-bit word size, cost in the neighborhood of $2,000,000, have extensive software and hardware support, and have many special hardware features, e.g., the input/output function is managed by an independent processor which interrupts the CPU when it has finished its previous assignment. The super computers are usually very large, specialized machines costing over $2,000,000 and frequently have architectural features not found in the other machines. A representative machine is the ILLIAC IV, which consists of many processors operating in parallel under the supervision of a master control unit.[13]

*Some writers place a medium class between midi and large.

In contrast to the other classes, the *microcomputer* is a relative newcomer, having been first introduced in 1971. It is one step down from the minicomputer in some ways; but the microcomputer has some special properties which give it a distinct advantage over the other computer classes for certain tasks. First, a microcomputer is distinguished from the other computer classes by having the four main units (control, ALU, input/output interface, and memory) all on a single LSI chip or on a small number of such chips. The corresponding physical size reduction in going from a minicomputer to a microcomputer is a factor of between 20 and 100. This size reduction is accompanied by a large reduction in cost (typically a factor of 5), and with a certain amount of reduction in performance. On the other hand, there are industrial, business, consumer, and research applications where full mini-computer capabilities are not required, e.g., less than a 16-bit word size is acceptable. In these cases, the reduced specifications permit corresponding reductions in the hardware size and cost. Moreover, there are a large number of industrial, business, consumer, and research applications where an order of magnitude reduction in the memory capacity and/or the input/output capabilities are permitted. As would be expected, the term *microcomputer* refers to a complete, programmable data-processing system. *Microprocessor*, on the other hand, is a more general term, which encompasses units ranging all the way from simple pocket calculators to full microcomputers, with the exact definition coming from the context. Nevertheless, the essence of all microprocessors is the LSI chip(s) encompassing the control section and the ALU, which means that it is equivalent to the CPU shown in Figure 3.1. Unless stated otherwise, when a manufacturer refers to a microprocessor, he means the CPU. Several such types of microprocessors are currently available for well under $100.[19]

At this point we should distinguish between a *macroprogram* and a *microprogram*. A macroprogram is written with ordinary machine-level instructions, *macroinstructions*, such as those presented in Table 3.2. On the other hand, a microprogram contains more primitive instructions, *micro-instructions* (each equivalent to one or two register transfer operations), from which macroinstructions can be implemented. For example, the multiply macroinstruction is implemented by means of a series of add and shift microinstructions.

An Elementary Microprocessor

Except for the fact that the CPU is implemented with LSI circuits, the architecture of a microprocessor is essentially the same as we have previously discussed in Figure 7.16. Thus, the programming concepts of Chapter 3 and the principles developed in the other chapters are applicable. Since it will be particularly valuable later and since it provides an application of the

ROM methods of Chapter 9, the elementary microprocessor shown in Figure 10.5(a) is based on the control memory concept.

Although it may not be obvious at first, Figure 10.5(a) is essentially the same as Figure 3.1 except that more details are provided concerning the way control signals are generated. The control memory is the source of signals (microinstructions) which actuate the various basic operations of the ALU. It also provides signals to other microprocessor units, e.g., the main memory receives its read/write signal from the CM. Of course, the control memory is addressed in the same way as it was in Chapter 9.

Figure 10.5(a) includes several classes of microprocessors.

1. The *nonprogrammable* class is one in which the user is given a fixed processor structure such as is present in most pocket calculators. About the only freedom the user has is in the selection of the available keyboard operations and the data. Since the unit is not programmable, the main memory and the macroinstruction bus is omitted in this type of processor. (The ALU in most microprocessors contains a few registers for the storage of intermediate results.) In addition, the control memory is loaded at the factory and end-user will not be able to modify it. Part of the microinstruction supplies the next-address function. Another part supplies the ALU signals for the various mathematical operations and for control of the input/output devices. For example, in a pocket calculator, the input is a small keyboard and the output is usually a LED, light-emitting diode, display. Surprisingly, a scientific calculator usually requires less than 1000 microinstructions for complete implementation including the keyboard decoding, arithmetic, and display functions. Within this space, it employs some intriguing algorithms to perform trigonometric calculations, logarithms, and other functions. In addition, the hardware contains some very interesting logic. The reader is directed to recent issues of the *Hewlett-Packard Journal*, particularly Reference 15 in the Bibliography at the end of this chapter, for details which are beyond the scope of this text.

2. The *macroprogrammable* class[14] is very similar to the standard computers we have been discussing throughout the text, with the main memory being user-programmable with conventional machine language instructions. As shown in Figure 10.5(a), each macroinstruction is decoded by means of the next-address logic; this usually results in a sequence of several microinstructions from the CM. (Details of this process will be discussed in Section 10.4.) Again the CM may not be modified by the end-user; but there usually is considerable freedom to interface various input/output devices with the processor. In fact, several microprocessor manufacturers have a series of LSI chips which have been especially designed for interfacing between their processor and external devices.

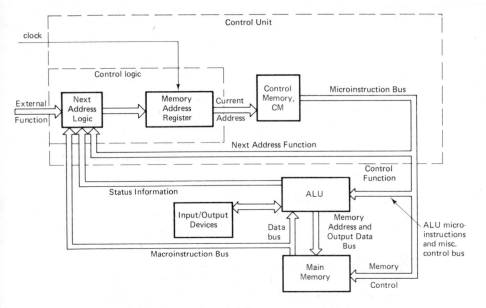

(a) General Configuration.

(b) Photomicrograph of the Motorola 6800 Microprocessor (Courtesy of Motorola Semiconductor Products Division).

Figure 10.5 A Typical Microprocessor.

Figure 10.5(b) shows a photomicrograph of the Motorola 6800 microprocessor, which is of the macroprogrammable class. The picture shows the entire CPU on a single-crystal silicon chip approximately 0.2 inch square, which has been enlarged about 300 times. The chip contains over 5000 transistors fabricated by the n-channel MOS technology. Black lines projecting outward are leads bonded to pins of the finished package. The 6800 is an 8-bit processor with a repertoire of 72 instructions. It can address 64K words of memory and performs addition in 2 microseconds.

3. The *microprogrammable* class is the most general and powerful of the three. Here the user is free to program the control memory (of course, the main memory also) which in turn allows direct management of the individual operations of the ALU (and of other parts of the processor). Through this method new macroinstructions may be defined, which means that the user has complete control of the instruction set employed in programming the main memory. Although we are concentrating on the application of microprogramming to microprocessors in this chapter, the basic ideas are applicable to all scales of digital computers. For example, extensive use was made of microprogramming in the implementation of both the IBM 360 and 370 families of computers,[8] and several minicomputers may be microprogrammed by the end-user. (Again, hardware details will be discussed in Section 10.4.)

Microinstructions and Microprogramming[8]

In Chapter 9 we saw how a control memory could be employed to implement sequential circuits (thus, also register transfer logic). From this, it is only a short step to visualize how a microprogram in the control memory of Figure 10.5 can implement the macroinstructions required by the processor. The ALU and the other units need a large number of control signals to manage the various register transfer operations. Over 200 separate signals is not an uncommon requirement for a large-scale computer (the IBM 360/65 employs about 250). The most straightforward method for realizing these signals is to assign one bit in the control memory's word for each needed signal. While this is a very general approach, it is extravagant in the number of bits required per instruction word. Moreover, many control signals are mutually exclusive, e.g., the same register would not be simultaneously shifted and cleared. Thus, the control word may be coded so that, in the extreme, n bits may manage 2^n microinstructions. In the first method, n operations are controlled simultaneously (parallel processing) but with a high cost in bits per word. In the second method, only one operation is controlled at a time, resulting in a low bit-per-word cost; but a decoder is required and the desired function may consume a series of several control memory cycles. For obvious reasons, the first approach is called *horizontal microprogramming* while the second is called *vertical microprogramming*. Most processors employ a combination approach, which is between these two extremes. As a simple example, if five bits are reserved for control, with three bits for vertical and two bits for horizontal microprogramming, one of eight ($2^3 = 8$) operations could be executed, through a decoder, with two others always available. This is a total of three simultaneous operations from a repertoire of $2^3 + 2 = 10$ instructions. Because it follows the trend of current microprocessors and for tutorial reasons, ours will be a combination approach leaning toward vertical microprogramming.

Establishing the Control Memory Address. Now let us consider the method of achieving the next control memory address. In Chapter 9, an elementary branching method was described in which a bit (or group of bits) was employed to control branching and the next-address logic was trivial. Here, a more powerful technique is achieved by employing an actual jump instruction which, together with some auxiliary conditions, establishes the memory address through a next-address logic circuit.

The control memory word format employed by our microinstructions is shown in Figure 10.6. As in the discussion in Chapter 9, the control word is broken into two parts—the next-address function and the control function. In order to achieve both conditional and unconditional branching with the next-address function, bits 8 through 11 code several different jump (branching) possibilities. These include jumps on flag conditions, jumps on status of groups of bits from the main memory, and random jumps. For convenience, the control memory is visualized as a square array of memory locations with bits 0 through 3 specifying the column and bits 4 through 7 specifying the row.

Note that bits 12–19 specify the ALU microinstruction, bits 20–23 specify flags, and bits 24 and above control other functions when required. These will be explained in more detail shortly.

Details of the next-address function are shown in Table 10.3. Note that each row of the table (particular address control function) contains the binary code of the next-address function (as read from the CM) and the resulting actual next memory address. Consider the Jump Random instruction, JR. In this case, a jump may be made to any place in the CM since there is a one-to-one correspondence between the next row-column in the function and the actual next memory address row-column. For example, the mnemonic instruction which is to cause an unconditional branch to row 5, column 7 would be JR 5,7. It would have the hexadecimal code 157_{16} for microinstruction bits 0 through 11.

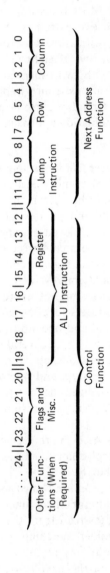

... 24 || 23 22 21 20 || 19 18 17 16 | 15 14 13 12 || 11 10 9 8 | 7 6 5 4 | 3 2 1 0

Other Functions (When Required)

Flags and Misc.

ALU Instruction

Control Function

Register

Jump Instruction

Next Address Function

Row

Column

(Bit numbers are shown divided into hexadecimal digits.)

Figure 10.6 Microinstruction Format.

364

Table 10.3 Address Controls

Branch Mnemonic	Next-Address Function 11 10 9 8 7 6 5 4 3 2 1 0												Next Memory Address Row	Column
JR ≡ Jump Random	0	0	0	1	d_7	d_6	d_5	d_4	d_3	d_2	d_1	d_0	$d_7 d_6 d_5 d_4$	$d_3 d_2 d_1 d_0$
FC ≡ Flag Column	0	0	1	0	d_7	d_6	d_5	d_4	d_3	d_2	d_1	–	$d_7 d_6 d_5 d_4$	$d_3 d_2 d_1 f$
MA ≡ Multiplexer Segment A	1	0	1	0	d_7	d_6	d_5	d_4	–	–	–	–	$d_7 d_6 d_5 d_4$	$A_3 A_2 A_1 A_0$
MB ≡ Multiplexer Segment B	1	0	1	1	d_7	d_6	d_5	d_4	–	–	–	–	$d_7 d_6 d_5 d_4$	$B_3 B_2 B_1 B_0$
etc.														
MF ≡ Multiplexer Segment F	1	1	1	1	d_7	d_6	d_5	d_4	–	–	–	–	$d_7 d_6 d_5 d_4$	$F_3 F_2 F_1 F_0$

Abbreviations: d_i ≡ bit i of data word.

           A_i ≡ bit i of macroinstruction word (main memory bus) segment A (similar for $B_i, C_i, \ldots F_i$).

           f ≡ current state of flag flip-flop.

           – ≡ bit is a don't care.

The Jump Flag Column function, FC, has the same one-to-one row-column correspondence except for the fact that the least significant bit in the next column address takes on the value of the flag, f. If the flag's value is established by the sign of some register, a Branch on Register Sign instruction would result. On the other hand, the Jump on Multiplexer Segment A, MA. instruction is somewhat different from the previous jump instructions. Here the next memory row address is the same as in the next-address function; but the column address is obtained from a four-bit segment of the main memory bus (part of the macroinstruction) selected by a multiplexer in the next-address logic. The various four-bit segments of the input bus are labeled A, B, C, etc. (with A being the most significant four-bit segment, B the next four-bit segment, etc.) resulting in multiplexer instructions with mnemonics MA, MB, MC, etc.

One major use of the multiplexer instructions is to decode macroinstructions from the main memory. The actual decoding process causes a jump to the proper starting location in CM for the realization of the macroinstruction by means of a microinstruction program.

As an example of address control, assume that the next-address function is $B53_{16}$, the main memory bus contains $9E8C_{16}$, and the multiplexer segments correspond, in sequence, with the hexadecimal digits starting with the most significant digit being segment A. Here, B, the most significant digit of the function, indicates that the branch operation is Jump on Multiplexer Segment B, MB; the next digit, 5, specifies the next memory address row; finally the next digit, 3, is neglected in this example. The resulting next memory address becomes 5E, where E is the digit found in the multiplexer Segment B.

ALU Microinstructions. As previously indicated, the ALU micro-instructions are the lowest-level, most primitive, operations defined for a computer. From a small group of these instructions the macroinstructions for *any* computer can be constructed. It is one of the major reasons for our interest in these primitives—they are the key to simulation (the more specific word in this context is *emulation*) of one computer by another. Emulation is employed by computer manufacturers to test a computer design while the actual hardware for the new machine is still underconstruction. In addition, it is frequently employed to represent an old machine by a newer one so that programs using the old instruction set can be run without any changes.

The ALU instruction set has some features in common with machine language instructions such as those discussed in Chapter 3. Again, the instruction is broken into operation and operand parts, with certain instructions like clear having the accumulator as an implied operand. Although the instructions under consideration refer only to the ALU, units such as the memory and input/output devices are controlled indirectly through registers manipulated by these instructions.

Table 10.4 summarizes the ALU microinstructions which are employed in this chapter. The reader should study the table, taking note of the RTL equuivalence and the several other features listed there which aid in understanding the instructions. The ALU instructions have been given two-letter mnemonics to emphasize their more primitive character and to distinguish them from the macroinstructions of Table 3.7. The following other features should be observed concerning the entries in Table 10.4:

1. Bits 16 through 19 in the control memory bit format indicate the operation, and bits 12 through 15 indicate the operand. Note that each part is expressed as a hexadecimal number; and to emphasize its base-16 nature, the bit pattern is divided into two groups by a vertical line.

2. There are eight registers within the ALU, which are denoted by 000 through 111 in bits 12 to 14. These are used to hold one operand during certain instructions and as temporary storage—R^i represents register i, $0 \le i \le 7$.

3. When bit 15 is zero, the register class of instructions is being considered. AD, AN, OR, LR, and several others belong to this class.

4. For bits 12 through 15, the following definitions are employed: $1000 \equiv$ accumulator and $1001 \equiv$ memory address register.

5. In the AD, CM*, and several other instructions, the quantity CI is the Carry Input. (A similar quantity is the Left Input, LI, found in the

---

*Any possible confusion between this mnemonic and the abbreviation for control memory should be resolved by the context.

Table 10.4 ALU Microinstructions

Mnemonic	Meaning	RTL Equivalent	Control Memory Bit Format 19 18 17 16 | 15 14 13 12 (Operation) (Operand)
1. CL	Clear accumulator	$A \leftarrow 0$	$0000 \mid 1000 = 08_{16}$
2. NO	No operation	—	$0000 \mid 1111 = 0F_{16}$
3. RR	Rotate Acc. right one bit (similar to LINC ROR)	$A \leftarrow rrA$	$0010 \mid 1000 = 28_{16}$
4. SR	Shift Acc. right one bit	$A \leftarrow srA$; $A_{n-1} \leftarrow LI$; $RO \leftarrow A_0$	$0011 \mid 1000 = 38_{16}$
5. SL	Shift Acc. left one bit	$A \leftarrow slA$; $CO \leftarrow A_{n-1}$; $A_0 \leftarrow CI$	$0100 \mid 1000 = 48_{16}$
6. AD R^i	Add $C(R^i)$ to Acc. results in Acc. R^i unchanged	$A \leftarrow A + R^i + CI$ (1's comp. result)	$0101 \mid OR_2^i R_1^i R_0^i = 5i_{16}$
7. ST R^i	Store Acc. in register R^i ; Acc. is unchanged	$R^i \leftarrow A$	$0110 \mid OR_2^i R_1^i R_0^i = 6i_{16}$
8. DR R^i	Decrement register R^i	$R^i \leftarrow R^i - 1$	$0111 \mid OR_2^i R_1^i R_0^i = 7i_{16}$
9. IR R^i	Increment register R^i	$R^i \leftarrow R^i + CI$ $(CI=1)$	$1000 \mid OR_2^i R_1^i R_0^i = 8i_{16}$
10. IM	Increment memory address register	$MA \leftarrow MA + CI$ $(CI=1)$	$1000 \mid 1001 = 89_{16}$
11. LR R^i	Load Acc. from register R^i	$A \leftarrow R^i$	$1001 \mid OR_2^i R_1^i R_0^i = 9i_{16}$
12. LM	Load memory address register, MA, from ACC.	$MA \leftarrow A$	$1001 \mid 1001 = 99_{16}$
13. LD	Load ACC. from data input bus	$A \leftarrow (Data\ Bus)$	$1001 \mid 1101 = 9D_{16}$
14. AN R^i	AND $C(R^i)$ with Acc. leave results in Acc. $C(R^i)$ unchanged	$A \leftarrow A \wedge R^i$	$1010 \mid OR_2^i R_1^i R_0^i = Ai_{16}$
15. OR R^i	OR $C(R^i)$ WITH ACC. leave result in Acc. $C(R^i)$ is unchanged	$A \leftarrow A \vee R^i$	$1011 \mid OR_2^i R_1^i R_0^i = Bi_{16}$
16. CM	Complement ACC.	$A \leftarrow A + CI$ $\left\{ \begin{array}{l} CI=0 \text{ is 1's comp.} \\ CI=1 \text{ is 2's comp.} \end{array} \right\}$	$1100 \mid 1000 = C8_{16}$
17. TR R^i	Test Register R^i for zero	IF $(R^i=0)$ $CO \leftarrow 0$ ELSE $CO \leftarrow 1$	$1111 \mid OR_2^i R_1^i R_0^i = Fi_{16}$

SR instruction.) It forms an LSB input for various functions much as the carry-in signal, C_{i-1}, does for the full adders shown in Figure 8.3. On the other hand, the SL and TR as well as some other instructions have a Carry Output, CO, signal. (A similar quantity is the Right Output, RO, found in the SR instruction.) This is analogous to the carry generated by the MSB in a parallel full-adder circuit. The exact role of the CI, LI, RO, and CO quantities for all instructions, where they are applicable, should be clarified by the RTL column in the table.

As an example of microinstruction operation, assume that the contents of register 3 and the accumulator for a six-bit processor are, respectively, 011011 and 110001. After execution of the microcode: 10100011, (bits 12–19) the contents of register 3 and the accumulator would be 011011 and 010001. These results may be checked from Table 10.4 which indicates that the mnemonic AN 3 corresponds to the above microcode. It requires the AND function between register 3 and the accumulator, with the results placed in the accumulator but leaving the register unchanged. (More extensive microinstruction examples will follow in the next section.)

10.4 MICROPROCESSOR SYSTEMS

The microword format, the ALU microinstructions of Table 10.4, and the system examples to follow have been somewhat idealized for tutorial purposes; but they are similar to those for existing microprocessors (e.g., the Intel 3000 series[19]). Once the material presented here is understood, the reader should be in a good position to master any of the commercial units.

An Elementary Application

Although the basic processor which we will consider in this section has the capabilities of handling both macroinstruction and microinstruction, for simplicity an elementary application requiring only microinstructions will be considered first. In this example, pressure and temperature are measured for a chemical plant. Next the hypothetical performance function:

$$F = T + \frac{P}{2} + 105$$

(where T is temperature and P is pressure) is calculated and displayed. Then the process is repeated. A flow chart for the problem is shown in Figure 10.7. It faithfully follows the above problem statement, but the need for the initialization step will be justified shortly.

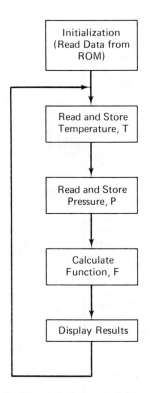

Figure 10.7   Method for Calculating the Performance Function.

The hardware to implement the flow chart for this problem is shown in Figure 10.8. It is a more specific version of the block diagram of Figure 10.5. While more details will be added to several blocks as the requirements develop, the basic form shown will be employed through the remainder of the text in examples and in the Exercises. The control logic, control memory, and ALU are the same major units that were discussed previously. Note that the carry-input (CI) terminal and the left-input (LI) terminal are both explicitly shown on the ALU; these signals are required by several of the ALU instructions of Table 10.4, e.g., AD and IM. Since only one of the two terminals receives information at one time, CI and LI share a control line. For the same reason, they also share a bit in the microinstruction.

The constants needed by the processor are stored in the data ROM, Figure 10.8, and the bracketed quantity at the ROM input is employed to represent the addresses accepted. If an address between 0 and $2F_{16}$ appears at the ROM input, the contents of that address are immediately placed on the data bus. On the other hand, when an address outside 0 to $2F_{16}$ appears at the ROM input, the output will be effectively disconnected from the bus. In all cases, the data bus contents will be determined only by the device

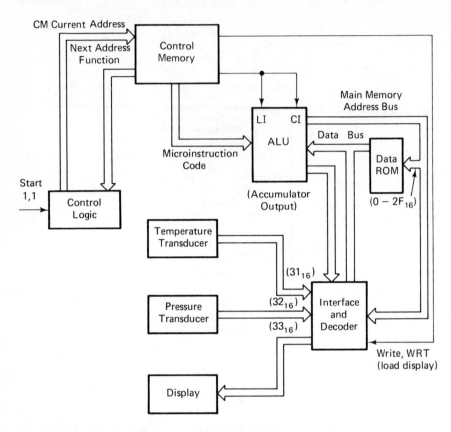

Figure 10.8   A Microprocessor System Employing Only Microinstructions.

whose address is on the address bus. The transducers and the display have addresses 31, 32, and 33, respectively, so that information may be exchanged with them much as is done in a RAM. (Note that the accumulator output serves as the source of data to be written, of loaded, by the display.)

Now consider Table 10.5 which contains the microprogram to implement the flow diagram of Figure 10.7. (The accompanying main memory contents are shown in Table 10.6.) The table is constructed in a manner very similar to those for the machine language programs given in Chapter 3. The leftmost column contains the control memory address, the next group of columns contains the microinstruction mnemonic, and the third group of columns contains the actual microinstruction code. The microinstruction mnemonic format is the same as previously shown in Figure 10.4. The address controls follow the scheme of Table 10.3, and the ALU Instructions follow the scheme of Table 10.4. In the present case, only two bits of the miscellaneous function

Table 10.5 Performance Function Program
(see block diagram Figure 10.5)

CM Address (Row, Col.)	( )	( )	$\left(\begin{matrix}W\\R\\T\end{matrix}\right)$	$\left(\begin{matrix}CI\\LI\end{matrix}\right)$	ALU Instruction	Next-Address Function	Microinstruction Code (Base 16)	Remarks	
1,1 (Start)	0	0	0	0	CL	JR 0,5	008105		
0,5	0	0	0	0	LM	JR 1,5	099115		
1,5	0	0	0	0	LD	JR 2,5	09D125	$A \leftarrow C(\text{Bus Add} \cdot 0)$ $=105_{10}$	Initialization
2,5	0	0	0	0	ST 2	JR 3,5	062135	$R^2 \leftarrow 105_{10}$	
3,5	0	0	0	1	IM	JR 4,5	189145		
4,5	0	0	0	0	LD	JR 5,5	09D155	$A \leftarrow C(\text{Bus Add} \cdot 1)$ $=31_{16}$	
5,5	0	0	0	0	ST 3	JR 6,5	063165	$R^3 \leftarrow 31_{16}$	
6,5	0	0	0	0	LM	JR 0,6	099106		
0,6	0	0	0	0	LD	JR 1,6	09D116	$A \leftarrow T;:T\equiv$Temperature	
1,6	0	0	0	0	ST 4	JR 2,6	064126	$R^4 \leftarrow T$	
2,6	0	0	0	1	IM	JR 3,6	189136		
3,6	0	0	0	0	LD	JR 4,6	09D146	$A \leftarrow P;:P\equiv$Pressure	
4,6	0	0	0	0	SR	JR 5,6	038156	$A \leftarrow \dfrac{P}{2}$	
5,6	0	0	0	0	AD 4	JR 6,6	054166	$A \leftarrow T + \dfrac{P}{2}$	
6,6	0	0	0	0	AD 2	JR 7,6	052176	$A \leftarrow T + \dfrac{P}{2} + 105_{10}$	
7,6	0	0	0	1	IM	JR 8,6	189186		
8,6	0	0	1	0	NO	JR 9,6	20F196	*Numeric Display* $\leftarrow$ *ACC*	
9,6	0	0	0	0	LR 3	JR 6,5	093165	Now repeat calculation	

Table 10.6 Data Controlled by the Main Memory Bus

Main Memory Address (Base 16)	Contents	Remarks	
0	105_{10}	Constant for performance function, F	Data ROM
1	31_{16}	Temperature transducer address	
2 through 2F	—	Not used	
31	(Temperature values)	Interface	
32	(Pressure values)		
33	(Display)		

and employed: $\begin{pmatrix} W \\ R \\ T \end{pmatrix}$, which loads the interface display from the accumulator

when $\begin{pmatrix} W \\ R \\ T \end{pmatrix} = 1$, and $\begin{pmatrix} CI \\ LI \end{pmatrix}$, which is the carry-input/left-input bit. The reader should now study the program. First consider the mnemonic group of columns and verify that the flow chart of Figure 10.5 is properly implemented; then check that the microcode column is a correct representation of the corresponding mnemonics. (The comments in the Remarks column should help to clarify the various steps of the program.)

A Macroinstruction Application

As pointed out in the previous section, our basic hardware is capable of supporting both macroinstructions and microinstructions. This facility will now be demonstrated by a design for implementing portion of the LINC computer, discussed in Chapter 3. Recognize that the proposed system is a member of the third microprocessor class, a microprogrammable processor. Here the LINC program (macroinstructions) and data will be placed in the main memory. The control unit, through the control logic and CM, will decode the macroinstructions and realize the necessary register transfer logic to implement each instruction. This will include the realization of the fetch and execution cycles and the maintenance (in the ALU) of such items as the memory address register and the program counter. The above process of realizing one computer by means of another computer is known as *emulation*.

Processor Configuration.   The block diagram for the current application is shown in Figure 10.9. It follows the basic pattern established by the general block diagram in Figure 10.5, but it emphasizes those details which are most appropriate in explaining the macroinstruction implementation. For example:

1. The macroinstruction word and the multiplexer are octal oriented while hexadecimal is employed throughout the rest of the system.

2. The multiplexer connections are defined so that it is easy to see which macroinstruction bits are sampled by each multiplexer segment, Seg. (Note that a 12-bit octal macroinstruction word is employed.)

3. The accumulator output is supplied directly to the RAM so that data can be stored (written) in the memory. Since the same memory address lines are employed both for data being retrieved from as well as for data being stored in memory, the write control bit is necessary. (When the write bit is 1, data are stored; when it is 0, data may only be read.)

4. The carry-output and right-output (CO and RO) bits from the ALU

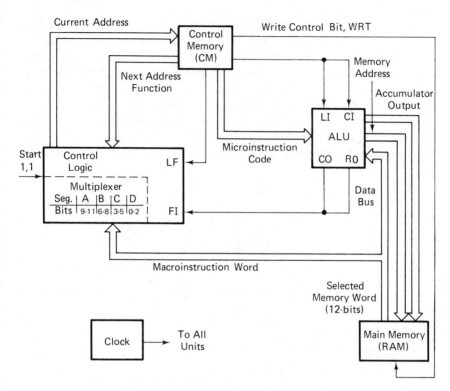

Figure 10.9 A Microprocessor Having Both Macroinstruction as Well as Microinstruction Capabilities.

are explicitly shown. The carry-output signal is familiar from Chapter 8—it goes to 1 whenever a carry would be produced from the MSB, e.g., when the accumulator's MSB is a 1 and a shift-left microinstruction is executed. The right-output signal is produced in a similar way, e.g., when the LSB is 1 and a shift-right microinstruction is executed. The presence of the *CO* or *RO* signal may be stored by setting the Flag Input bit to 1 in the control logic (through the *FI* terminal).

5. It is assumed that the Load Flag logic is such that, when $LF = 1$, the flag flip-flop is connected to the ALU through the *FI* and *CO/RO* terminals in such a way that it looks like a bit added to one end of the accumulator. For example, when the SL instruction is executed and $LF = 1$ in the microinstruction, the flag will look like it is a bit added to the left end of the accumulator, forming a large shift register.

The flag must be set on a previous step to effect a branch instruction like FC. (It is the state of the flag before the clock pulse which is transmitted to the FC logic and which determines the next memory address.)

CM Programs for the LINC Instructions. Once the basic units of the microprocessor, shown in Figure 10.9, have been properly connected, all that is required to complete the LINC design is to deposit the required programs in the control memory. Before actually writing any of the microprograms, however, it is desirable to construct a map of the CM and plan the arrangement of the various programs within the memory. (As programs are completed, Figure 10.9 can be updated to provide a map of available memory locations.) Figure 10.10 shows the map for our emulation of the LINC. Since we do not know beforehand exactly how many locations are required

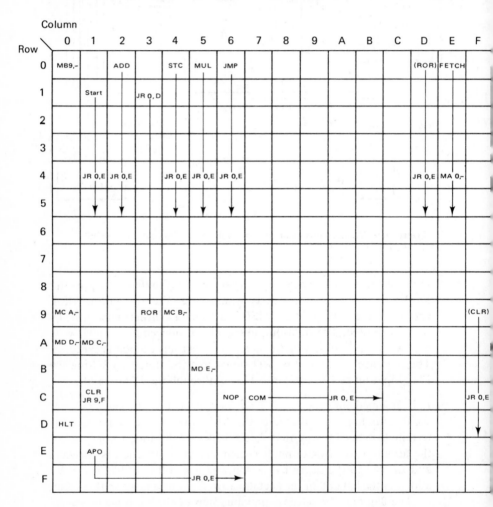

Figure 10.10   Control Memory Map for the LINC Emulation.

to implement each macroinstruction, a line with an arrowhead is employed to represent the intended sequence of locations. The line is broken with an unconditional branch symbol, e.g., JR, to indicate the approximate location of the jump to the next program. When no arrow is associated with an unconditional branch, such as MA or MB, the instruction will be placed in that exact location. Brackets around a macroinstruction, e.g., (ROR), denote the continuation of a program.

Figure 10.10 shows that the start signal initiates the CM at address 1,1. A small initialization program begins at that point, and it is followed by an unconditional branch to begin the first fetch-cycle program, at address 0,E. The fetch program ends each time with the next-address function MA 0,–. The purpose of this instruction is to cause a jump to row 0, with the column determined by the multiplexer sampling of segment A (bits 10–12) of the macroinstruction word. Thus, if the macroinstruction word is 6213_8, the new CM address will be 0,6, which is indicated by the LINC mnemonic JMP in the corresponding position on the CM map. (Once a multiplexer branch instruction entry is made in a row, we reserve the remaining columns, 0 through 7, in that row so that other macroinstructions, with different machine codes, can be detected there.)

The reader should now verify that ADD, STC, and MUL are in their proper positions on the map. If segment A of the multiplexer yields 0_8, the 0,0 location shows that the jump MB 9,– would be executed, which means that the next instruction would come from row 9. From Table 3.2 it can be seen that, for our pseudo-LINC instructions, the only possibilities for segment B are 0, 3, and 4. If it is 3, the ROR instruction sequence begins. If it is 4, an APO instruction is probably required, but the other bits will also need to be checked. Finally, if it is a 0, a branch to row A is indicated for further decoding. The reader should continue this process and verify that the programs for the other instructions start at the proper locations as indicated on the map. (Note that the only code allowed for the instruction NOP is 0016; but the YYYY from of Table 3.2, will be considered in the Exercises.)

The implementation of fetch, STC, and ROR are now presented since they are typical of a large class of microprograms. After studying them, you should be able to justify all the entries in each column. (The RTL programs in Figure 7.17 and Table 7.6 may serve as a useful guide to the required algorithms.) The fetch program is shown in Table 10.7. As predicted, the MA 0,– next-address function is contained in the program as a first step in decoding the macroinstruction. In this and the following two programs, it will be observed that register 1 is employed for the program counter, and register 2 is employed to store the macroaccumulator. The term *macroaccumulator* emphasizes the fact that the accumulator for the machine being emulated is different from the microaccumulator, or accumulator for short, of

Table 10.7 Microprogram for the Fetch Cycle

CM Address (Row, Col.)	( )	( )	$\begin{pmatrix} W \\ R \\ T \end{pmatrix}$	$\begin{pmatrix} CI \\ LI \end{pmatrix}$	ALU Instruction	Next-Address Function	Micro-instruction Code (Base 16)	Remarks
0,E	0	0	0	0	LR 1	JR 1,E	09111E	$A \leftarrow PC$
1,E	0	0	0	0	LM	JR 2,E	09912E	$MA \leftarrow A$
2,E	0	0	0	1	IR 1	MA 0,–	181A0–	$PC \leftarrow PC+1$; then to execute cycle

(header spanning: Microinstruction Mnemonic)

the microprocessor itself. Because this distinction is made, the microaccumulator can be employed for other tasks without disturbing the macroaccumulator.

Next consider the microprogram for the STC instructions, shown in Table 10.8. A small problem develops here when the first instruction loads the accumulator from the memory. Only the three least significant digits of the octal number are valid address information—the fourth digit is the code of STC, which is a 4. The MSB of this digit, however, is the only bit differing from zero, and the SL and SR microinstruction sequence nicely eliminates the objectionable bit and restores the desired address. Another feature of the program worthy of mention is the way the data are stored in the main memory by the $\begin{pmatrix} W \\ R \\ T \end{pmatrix} = 1$ microinstruction at address 5,4.

Table 10.8 Microprogram for the STC Instruction

CM Address (Row, Col.)	( )	( )	$\begin{pmatrix} W \\ R \\ T \end{pmatrix}$	$\begin{pmatrix} CI \\ LI \end{pmatrix}$	ALU Instruction	Next-Address Function	Micro-instruction Code (Base 16)	Remarks
0,4	0	0	0	0	LD	JR 1,4	09D114	$A \leftarrow (4XXX_8)$
1,4	0	0	0	0	SL	JR 2,4	048124	Remove leading bit to eliminate "4" in $4XXX_8$
2,4	0	0	0	0	SR	JR 3,4	038134	Restore address
3,4	0	0	0	0	LM	JR 4,4	099144	$MA \leftarrow XXX_8$
4,4	0	0	0	0	LR 2	JR 5,4	092154	$A \leftarrow (Macroacc.)$
5,4	0	0	1	0	NO	JR 6,4	20F164	Store data
6,4	0	0	0	0	CL	JR 7,4	008174	$A \leftarrow 0$
7,4	0	0	0	0	ST 2	JR 0,E	06210E	$(Macroacc.) \leftarrow A$; then return to Fetch cycle

(header spanning: Microinstruction Mnemonic — Misc. Function)

Finally, we consider the ROR microprogram shown in Table 10.9. The machine code itself, i.e., the 3 in $0300+n$, is eliminated this time by what is called *masking*. Since 0 through 5 are the only valid bits, a mask formed by the number 0077_8 when ANDed with $0300+n$ will yield, n, the required number of rotate steps. Another point to emphasize here is the loop implemented by the instructions at addresses 3,D through 7,D. Of particular interest is the way the loop counter is decremented at address 4,D and the way the test for completion is made at address 5,D.

Table 10.9 Microprogram for the ROR Instruction

CM Address (Row, Col.)	( ) $\begin{pmatrix}L\\F\end{pmatrix}$ $\begin{pmatrix}W\\R\\T\end{pmatrix}$ $\begin{pmatrix}CI\\LI\end{pmatrix}$				ALU Instruction	Next-Address Function	Remarks
9,3	0	0	0	0	CL	JR 8,3	⎫
8,3	0	0	0	1	SL	JR 7,3	⎪
7,3	0	0	0	1	SL	JR 6,3	⎪
6,3	0	0	0	1	SL	JR 5,3	⎬ Create mask, which is
5,3	0	0	0	1	SL	JR 4,3	0077_8
4,3	0	0	0	1	SL	JR 3,3	⎪
3,3	0	0	0	1	SL	JR 2,3	⎭
2,3	0	0	0	0	ST 3	JR 1,3	$R^3 \leftarrow A$
1,3	0	0	0	0	LD	JR 0,D	$A \leftarrow (0300+n)$
0,D	0	0	0	0	AN 3	JR 1,D	$A \leftarrow n$; by masking
1,D	0	0	0	0	ST 3	JR 2,D	$R^3 \leftarrow n$
2,D	0	0	0	0	LR 2	JR 3,D	$A \leftarrow (Macroacc.)$
3,D	0	0	0	0	RR	JR 4,D	$A \leftarrow rrA$
4,D	0	0	0	0	DR 3	JR 5,D	$R^i \leftarrow R^i - 1$
5,D	0	1	0	0	TR 3	JR 6,D	Test Reg. 3 for 0 (finished)
6,D	0	0	0	0	NO	FC 7,D	If finished jump 7C otherwise 7D
7,C	0	0	0	0	ST 2	JR 0,E	Finished; $(Macroacc.) \leftarrow A$; then to execute cycle.
7,D	0	0	0	0	NO	JR 3,D	Not finished; return to 3,D

Summary. The hypothetical microprocessor we have been discussing is similar to several commercial units,[14] but it has been simplified for instructional reasons. The objective has been to present the basic principles of microprocessor hardware and some of the main design techniques for their use without getting tied up in the idiosyncrasies of particular types. In addition, an effort has been made to minimize routine details, e.g., both the CM and the microinstruction set are much smaller than most commercial units. For generality, the emphasis here has been on user-microprogrammable

units. It should be understood, however, that microprocessors which do not
have this capability are also quite popular, and as a matter of fact, are often
easier to apply.[14] By having an understanding of the concepts upon which
all microprocessors are built, it is felt that the reader should be able to quickly
learn how to apply any of the commercial types to practical design problems.

Most microprocessor manufacturers supply computer-aided design software packages, which can be very helpful in applying their products. For
example, simulation programs allow a microprocessor system to be tested
so that most design errors can be corrected even before a breadboard is
actually wired. Moreover, programs called *cross assemblers* are available
permitting the designer to employ the more extensive facilities of a larger
scale computer for writing and debugging machine language programs,
which are later to be run on a particular microprocessor. Since these software
packages are tailored for a specific microprocessor, the reader is referred to
the manufacturers' literature for details. Many of the required fundamentals,
however, have already been developed in Chapters 3 and 7.

‡10.5 BASIC ARCHITECTURAL FEATURES OF MODERN COMPUTERS[2,6,9,18]

It is the purpose of this section to give the reader further insight by
presenting samples of various architectures which are currently being
employed. The treatment is not exhaustive, but the Bibliography should provide most additional sources of information which may be required.

An appreciation for the variety of architectural decisions which must be
made by the computer designer may be obtained from Table 10.10. As indicated by the table's title, the classification is by no means complete but it
does present a reference frame for the remainder of this chapter. Concerning
low-level features, busing has already been discussed at length in an earlier
section. Likewise, the distinction between serial and parallel operations was
made in Chapter 8 when methods for addition were compared. The concept
of synchronous and asychronous operation was mentioned in connection
with various types of counters in Chapter 5, yet a broader development of
the subject is appropriate here.

Synchronous operation, as the name implies, requires that all circuit
actions occur in coincidence with a clock pulse. The synthesis method of
Chapter 5, except for the ripple technique, leads to a synchronous circuit.
The reader should verify this by referring to Chapter 5 and noting that the
input pulse(s) provides the required synchronization signal. The clocked
J-K flip-flops used in that chapter make the point still more obvious. An
important fact to observe concerning synchronous circuits is that the output

Table 10.10   Typical Architectural Decisions

Organizational Level	Alternatives
Low	Series—Parallel Synchronous—Asynchronous Individual connections—Bussing
Intermediate	Microinstruction—Macroinstruction Single address—Multiple address Fixed-point arithmetic—Floating point arithmetic
High	Uniprocessor—Multiprocessor—Multicomputer—Computer network—Pipeline processor—Array processor—Special-purpose processor
Memory	Conventional—Cache—Virtual—Interleaved—Scratch pad—Associative
Input/Output	Byte multiplexing—Block multiplexing CPU transfers—Data channel transfers—Peripheral computer CPU interrupts—Cycle stealing—Programmed input/output

of any flip-flop can be used in any gating circuit to drive other flip-flops without concern about timing. Moreover, no matter how complicated the circuit is, one need only wait the gating delay plus one flip-flop delay before bringing in the next pulse or before reading the state of the flip-flops. Asynchronous systems, of course, are not so simple to use. For example, in a ripple counter we have already seen that the delay before all flip-flops have settled depends on the state the circuit was in before the input pulse occurred, and the maximum delay is proportional to the number of bits in the counter.

The preceding discussion suggests two ways of insuring that an asynchronous system has settled and is ready to accept the next command (input pulse) or to transfer its results:

1. Provide sufficient time between commands to allow for the worst-case delay.

2. Have the system automatically develop a signal indicating the completion of circuit action.

These two methods are represented by the system shown in Figure 10.11, where three operations are arranged in series and each requires the finished data from the previous process before carrying out its own task. Note that in both methods a signal must be received at the start input before the incoming data will be processed. Which of the two methods is best depends on the

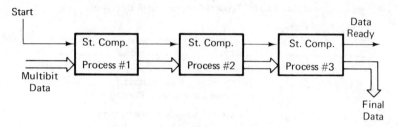

(a) Internally Generated Completion Signals

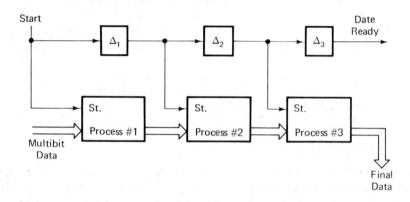

(b) Delay Method for Generating Completing Signals

st. ≡ start, comp. ≡ completion, Δ_i ≡ maximum delay for processor i
$\Longrightarrow$ ≡ multibit data.

Figure 10.11   Two Methods for Coupling Asynchronous Systems.

particular situation. However, the delay method is often cheaper to implement, but usually it is slower because each delay Δ_i must be set for the worst
combination of requirements possible for any run with that processor. On
the other hand, when the completion signal is generated internally, it can
automatically adjust to the exact processing time needed for each run.

One of the most common types of asynchronous systems is composed
of two processors, each driven by its own independent clock. Such a system
in which the outputs of the two processors are combined in a third processor
is shown in Figure 10.12 Note how the start signal for processor 3 is generated
by ANDing the completing signals from the other two processors.

In order to make the system of Figure 10.12 more specific, assume that
both processors 1 and 2 consist of a switch buffer (register) feeding into a

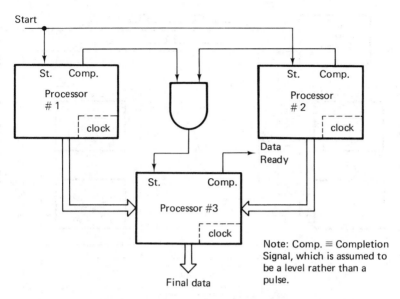

Figure 10.12 Generalized Diagram for the Coupling of Three Asynchronous Processors.

shift register (performing division by 2)* and that processor 3 adds the contents of the two shift registers, with the results being transferred to a buffer register.

For a synchronous system, the above configuration colud be represented as shown in Figure 10.13. The clock employed in the figure is a so-called two-phase clock because it has two outputs which are in synchronization, but output 2 is delayed by exactly one-half clock period† from output 1. The waveforms above the clock output lines show the delay, and the arrow marks the beginning of the first pulse, which is controlled by the clock start switch. Note that both registers 1 and 2 possess a transfer input, tr., and a shift-right input, shr. When the tr. input receives a pulse, it causes data from the switch buffer to be transferred into the main register; when the shr. input receives a pulse, the data in the main register are shifted right by one bit. Thus, during each clock period, the number in each buffer is divided by 2 and added. Then, with the occurrence of the next clock pulse at output 1, the results are transferred to register 3.

An asynchronous version of Figure 10.13 is shown in Figure 10.14 with the same problem being solved as before. In this instance, however, the solu-

---

*In order to simplify the circuitry for this example, each input is divided by 2. However, in a more practical case the divisors would be different.

†The time between two clock pulses at *one* output is defined as the *clock period*.

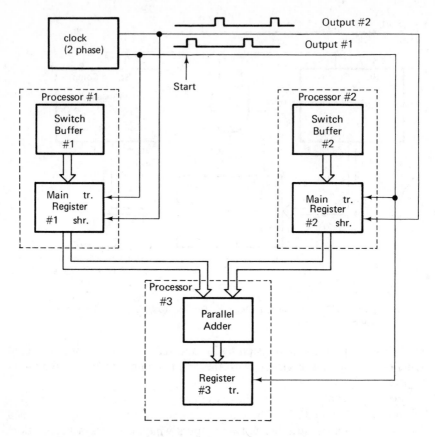

Figure 10.13   Synchronous Coupling of Three Arithmetic Processors.

tion is a good deal more complicated because of the asynchronous requirement. Carefully examine Figure 10.14 and, in particular, observe how the
completion signals for processors 1 and 2 are generated. In addition, note
that in Figure 10.13 the switch buffers may be changed and the new results
will appear in register 3. On the other hand, the circuit of Figure 10.14 does
not possess the same capability in that, once flip-flops 1 and 2 are set, there
is no circuitry to reset them. Thus, only one group of data is processed,
regaraless of the number of times the switch buffers are changed.

Although the input/output units are often asynchronous, the CPU and
most other units of a digital computer usually employ synchronous circuitry.
On the other hand, there are systems which are based entirely on asynchronous
circuitry.

Returning to Table 10.10, we see that the intermediate-level features
have already been considered either in the beginning of this chapter or in an

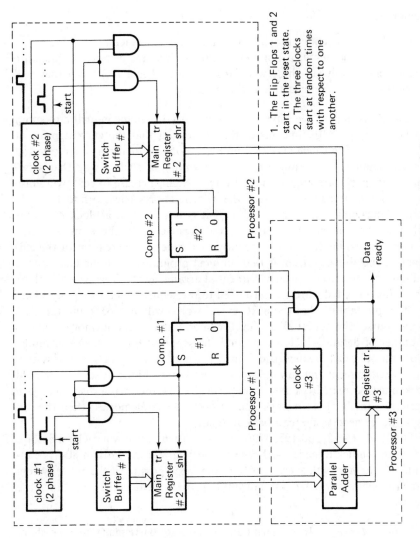

Figure 10.14 Asynchronous Coupling of Three Arithmetic Processors.

1. The Flip Flops 1 and 2 start in the reset state.
2. The three clocks start at random times with respect to one another.

383

earlier part of the text. Moreover, the input/output features are considered at length in Chapter 11. Thus, we are left with the memory and high-level features, which will be the subjects of the next two sections of this chapter.

Memory Configurations

In addition to conventional core and solid-state memories, which were discussed in Chapter 9, a number of innovations in memory architecture have recently evolved. These were mentioned in Table 10.10, but they will be considered in greater detail at this point.

Cache Memory. We have already seen that a semiconductor memory *can* be more than 10 times faster than core, but that the current price of core is lower. The cache system employs a conventional core memory and a much smaller semiconductor, cache, memory in such a way that semiconductor speeds are approached at only a slight increase in cost over core. As a program is being executed, the system always tries to find the next word in the semiconductor memory. If it is not present, a block of words including the one needed is transferred from core to the semiconductor memory, displacing the least frequently used block.* The reason for the efficiency of the above concept is that, in most programs, both the instructions and data are found in a nearly sequential arrangement. Thus, when a block containing a needed word is transferred to the semiconductor memory, there is a high probability that the next several words will also be found there.

Of course, the actual algorithm for operating a cache memory contains a number of details which are beyond the scope of this text. For example, before an old block is displaced from the semiconductor section, the system must chack to see if any new data were stored in the block. If so, then the old block must be read bvck into core instead of merely being displaced. (An alternative to this is to always update both of the memories; then the old blocks never need to be read back into core.)

The IBM 360/85, 360/195, 370/155, and 370/165 employ a cache memory, and studies indicate that a 1-μs core memory combined with an 80-ns semiconductor memory will average 80% of the speed possible if the entire system were fast memory, which is a significant improvement at a fairly small increase in cost.

Virtual Memory. In modern data processing, particularly in time-share applications, it frequently happens that the instructions and data occupy more space than is available in the main memory. There are two possible approaches to the solution of this problem (both of which employ an auxiliary store, usually a disk, to hold a copy of all programs/data):

---

*Other criteria have also been successfully employed.

1. *Static scheduling,* where the programmer is responsible for keeping track of the location of all routines and blocks of data and for swapping these in and out of the main memory at the proper times.

2. *Dynamic scheduling* (or *virtual memory*), in which the main memory appears to the programmer to be much larger than its actual physical size, with an automatic hardware/software system performing the swapping and "bookkeeping" functions.

When the CPU tries to fetch a particular word in a virtual memory system, it first looks in the main memory. If it is not present there, a block of words including the needed one is transferred into the main memory much as is done in a cache memory system. There are other similarities between these memory systems, but there are also major differences. The most obvious difference is in the hardware composing the two (semiconductor and core versus core and disk). Another more fundamental difference resides in the fact that the virtual memory, by means of a series of pointers, automatically keeps track of the programs/data which may be scattered about the main and auxiliary memories. The RCA computers were among the first to employ the virtual memory concept, but the IBM 360/67 and IBM 370 also use it.

Memory Interleaving. As indicated in Chapter 9, a complete memory cycle typically requires on the order of 1 μs. When this is contrasted with a typical CPU cycle time of 100 ns, we see that a substantial mismatch exists. One way to improve the efficiency is to divide the memory into independent segments, each of which has an independent read/write cycle. The simplest interleaving scheme employs two memory segments with the odd addresses assigned to one and the even addresses assigned to the other. Since instructions for the most part occupy sequential memory locations, while one instruction is being processed and before its read/write cycle is completed, the next instruction can be fetched. In this way a speedup factor approaching 2 may be obtained.

Even larger speedup factors are obtainable by segmenting the memory further, e.g., by dividing it into an independent instruction half and a data half, each of which is also interleaved into odd and even addresses. Then two instructions and two data words can be accessed concurrently. Many small- and large-scale computers employ the interleaving concept, e.g., the DEC PDP-11/45 and the UNIVAC 1110.

Summary of Memory Architectures. Table 10.11 presents a concise summary of representative memory architectures, and the reader may want to use it to review the material in this section. It should be noted that the scratch pad memory, a rather old concept, has been added for completeness. This memory type is of more limited applicability than the others listed since it is used mainly to store intermediate results.

Table 10.11 Summary of Memory Architectures

Memory Type	Mechanization	Main Advantages	Main Disadvantages	Special Distinguishing Features
Conventional	All core or all semi-conductor	1. Simplicity 2. Low cost	1. Slow if core 2. More expensive if semiconductor	This is the norm with which more innovative memory organizations must be compared.
Cache	Mostly core but also a small section of semiconductor	$\approx$ 80% of semi-conductor speed at near core prices	Added complexity (circuitry)	Small semiconductor working section which automatically swaps in currently needed instructions/data from core.
Virtual	Main memory: usually core Auxiliary: usually disk	1. Increase in apparent size of memory. 2. Automatically keeps track of programs scattered throughout memory	Added complexity (circuitry and software)	The programmer is not aware of the presence of the auxiliary memory.
Interleaved	Always core	Speed improvement over normal core	Added complexity (circuitry)	More than one independent memory cycle.
Scratch pad	Semiconductor	1. Increase in speed over normal core 2. Simple 3. Fairly inexpensive	1. Requires extra programming 2. Not applicable to all memory problems	Usually employed as a small, fast store for intermediate results.

Processor Configurations—Higher-Level Features

Why are there so many different computer configurations and by what criteria can the performance of a computer be judged? Let us consider the following typical criteria:

1. Speed (Number of tasks which may be completed per second.)

2. Reliability

3. Versatility (How easy is it to adapt the basic configuration to a variety of processing tasks?)

4. Programming convenience (How simple can the programmer's task be made? For example, a few special instructions may significantly reduce programming time for certain tasks.)

5. Computing Power (How many bits per word? How large is the main memory? etc.)

6. Cost (Hardware, software, and operating.)

Special test problems, called benchmarks, can be employed to compare (develop numerical scores for) various architectures on the basis of the above criteria, and this certainly is the objective way to perform computer systems analysis. However, this is currently a fairly complicated procedure, which is beyond the scope of our text. (The interested reader is referred to the Bibliography at the end of this chapter.)

Of course, all of the above criteria could eventually be assigned a dollar cost; yet if each of these is considered as an entity, it appears that many of the innovations in computer architecture are concerned with processor speed. Since most of these involve some type of parallelism, there is also an amount of reliability through redundancy in many of the systems; yet in the final analysis it appears that speed itself has been the main motivation in many recent computer developments.

The type of computer we have considered thus far, which employs a single instruction stream to operate on a single data stream in sequential fashion, is called a *uniprocessor* or *von Neumann machine* (after the man who contributed much to the development of the digital computer). Most early systems and contemporary minicomputers are uniprocessors, which is not surprising since it is the most elementary machine to design and build. On the other hand, modern large-scale digital computers frequently employ simultaneous (or concurrent) processing in various portions of the system; thus, we will label this large class of systems *concurrent processors*.

Concurrency can be obtained in many different ways; some of the most popular are:

1. *Multiprocessor*—a group of computing units, each with its own instruction and data stream, sharing a common memory and control unit.

2. *Multicomputer*—independent computers often with one acting as a supervisor in performing a common task at a single location. (A common though somewhat limited application consists of a mini-computer acting as a preprocessor for a large-scale computer.)

3. *Computer Network*—independent computers at different geographical locations connected by a communication channel so that the unique resources available at a particular site can be made available to all members of the network.

4. *Pipeline Processor*—a single computer which simultaneously performs operations on several sections of a calculation, in various stages of completion, much as is done on an assembly line.

5. *Array Processor*—a group of computing units each of which simultaneously performs the same operation on a different data stream.

All of the above systems have variations and of course a combination of types is also possible, but this discussion will be limited to the pure types listed above. The first thing that should be observed is that the name of each system is fairly descriptive of its basic nature.

In the discussion of the various methods for obtaining concurrency, first more details will be presented concerning some of them; then a table summarizing the characteristics of each will be presented.

Multiprocessor.   Many problems can be divided into several independent tasks, and the possibility of capitalizing on this characteristic to speed the completion of the overall problem is the motivation for the type of concurrency called *multiprocessing*. Here there are several identical processors, each of which employs its own instruction stream to complete one of the tasks composing the total problem. The Control Data 6600 is a member of this class. It employs 10 such identical units called *peripheral processors*. Of course, this is a simplified description in that actual implementation results in several interesting situations, e.g., the need for a supervising unit to break the original problem into sections and to schedule the sections to the available processors.

One variation of the multiprocessing concept employs nonidentical processors specialized to common types of calculations, e.g., floating-point arithmetic, evaluation of logic expressions, fixed-point arithmetic, etc. Again, each of these operates on an independent section of the problem.

Although not operating as true multiprocessors, even minicomputers have been able to capitalize on some of these ideas. For example, the Digital Equipment Corporation PDP-11 speeds up the completion of floating-point calculations with a special processor. This device is controlled by the CPU, but after a delay time required for initiating the calculation, the CPU can turn to other tasks while the floating-point calculation is being completed. Thus a certain amount of concurrency is achieved.

Not a lot of emphasis has been placed on possible reliability improvements, but with identical processors and the proper design of the other equipment, it should be possible to operate a multiprocessor with reduced performance when some of the processors have failed.

Most large-scale computers employ an elementary type of multiprocessing in that they have a special unit to handle input/output tasks which is completely independent of the central processor. By this technique, the CPU is not burdened or interrupted by what often is a very time-consuming but, in reality, a rather simple task.

Pipeline Processor.[17]   Although circuit speed has increased by orders of magnitude since the first computer was built, it sometimes is necessary to further increase the rate at which calculations are performed beyond what the state of the art allows with a uniprocessor. If very similar operations are required on each set of data such as is true in making payroll calculations, an "assembly-line technique" called *pipelining* can be used. The basic philosophy here is to divide the overall task into several suboperations which are performed in sequence just as on an assembly line. Also as on an assembly line, each station operates simultaneously so that solutions closer to completion are present as the end of the line is approached. Although the time required for any individual problem to get completely through the pipeline may be the same or slightly longer than with a uniprocessor, the fact that several parts of the problem are being worked on simultaneously provides a significant multiplying factor for the number of problems which may be completed per second, and speedup factors of more than two orders of magnitude are obtainable. It should be noted that pipelining achieves improved performance through a serial rather than a parallel approach. The Control Data STAR is one example of a computer which employs the pipeline concept.

Like so many architectural innovations, pipelining was conceived as a design approach for an entire system; but its greatest application is likely to be for individual computer units. A typical example is the arithmetic unit which has recently received considerable attention, with the pipeline concept being found particularly effective in improving the speed of multiply operations. The IBM 370–195 employs a pipeline floating-point multiply unit.

Array Processor.[17]   In contrast to a pipeline processor which achieves concurrency by serial means, the array processor shown in Figure 10.15 takes a parallel approach. This system employs a single control unit and a single instruction stream to service a large number of identical computing units or cells, each of which operates on its own data stream. The simultaneous operation of each instruction on wholesale amounts of data is, of course, the secret to the success of this system. On the other hand, operations such as input/output and compiling are usually performed by a standard

Independent Results

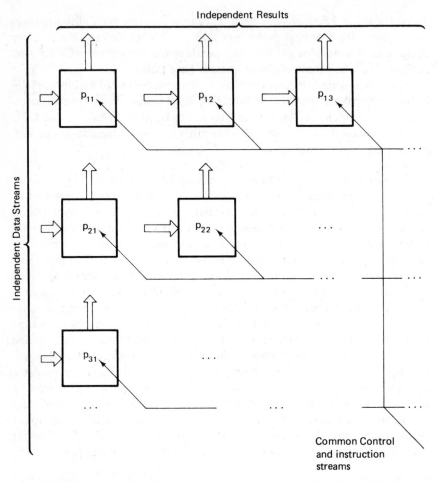

Figure 10.15   The Basic Array Organization (where P_{ij} is the processor in the ith row jth column of the array).

uniprocessor. The array processor has an advantage with those problems in which the data can be treated in blocks, whether this be due to their physical geometric arrangement or merely because the data allow an artificial arrangement of this type. Matrix operations, solution of partial differential equations, the Fourier transform, weather modeling, and many other practical problems would appear to benefit from this technique. Moreover, it does seem that most problems which require very high performance can eventually be configured so that the array approach is valid. Typical computers of this type are the ILLIAC IV and the Goodyear Aerospace Corporation STARAN.

Conclusions

Table 10.12 is presented as a summary of the discussion of architectural styles. The reader may now want to review the material in the previous few

Table 10.12 Summary of Representative Computer Architectural Styles

System	Instruction Stream	Data Stream (Per Instruction Stream)	Typical Number of Processors in System	Memory	Main Advantages	Main Disadvantages	Special Distinguishing Features
Uniprocessor	Single	Single	1	Central	Simplicity and low cost	CPU must do all operations including input/output.	This is the norm with which other systems should be compared.
Multiprocessor	Multiple (central control)	Single	8	Central (shared)	Speed, reliability, versatility	Higher cost; supervision requires overhead.	
Multicomputer	Multiple (local control but may be slaved)	Single	≥ 2	Local	Speed, reliability, versatility	Overhead when in slaved mode.	The several computers are part of a local facility; one computer may become master.
Computer Network	Multiple (local control but may be slaved)	Single	10	Local	Availability of unique resources	Communication channels and interconnections are expensive.	Typically, computers are spread across the entire country.
Pipeline Processor	Single	Single	1	Central	Speed	Higher cost; not applicable to all problems.	Several simultaneous operations in assembly-line fashion.
Array Processor	Single	Multiple	≥ 64	Part central and part local	Speed on "matrix-like" problems	Higher cost; not applicable to all problems.	

pages by employing the table as a guide. Note the way the Instruction Stream and Data Stream columns are defined. For example, the multicomputer which employs one instruction stream and one data stream for each of several computers is classified as a *multiple* instruction stream but *single* data stream system because the instruction stream is quoted per system and the data stream is quoted per instruction stream.

The classification and analysis of various computer configurations is an aspect of engineering that is still in its infancy but one that is beginning to arouse considerable research and developmental interest. It seems that the opportunities for improved computer performance are limited only by one's imagination, e.g., it has recently been suggested that a system be designed which will automatically adjust its architecture in conformance to the requirements of individual problems.

The reader should now be prepared to study journals and more advanced texts on computer architecture. (The Bibliography is a good place to start.) This will no doubt continue to be a profitable area for research and development for a long time to come.

10.6 BIBLIOGRAPHY

1 BEIZER, BORIS, *The Architecture and Engineering of Digital Computer Complexes*. New York: Plenum Press, 1971.

2 BELL, GORDON C., and ALLEN NEWELL, *Computer Structures: Readings and Examples*. New York: McGraw-Hill Book Company, 1971.

3 CHU, YAOHAN, *Computer Organization and Microprogramming*. Englewood Cliffs, N.J.: Prentice-Hall, Inc., 1972.

4 DAVIES, P. M., "Readings in Microprogramming." *IBM Systems Jour.*, Vol. 11, 1972, pp. 16–40.

5 FLORES, IVAN, *Computer Organization*. Englewood Cliffs, N.J.: Prentice-Hall, Inc., 1969.

6 FOSTER, CAXTON C., *Computer Architecture*. New York: Van Nostrand Reinhold Company, 1970.

7 HILL, F. J., and GERALD R. PETERSON, *Digital Systems: Hardware Organization and Design*. New York: John Wiley & Sons, Inc., 1973.

8 HUSSON, S. S., *Microprogramming—Principles and Practices*. Englewood Cliffs, N.J.: Prentice-Hall, Inc., 1970.

9 KATZAN, HARRY, *Computer Organization and the System/370*. New York: Van Nostrand Reinhold Company, 1971.

10 LEWIN, DOUGLAS, *Theory and Design of Digital Computers*. New York: John Wiley & Sons, Inc., 1972.

11 LEWIN, MORTON H., "Integrated Microprocessors." *IEEE Trans. Circuits and Systems*, Vol. CAS-22, 1975, pp. 577–85.

12 *PDP 11/45 Processor Handbook*. Digital Equipment Corporation, Maynard, Mass. (Consult most recent edition.)

13 SLOTNICK, D. L., "The Fastest Computer." *Sci. Am.*, Vol. 224, 1971, pp. 76–87.

14 SOUČEK, BRANKO, *Microprocessors and Microcomputers*. New York: John Wiley & Sons, Inc., 1976.

15 "Special Issue on the HP-65 Programmable Pocket Calculator." *Hewlett-Packard Journal*, Vol. 25, May 1974, pp. 2–17.

16 "Special Issue on Microprogramming." *IEEE Trans. Comput.*, Vol. C-23, 1974, pp. 753–837.

17 "Special Issue on Minicomputers." *Proc. IEEE*, Vol. 61, 1973, pp. 1524–40.

18 STONE, HAROLD S. (ed.), *Introduction to Computer Architecture*, Chicago, Ill.: Science Research Associates, Inc., 1975.

19 THEIS, D. J., "Microprocessor and Microcomputer Survey." *Datamation*, Vol. 20, December 1974, pp. 90–101. (This article is now mainly of historical interest, but it is typical of the type of summary data that are currently available in the literature.)

10.7 EXERCISES

10.1 Draw a diagram of the instruction format for a two-address, one-instruction per word computer; allocate space in bits for each part of the basic word. (Assume: 10 index registers, 16 instructions, and a 4096-word memory.)

10.2 In a single-address machine which employs 95 instructions, how many bits are required in the instruction word so that all 4096 words in the memory can be directly addressed?

10.3 Using Reference 12, prepare about a two-page, basic explanation of the instruction set employed in the PDP 11/45.

10.4 Write a report discussing the features of the computer presented in Reference 13.

10.5 A microprocessor does not have direct capabilities for loading constants, which may be required by its microinstructions. List, and discuss in a couple of sentences each, at least three good ways to solve this problem.

10.6 It is desired to create a new next-address function which jumps to a new row on the basis of the contents of the flag flip-flop. Following the scheme of Section 10.3 *as closely as possible*, prepare the required information for a new line in Table 10.3 which describes this instruction.

10.7 It is desired to create a new ALU microinstruction which increments the accumulator. Following the scheme of Section 10.3 *as closely as possible*, prepare the required information for a new line in Table 10.4 which describes this instruction.

10.8 In addition to the formally defined NO instruction ($0F_{16}$), list at least two other instructions which will perform no operation under certain conditions.

10.9 Create a new column for Table 10.9 which shows the microinstruction codes (in base 16) corresponding to the microinstruction mnemonics.

10.10 (a) In constructing Figure 10.10, it was assumed that only the LINC machine code 0016 would be employed as the NOP instruction. If the additional YYYY code of Table 3.2 is also allowed, insert NOP in the resulting locations on the CM map.

(b) What microinstruction should be placed at these locations?

10.11 Using a format like Table 10.7, write the start (initialization) microprogram, which has its first location at CM address 1,1 (Figure 10.10). Assume that all macroprograms start at location 20 in the main memory.

10.12 Write a microprogram to implement the APO instruction of Table 3.2. (Present your solution in a form similar to Table 10.8.)

10.13 Repeat Exercise 10.12, but implement the JMP instruction.

10.14 Using the microprocessor system of Figure 10.8, including the data ROM contents of Table 10.6, determine the CM address where the following microprogram stops and also the contents of register 2 at that time. All CM, ROM, and register locations, not originally specified, initially contain zero. Assume a 12-bit ALU; and any instruction not defined by Table 10.4 is a NO instruction, except that a halt instruction is added with mnemonic HT and code 00_{16} which halts the processor at the current CM location.

CM Address (Row, Col.)	Microinstruction Code (Base 16)
(Start) 1,1	008117
1,2	062118
1,3	028112
1,4	1C8116
1,5	148113
1,6	06211A
1,7	148115

11 | COMPUTER INTERFACE DESIGN

As digital computers become increasingly popular, there is a growing demand for specialized interfaces which allow them to be *directly connected* to an ever-wider variety of peripherals. These external devices can be standard computer units such as disks, tape drivers, CRT displays, and printers; they can be various electrical, medical, or other instruments that employ the computer for *on-line* data processing; or they can be industrial equipments that employ the computer to supervise operations such as an automatic oil refinery or a computerized steel-rolling mill.

The process of selecting an interface for a particular application can be grouped into the following classes:

1. Purchase of a complete interface system from the computer manufacturer. (Even here there is often a significant engineering problem in selecting the proper system from the many that are available.)

2. Assembly of the interface from commercially available printed circuit cards.

3. "From-scratch" design and construction of the interface with individual integrated circuits.

Typical sources for these three levels of interface design are found in the Bibliography.[5,8,9] Although further study and experience will be required for the engineering of complex interfaces,* it is our purpose, in this chapter, to employ the tools previously developed (particularly from Chapters 7 and 10) to describe the basic principles of interface design and to prepare the reader for more advanced work.

11.1 DATA TRANSFER TECHNIQUES[11]

There are several methods for transferring data between a computer and the outside world. This area has been one of intensive development, and there is every indication that this activity will continue into the foreseeable future.

---

*This is particularly true for item 3 in the above list.

Fortunately, the subject consists of a few basic ideas, and when the reader masters these, he should be in a position to understand the complexities of both present and future systems.

Data transfer methods may be divided into the following three major areas:

1. Programmed transfers

2. Interrupts

3. Direct memory access

In the *programmed transfer* method, the peripheral is serviced at the convenience of the computer, i.e., whenever the I/O instructions happen to be reached in the *normal* program sequence. By contrast, the *interrupt* method places the data transfer at the convenience of the external device, i.e., service is initiated through what amounts to an I/O subroutine *as soon as the peripheral requests it* instead of when the computer gets around to it.

Direct memory access, DMA, is a very quick and efficient method for transferring large blocks of information such as when data is passed to or from a disk. It is similar to the interrupt method in that *service is initiated immediately*; but instead of having a moderate-sized software subroutine supervise the transfer, it is done entirely by special hardware. Thus, the data are transferred much quicker at the expense of more complicated and costly hardware.

11.2 PROGRAMMED TRANSFERS

Let us now consider programmed data transfer from a peripheral into the accumulator. For I/O operations we divide the 12-bit instruction word into the three sections shown in Figure 11.1. As usual, the three most significant bits are reserved for the operation code, with 7XXX having been selected to represent the entire class of I/O instructions. The command section of the word indicates the type of I/O operation being performed, e.g., 4_8 signifies the data is to be read by the accumulator from the device selected. Finally, the least significant six bits provides coding capability for selecting one of $2^6 = 64$ peripheral devices. For example, the instruction word 7410 indicates

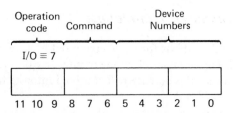

Figure 11.1 I/O Instruction Word Format.

that information is to be transferred from the *data buffer*, DB, of device number 10 onto the data bus where it will be read by the accumulator. R Ead data from Device number n, 74 n, will be given the mnemonic RED n. (For those I/O instructions, to be described shortly, which do not require a device number, bits 0–5 will be held at 00_8. Here the machine code will be of the form 7c00, where c is one of eight possible commands represented by a single octal digit. Thus, there are available eight commands employing a device number; eight different commands without a device number; and $64 - 1 = 63$ possible devices.)

The logic for implementing the RED instruction is shown in Figure 11.2. There are three separate buses connecting the various interfaces to the com-computer. The six-bit word on the device decoder bus (through the device decoder, DD) controls which one of the 63 possible interfaces is to be connected to the computer through the command bus and the data bus. With the device decoder actuated, the command decoder, CD, is free to take the information from the command bus and, in this case, place the word contained in the data buffer register onto the common data bus so that registers from many other interfaces may be simultaneously attached to the bus. (Only a single register, however, is actually delivering information to the bus at one time.)

What type of RTL program is required by the RED instruction? Assume it is known that the device is ready to deliver the needed data word. Then all that the I/O instruction need do is place the word from the device onto the data bus and then actuate the transfer from the bus into the accumulator. Figure 11.2(b) shows the execution portion of the RTL program for accomplishing this. It should be understood that once the transfers are made from the *MB* to the decoders (DD and CD), the levels will be held until the end of the $T = 2$ state. The RTL for the RED instruction in Figure 11.2(b) should be compared with that for the other instructions shown in Table 7.6.

The minor changes required in Figure 11.2 to implement the companion instruction WED n, WritE the information from the accumulator into Device n, is left as an exercise. (WED n will have the numerical code 73 n.)

Implementation of Flags and Branch Instructions

In describing the RED instruction, it was assumed that the peripheral was always ready to transmit the needed information. Frequently, this is not the case, and the computer must wait until the device is ready; otherwise, erroneous data will be received. A typical case where the data are not immediately available is just after a tape controller has been instructed to read from a remote section of tape. Here, the computer must wait until the tape drive has found the data before the ready signal can be given. Often device-readiness is indicated by a flag flip-flop, which is gated onto the common flag but by a level from the command decoder. (Since wired-OR logic is employed where connection is made to the bus, only one flag line is required

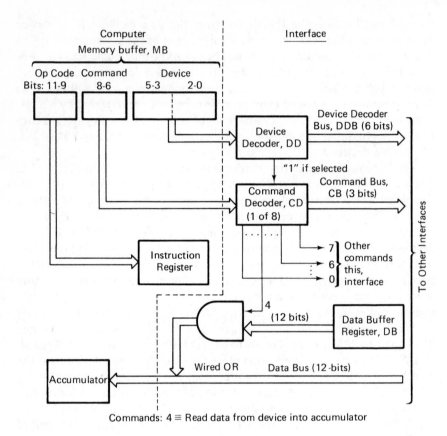

Commands: 4 ≡ Read data from device into accumulator

(a) Required Logic

$$T = 0 \quad DDB \leftarrow MB_{0-5}; CB \leftarrow MB_{6-8};$$
$$T = 1 \quad A \leftarrow DB;$$
$$T = 2 \quad S \leftarrow 0; T \leftarrow 0; DDB \leftarrow 0; CB \leftarrow 0;$$

(b) RTL Program

Figure 11.2   Implementation of the RED Instruction.

for all 63 interfaces.) Figure 11.3(a) shows the logic required to implement the flag instruction. The reader should carefully study the diagram noting such things as:

1. The method of setting and resetting the flag.

2. The fact that command 5 controls both the flag's presence on the bus and its reset line.

3. The way that flag signals enter the computer logic.

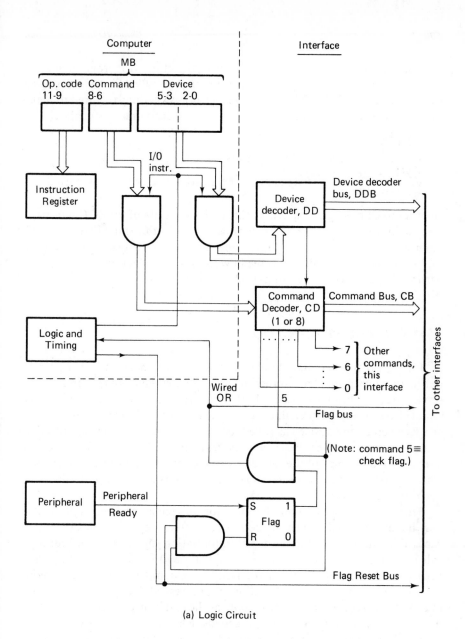

(a) Logic Circuit

T = 0 $DDB \leftarrow MB_{0-5}$; $CB \leftarrow MB_{6-8}$;
T = 1   IF (Flag Bus = 1) $PC \leftarrow PC + 1$;: Flag high therefore skip;
T = 2 $S \leftarrow 0$; $T \leftarrow 0$; (Flag of selected device) $\leftarrow 0$;
        $DDB \leftarrow 0$; $CB \leftarrow 0$;
T = 5 $M(MA) \leftarrow MB$; $PC \leftarrow 1$; $S \leftarrow 0$; $T \leftarrow 0$;

(b) Execution Cycle for SOH Instruction

Figure 11.3 Implementation of the Flag Instruction.

Usually the flag concept is employed in conjunction with program branching; for this reason, we will define a new instruction called S̲kip O̲n flag H̲igh (from device n), SOH n. Thus, the sequence of instructions required to make the computer wait until the flag of device 15 produces a ready signal is:

SOH 15

JMP $p - 1$:   Flag is low; therefore, go back and try again.

$\begin{Bmatrix} \text{Device Service} \\ \text{Routine} \end{Bmatrix}$:   Flag is high; therefore, service routine is started.

In order to gain insight concerning the implementation of the SOH n instruction, refer to Figure 11.3(b). In particular, study the similarities and differences between this RTL program and the one for RED shown in Figure 11.2(b). (It is also instructive to see how the RTL productions in part (b) of Figure 11.3 are mechanized in part (a) of the figure.)

11.3 INTERRUPT TRANSFERS

Frequently it is necessary to concurrently service one or more peripherals during the running of a main program, or a peripheral may require immediate service once it becomes ready. Both of these situations are conveniently handled by means of interrupt logic which, at the end of each fetch cycle, checks whether any device has called for service. If one has, the current value of the program counter, PC, is automatically stored in a specified location, and the starting point of the service routine is transferred to the PC. (Somewhat arbitrarily, we choose memory location 0 as storage for the PC and begin the service routine at location 1.)

Figure 11.4(a) shows how the logic for an elementary interrupt facility may be implemented. When the peripheral requires service, it sets the interrupt call flip-flop in the interface. This information is placed on the interrupt call bus; and if the computer's interrupt allowed flip-flop has previously been set, the interrupt request is received by the computer's logic and timing unit. In order to save the calculation in progress at the occurrence of the interrupt, the request is recognized only at the end of each execution cycle—usually producing only a few microseconds delay. At that point the current value of the program counter is stored and a software service routine is initiated. The service routine must perform the following tasks:

1. Disable the interrupt flip-flop (with the I̲nterrupt O̲Ff, IOF $\equiv$ 7100, instruction) so that the computer is able to finish servicing one device before being interrupted again.

2. Save the register information from the original program so that it may be properly started again.

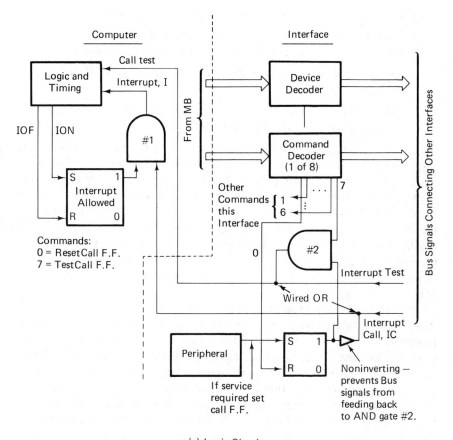

(a) Logic Circuit

T = 0 $MA \leftarrow MB_{0-8}$;
T = 1 $MB \leftarrow M(MA)$;
T = 2 $A \leftarrow A+MB$; $M(MA) \leftarrow MB$;
         IF (Interrupt = 1) $MA \leftarrow 0$ ELSE $S \leftarrow 0$, $T \leftarrow 0$;
T = 3 $MB \leftarrow M(MA)$; :Must read core location before writing;
T = 4 $MB \leftarrow PC$;
T = 5 $M(MA) \leftarrow MB$; $PC \leftarrow 1$; $S \leftarrow 0$; $T \leftarrow 0$;

(b) ADD Instruction — Modified Execution Cycle

Figure 11.4   Interrupt Implementation.

3. Determine which interface caused the interrupt. One way to do this is
   to poll the devices by means of the Skip if Interrupt call flip-flop of
   device n is Low $\equiv$ SIL n $\equiv$ 77 n.

[From Figure 11.4(a) it can be seen that the result of the test is wired-OR
connected to a bus; but with only one of the gates selected at a time, a unique
determination of the calling device is made.]

4. Carry out the required data exchange with the interface.

5. Reset the call flip-flop (with the instruction flip-flop n $\underline{R}\underline{E}$set $\underline{C}$all $\equiv$ REC $n \equiv 70\ \underline{n}$), thereby indicating device service completed.

6. Set the interrupt flip-flop (with the $\underline{I}$nterrupt $\underline{O}\underline{N}$, ION $\equiv$ 7200, instruction).

7. Restore the program counter, the accumulator, and any other registers so that the original program can continue.

In order to recognize the interrupt request during execution of a program, some statements must be added to the original RTL description of each machine language instruction. Since ADD is a typical instruction, its RTL implementation (originally found in Table 7.6) has been modified as shown in Figure 11.4(b). Attention should be given to the way original information in the PC is saved and how the computer program is forced to branch when the interface calls for service. The production at $T = 3$ is necessary since core memory must be read before writing into it. Moreover, it is not possible to combine $T = 3$ and $T = 4$ because they produce conflicting changes in the MB.

The interrupt service routine is presented in Table 11.1. The reader should study the way the six device-service tasks listed above are accomplished by the program. For brevity, only two interrupting devices are checked in this program; but the extention to additional devices should be obvious. Since the interrupt allowed flip-flop is disabled at the beginning of the program and not set until the end, we see that devices, perhaps of higher priority, requiring attention during the service routine are not recognized until service to the first device is complete. Moreover, our scheme always tests the devices in numerical sequence and does not take into account device priority or length of time a device has been waiting. These requirements are properly handled, at the expense of greater complexity, in many contemporary systems. More will be said on this subject in Section 11.5.

Another problem concerning more than one interrupt call existing at a time is that a new interrupt must *not* be recognized until the program counter again points to location X. If an interrupt were recognized by our program during execution of the instruction in location 14, the hardware would automatically write 15 into location 0, destroying X, which is where the original program should be restarted. Then it would not be possible to return to the original program. One solution, which is employed in some contemporary computers and the one we will adopt, is to prevent ION from taking effect until execution of the next instruction following it. Thus, in the program of Table 11.1, ION is executed at 14 but JMP X will be in the process of completing execution and PC $= X$ before the interrupt is recognized. For this reason, the return to the original program will *not* be lost no matter how many interrupt service cycles are completed.

Table 11.1 Interrupt Service Routine

Memory Contents

Memory Address	Mnemonic	Machine Code	Remarks
0	—	(X)	Address where the interrupt occurred
1	IOF	7100	Interrupt O Ff. Disable interrupt allowed flip-flop.
2	STC 600	4600	Store accumulator contents in 600
3	SIL 1	7701	Skip If device 1 flip-floop is Low
4	LMP 701	6701	Jump to device 1 service routine at 701
5	SIL 2	7701	Skip if device 2 flip-flop is low
6	JMP 711	6711	Jump to device 2 service routine at 711
7	CLR	0011	← Return here after calling device is serviced
10	ADD 0	2000	$A \leftarrow X$
11	ADD 601	2601	$A \leftarrow$ JMP X
12	STC 15	4015	$M(15) \leftarrow$ JMP
13	ADD 600	2600	$A \leftarrow$ original C(A)
14	ION	7200	Set interrupt allowed flip-flop
15	(JMP X)	(6 $\underline{X}$)	Jump back to original program
. . .			
. . .	—	—	← (Interrupt occurs immediately before this instruction. Hardware stores X in location 0 and sets PC for address 1.)
. . .			
. . .			
600	—	—	Accumulator contents stored here
601	JMP	6000	
. . .			⎫ Device service routines. At the end
701			⎬ of each are the instructions: REC n
. . .			⎱ followed by JMP 7, where n is the
777			⎭ device number.

Hardware produces this jump

Interrupt Occurs Here

Original Program

11.4 DIRECT MEMORY ACCESS

When large amounts of data are to be transferred to or from sequential memory locations in concurrence with the running of a main program, it is not feasible to use an interrupt system because it requires too much time for service by the software subroutine. This problem is solved in many computers through the use of *Direct Memory Access, DMA*. Here, each time the peripheral is ready to transfer a data word, it merely signals the computer; and without disturbing the PC or the accumulator, the main program instruction stream is *delayed one memory cycle* while hardware logic completes the data transfer. (For this reason, the method is frequently called *cycle stealing*.) Thus we see that, once DMA is initiated, the entire full data transfer will take place, whether it be 100 or 10,000 words, independently of the main program and without use of software. Although all three types of data transfers—programmed, interrupt, and DMA—can be active concurrently in a modern computer, for simplicity we will draw logic diagrams and write RTL programs here as if only DMA exists in our computer. Through this assumption, the principles of DMA can be quickly grasped; and after that it will be easier to understand systems composed of two or more transfer methods operating simultaneously.

Since DMA does not employ a software subroutine, the following aspects must be implemented entirely in hardware:

1. An address register, AR, pointing to the next memory location which is to receive or transmit data.

2. A method for incrementing AR so that a sequential group of memory locations can take part in the data transfer operation.

3. A method of determining when the last memory location has been served.

4. A data buffer, DB, to serve as a temporary storage register for data as they are transferred between the memory buffer, MB, and the peripheral.

5. Control signals to let the computer know when device service is required and to indicate whether data are being transferred in or out.

The circuitry required to implement this list is presented in Figure 11.5. Note how each item on the list is realized by one or more logic blocks and how these interact with the major computer circuits on the left side of the figure. A comparator is included within the interface to determine when the address register is pointing to the final memory address so that the interface logic will know when the DMA operation is complete. (For simplification. the methods of loading the address and FA registers, as well as several other details, have been omitted.)

The RTL program required to implement DMA within the ADD instruction is shown in Table 11.2. (Refer to Table 7.6 for the RTL of the normal ADD instruction.) The reader should study the correspondence between the information given in the table and that in Figure 11.5, Note that, when no service is required (SR $= 0$), the ADD execution cycle ends, as previously implemented, at $T = 2$. On the other hand, when service is required, states $T = 3$ through $T = 5$ are executed and the IF statement, at $T = 4$, controls whether the computer receives or sends data. Thus, $T = 3$ through $T = 5$

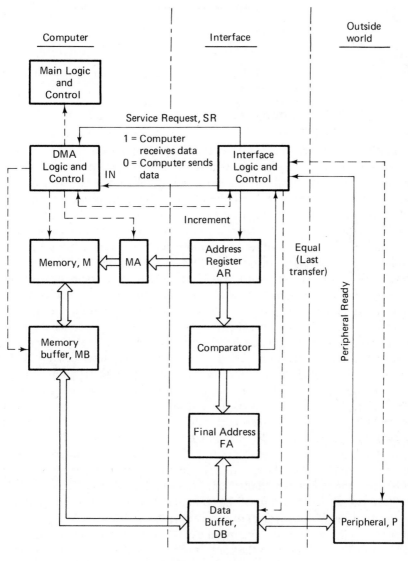

Notes: 1) Except where stated otherwise all blocks are registers
2) The connecting lines employ the following notation:
)--→ General timing and control signals
)—→ Specific control signals (labeled)
)⇨ Data lines
3) Only DMA Logic shown

Figure 11.5   Direct Memory Access System.

Table 11.2 ADD Instruction—Execution Cycle for DMA

$T = 0$ $MA \leftarrow MB_{0-8}$;
$T = 1$ $MB \leftarrow M(MA)$;
$T = 2$ $A \leftarrow A + MB$; $M(MA) \leftarrow MB$;
IF $(SR = 0)$ $S \leftarrow 0$, $T \leftarrow 0$ ELSE $MA \leftarrow AR$;
$T = 3$ $MB \leftarrow M(MA)$;
$T = 4$ IF $(IN = 0)$ $DB \leftarrow MB$ ELSE $MB \leftarrow DB$;
$T = 5$ $M(MA) \leftarrow MB$; $S \leftarrow 0$; $T \leftarrow 0$;

essentially constitutes the memory cycle which is stolen from the main program in order to transfer one data word. The significant amount of time saved by DMA compared to the interrupt method should now be obvious. (Observe that in attempting to eliminate a state from the table, it is not possible to move the MB ← DB production from state $T = 4$ back into state $T = 3$ because there would then be a contradiction on the contents of MB.)

Frequently there is a millisecond or more delay between the time one data word transfer takes place, through the DMA channel, and the time the peripheral is ready for the next word, which means that 500 or more instructions from the main program could be executed. Thus, the increase in efficiency by using DMA to run a program concurrently with data transfer, rather than doing the two operations separately, presents a tremendous saving—one much larger even than the savings between DMA and the interrupt method.

Can the logic required by the DMA interface be described by RTL? Certainly, it is feasible, and as a matter of fact, the required program is shown in Table 11.3. The symbol T^* is used to represent the program states of the interface in order to distinguish them from the T states of the computer (Table 11.2); but it can be seen that synchronization is established between the two state generators at $T^* = 0$ and $T^* = 2$.

It should be obvious that $T^* = 0$ is the state where the interface waits until the peripheral is ready, and that the computer must be starting the fetch state before the interface actually begins the next DMA cycle. The reason the interface pauses at $T^* = 2$ until the computer is at $T = 4$ is to prevent premature transfer of data from DB to the peripheral. This, as well as several other principles, will be clarified by an examination (in sequence) of the productions in Table 11.3 within the context of both Table 11.2 and Figure 11.5.

Table 11.3 Interface Control Signals

$T^* = 0$ IF (Peripheral Ready AND $S,T = 10$) $SR \leftarrow 1$, $AR \leftarrow AR + 1$ ELSE $T^* \leftarrow 0$;
$T^* = 1$ If (Computer Receives Data) $IN \leftarrow 1$, $DB \leftarrow$ Peripheral Data;
$T^* = 2$ IF $(T \neq 4)$ $T^* \leftarrow 2$;
$T^* = 3$ IF $(IN = 0)$ Peripheral $\leftarrow DB$;
IF $(AR \neq$ Final Address$)$ $T^* \leftarrow 0$;
$T^* = 4$ HALT;

Since the content of the address register is incremented before it is sent to MA, the initial value of AR should be one less than the starting address of the data in memory.

11.5 INPUT/OUTPUT SYSTEM VARIATIONS

The previous discussion focused on single examples of each main type of data transfer method. We will now present some of the many variations which can be employed. No attempt will be made to be exhaustive in this treatment since there are a vast number of alternate schemes in existence and many more likely to be developed.

Interrupt Polling Revisited

In the discussion of interrupt data transfers, a single interrupt bus connected all interfaces to the CPU. Thus, the computer had to poll the individual interfaces until the first one requiring service was identified; this can take a long time if there are many interfaces. There are several ways to solve this problem; but we will choose a solution that services the interrupts on a priority basis under program control.

The system is based on the following principles:

1. Each interface interrupt line is separately connected to the computer rather than being combined on a bus. Thus, the requesting device's identity is not lost.

2. A decoder receives the interrupt lines and produces an output, in binary, which is the number (address) of the device requiring attention. If two or more devices are requesting service, the one with the lowest device number is given priority.

3. When the computer executes the instruction Load Accumulator from Decoder $\equiv$ LAD $\equiv$ 7300, the original content of the accumulator is replaced by the decoder's output. This number may then be employed as the address part of various instructions to perform the actual data transfers between the computer and the device. For example, if 5 were the interrupting device with its binary number in the accumulator, and if location 200 contained 7400, execution of the instruction ADD 200 would yield 7405 in the accumulator, which is the instruction RED 5. (In order to execute RED 5, ADD 200 would be followed by STC $p + 1$.)

4. The interrupt call signal is developed from an OR combination of all the service requests. The CPU is then interrupted much as was done in Figure 11.4.

5. The ordering of the device signals entering the decoder gives priority to the lowest-number device requesting service. To provide program control over the priority, however, the bits of a *mask register* are ANDed with the corresponding device lines before they reach the decoder. When a particular mask bit output is zero, that interrupt is inhibited and the next interrupt in sequence will be selected by the decoder's output. The contents of the mask are controlled by the instruction: Transfer data from the Accumulator to the Mask $\equiv$ TAM $\equiv$ 7400. (Since n is never equal to zero, RED n = 74 n is easily distinguished from TAM.)

A logic diagram for the interrupt identification and priority system is shown in Figure 11.6. The reader should carefully study how each of the above features is realized by a specific part of the circuit. It should be clear that the mask register has no effect on the interrupt itself since AND gates 0 through 11 occur after the interrupt call signals is developed in the OR gate. Note that the O-output of each mask bit is employed as an input to an AND gate; thus, a cleared mask register, has 1's on the mask output lines and allows *all* interrupt requests to reach the decoder.

How is the decoder implemented in order to select the lowest-numbered calling device and to produce a corresponding binary code? In Figure 11.6 there are 12 devices, which means that the decoder must have four output bits ($2^3 < 12 < 2^4$). For simplification, however, the design of a three-device decoder is presented in Figure 11.7. The first step, part (a) of the figure, requires that all possible combinations of the device inputs be expressed in a truth table. Then, it is only necessary to scan each line of the table from right to left and determine the bit number (in binary) of the first logical 1. This number becomes the decoder output in the second half of the table. Having completed the truth table, the information is transferred to Karnaugh maps, and the output functions IA_i (for all i) are read.

A Multilevel Interrupt System

Although the circuit in Figure 11.6 has removed some of the deficiencies of the previous interrupt system, it does not permit higher-priority devices to interrupt the service routine of lower priority devices. Thus, an urgent interrupt (such as a power failure warning occurring just after the service initiation of a low-priority device) would be forced to wait until the computer was finished with the first device before the new interrupt could be considered. Frequently such a delay would be unacceptably long. Fortunately, this problem can be solved by a modification of the interrupt logic. The major change is a circuit on the output of the decoder which determines whether a new device, of higher priority, is requesting service; and if it is, a new interrupt is immediately generated. One way to accomplish this is to

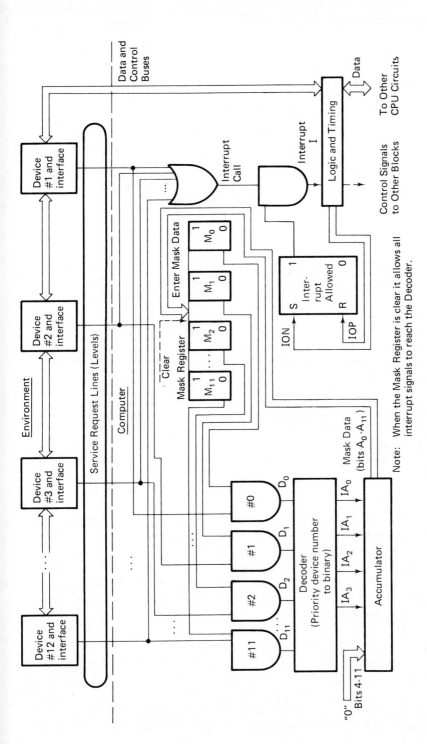

Figure 11.6 Interrupt Identification and Priority System.

Note: When the Mask Register is clear it allows all interrupt signals to reach the Decoder.

D_2	D_1	D_0	IA_1	IA_0
0	0	0	0	0
0	0	1	0	1
0	1	0	1	0
0	1	1	0	1
1	0	0	1	1
1	0	1	0	1
1	1	0	1	0
1	1	1	0	1

(a) Decoder Specification
(Priority given to least significant on-bit)

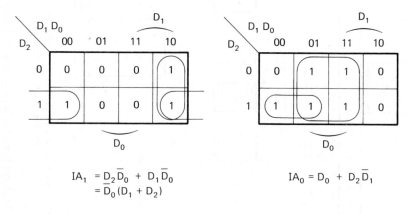

$$IA_1 = D_2\overline{D_0} + D_1\overline{D_0}$$
$$= \overline{D_0}(D_1 + D_2)$$

$$IA_0 = D_0 + D_2\overline{D_1}$$

(b) Interrupt Address Functions

Figure 11.7   Decoder Design.

store the number of the device currently being served in a register and continuously compare the register's contents with the decoder's output. When the comparator generates a not-equal signal, a new service request has occurred with higher priority than the one currently being handled. (Recall tha the decoder, similar to the one shown in Figure 11.7, develops a binary output which is the lowest-number device currently requesting service; hence, it is the device of highest priority.)

A block diagram of the multilevel priority interrupt system is shown in Figure 11.8; note how the interrupt is generated by the comparator from the decoder and current device register signals. The similarity between this circuit and the one in Figure 11.6 will be quickly recognized. Actually, the only major difference is that, in Figure 11.8, a new interrupt is produced whenever a request for service is received from a device of higher priority than the one currently being processed. In contrast, the system in Figure 11.6 disables the

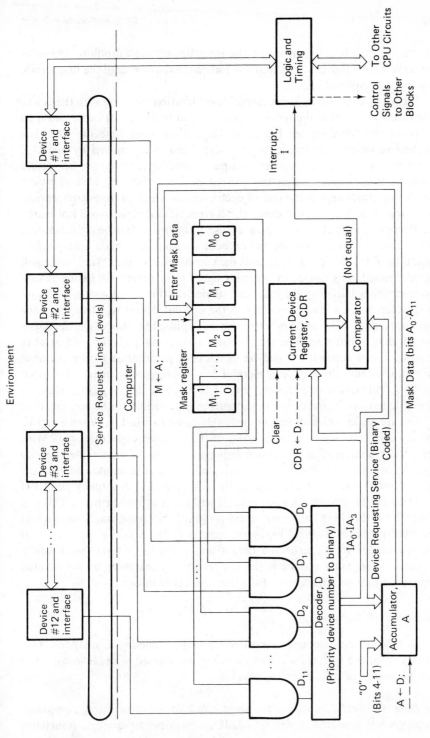

Figure 11.8 A Multilevel Priority Interrupt System.

411

interrupt capability, by means of the interrupt allowed flip-flop, whenever the service for a device starts, and it is not enabled again until the first service request is completed.

Figure 11.8 should be given careful consideration. Starting with the device interfaces, we see that the service request signal is a level. [It comes from an interrupt call flip-flop; see Figure 11.4(a).] Thus, each time the CPU has finished servicing a device, its request signal must be removed by the REC n instruction. Since the decoder's output is zero when there are no service requests, it is necessary that the current device register be cleared before processing starts and at the end of each cycle in which *all* interrupt services are completed. If this is not done, the current device register will not match the decoder's output, resulting in a false interrupt signal being generated. When a service request occurs, the decoder's output becomes a binary representation of the device number, and the comparator's output (i.e., not-equal signal) immediately causes an interrupt signal to be generated. Included with the electronics initiating the interrupt service routine is automatic hardware to transfer the new device number to the current device register. Thus, the interrupt signal, I is turned off after service begins; but, should the decoder's output change during the service routine, indicating a higher-level priority request, a new interrupt would be developed. In contrast to the previous interrupt systems, the multilevel method does not require an interrupt allowed flip-flop; but similar to previous systems, a machine language instruction is available to transfer the device number from the decoder to the accumulator.

One question which should immediately come to mind is: When a service routine is interrupted, how is the value of the program counter stored? With the proper combination of service requests, a chain of interrupted service routines would exist and a whole series of PC, accumulator, and perhaps other register values would need to be stored so that eventually the CPU could correctly accomplish each request within its proper priority. This can produce a rather long and complicated program; but some machines, such as the PDP-11, have a special hardware register called a *push-down stack* that stores numbers in sequence and then allows them to be removed in last-in first-out order. Such a stack is the key to quickly matching an interrupted service routine with its correct PC value and other information.

Advanced I/O Systems

Significant improvements are currently being made in I/O methods. Two typical advanced systems will now be briefly mentioned in order to complete our discussion of data transfer methods.

Data Channels. Large-scale, third- and fourth-generation computers employ what amounts to several small independent processors, sometimes called *data channels*, to handle I/O transfers with the outside world. In this scheme, the main CPU is never burdened with routine I/O operations, but it is free to devote its full time to computational tasks. While the CPU is finishing computational work currently in memory, the data channels load new jobs into the main memory from disk or other sources. When the CPU finishes a job, it signals the I/O controller which then assigns a free data channel to output the results on the specified peripheral. Thus, parallel processing is achieved with several I/O jobs simultaneously in progress and overlapping in time with the actual computation work.

The Control Data CYBER-70 series of computers is representative of this approach. It contains as many as 20 of these independent computers called *peripheral processors*. Each has its own small memory, which can execute programs independently of the main processor and can communicate with the main memory as well as with the CPU.

Certain models of the IBM System 370 (also System 360) provide for high-speed peripherals (such as disk and tape) on *selector channels* and for low-speed peripherals (such as printers and card readers) on *multiplexer channels*. Because of the speed requirement, selector channels are devoted exclusively to servicing a single peripheral at one time. As the name implies, however, multiplexer channels are designed to work with several peripherals at once by rapidly going through an operation cycle in which each is served for a short time in sequence. Of course, special precautions must be taken to insure that the data for the various peripherals are not mixed.

Recent developments in microcomputer technology will certainly lead to even more efficient ways of handling data transfers; and they will no doubt allow parallel I/O processing to be achieved on much smaller-scale systems.

Memory-Increment and Add-To-Memory. Two features which are proving to be a very popular addition to computer DMA units are *memory-increment* and *add-to-memory*. Memory increment is an operation in which the current contents of a selected location are read, 1 is added, and the result is written back into the original location. This process is very handy for producing histograms from statistical information, e.g., a voltage frequency spectrum from a noise source. The main additions required to our original DMA units are a circuit to do the incrementing and one to set the address register in the interface to the required histogram bin number. (Usually this number will be from an A/D converter, which is sampling at a rate specified by a clock in the interface.)

Add-to-memory is similar to memory-increment except that other number instead of just 1 may be added to the original memory contents. One of the main uses for this process is to average repetitive signals which are corrupted by noise. Here sequential memory locations are employed to store increasing time samples of the signal. The address register in the interface is always reset to the first sample's memory location before each new signal occurs. When many corresponding (in time) measurements of the noisy signal are added at each sample point, the random noise cancels, leaving the uncorrupted signal.

A detailed discussion of memory-increment and add-to-memory is found in Reference 7. Moreover, References 4, 6, 7, 10, 11, and 12 are a broad and rich source of material on all facets of computer interfacing.

11.6 BIBLIOGRAPHY

1 BLACK, W. WAYNE, *An Introduction to ON-LINE COMPUTERS*. New York: Gordon & Breach, 1971.

2 DAVIES, DONALD W., and DEVEK L. A. BARBER, *Communication Networks for Computers*. New York: John Wiley & Sons, Inc., 1973.

3 ENSLOW, PHILIP H. JR. (ed.), *Multiprocessor and Parallel Processing*. New York: John Wiley & Sons, Inc., 1974.

4 FALK, HOWARD, "Linking Microprocessors to the Real World." *IEEE Spectrum*, 1974, pp. 59–67.

5 *Integrated Circuits: Digital-Linear-MOS*. Sunnyvale, Ca.: Signetics Corporation. (Consult most recent edition.)

6 JURGEN, RONALD K., "Minicomputer Applications in the Seventies." *IEEE Spectrum*, 1970, pp. 37–52.

7 KORN, G. A., *Minicomputers for Engineers and Scientists*. New York: McGraw-Hill Book Company, 1973.

8 *Logic Handbook*. Maynard, Mass.: Digital Equipment Corporation. (Consult most recent edition.)

9 *PDP-11 Peripherals and Interfacing Handbook*. Maynard, Mass.: Digital Equipment Corporation. (Consult most recent edition.)

10 SCHOEFFLER, JAMES D., and R. H. TEMPLE, *Minicomputers: Hardware, Software and Applications*. New York: IEEE Press, 1972.

11 SOUCEK, BRANKO, *Minicomputers in Data Processing and Simulation*. New York: John Wiley & Sons, Inc., 1972.

12 ZISSOS, D., and F. G. DUNCAN, *Digital Interface Design*. London: Oxford University Press, 1973.

11.7 EXERCISES

11.1 Propose a logic circuit to be contained in the device decoder block of Figure 11.2(a).

11.2 In a manner similar to that in Figure 11.2, show how the WED n instruction may be implemented.

11.3 Instead of employing a skip instruction, it is desired to write a new instruction, FAC, to cause the Flag to make the ACcumulator positive for flag set and negative for flag not set. Write the RTL program to accomplish this.

11.4 If everything is the same except a SOL $n \equiv$ Skip On flag Low instruction, instead of SOH n, is the only flag test available, write the sequence of instructions to test flag 15.

11.5 Twelve interrumpt interfaces are to have a flag which can be identified quickly without polling. Explain your method and draw the logic diagram for the flag portion of the system.

11.6 Rewrite the RTL program in Figure 11.4(b) assuming that memory location zero is a semiconductor register with none of the limitations of core but which is controlled by the MA and MB registers.

11.7 Describe the logic employed by the ION instruction so that an interrupt is not recognized until the instruction following it is being executed.

11.8 Estimate the approximate percentage of savings in time achieved by using DMA instead of an interrupt to read data into memory.

11.9 With a computer having a 12-bit accumulator, is it feasible to employ the general technique of Figure 11.6 to service 24 or more interrupts? Explain in detail.

11.10 Consider a decoder for an interrupt identification system with four interrupt lines and with highest priority on the most significant bit. Find the logic function which bit *IA* must satisfy.

11.11 If a large number of interfaces exist in a system, the decoder technique of Figure 11.7 can become unwieldy. Present a heuristic approach for decoder design which does not have this limitation. Break your design into a priority chain circuit having:

(a) The same number of outputs, n, as it has inputs.

(b) A logical-1 output *only* for the lowest-number device requesting service. Follow the priority circuit by logic which accepts n input lines and produces the binary code of the device requesting service. (If no device is requesting service, output the number 0.) Demonstrate your method with a system having four devices.

11.12 For the interrupt system in Figure 11.6, write a pseudo-LINC program for the device identification portion of an interrupt service routine having the following requirements: If the device requiring service has high priority (as defined by the original mask register contents), be sure the device number is in the accumulator and branch to location 300 (service routine 1). If the device has low priority, place the device number in the accumulator and branch to location 350 (service routine 2). Assume that the mask register was originally loaded with a nonzero value. (You may want to employ the instruction Skip if Accumulator is ZEro $\equiv$ AZE $\equiv$ 0450.)

FORTRAN SUBROUTINES FOR THE LOGICAL MODEL

```
1           SUBROUTINE ADD (X,Y,S,N)
     C***   ONES COMP. ADD
2           LOGICAL*1 X(N),Y(N),S(N),CC(12),C
3           CALL TADD(X,Y,S,C,N)
4           CALL FALSE(CC,N)
5           CC(N)=C
6           CALL TADD(S,CC,S,C,N)
7           RETURN
8           END


9           SUBROUTINE AND (A,B,C,N)
     C***   A&B ARE ADDED WITH RESULT STORED IN C
10          LOGICAL*1 A(N), B(N), C(N)
11          DO 10 I=1,N
12    10    C(I)= A(I) .AND. B(I)
13          RETURN
14          END


15          SUBROUTINE COM (A,B,N)
     C***   B=A'.N=BITS. A IS UNCHANGED
16          LOGICAL*1 A(N),  B(N)
17          DO 10 I= 1,N
18    10    B(I) =.NOT. A(I)
19          RETURN
20          END


21          SUBROUTINE EQ (B,A,N)
     C***   A(I)=B(I) FOR ALL I. B IS UNCHANGED
22          LOGICAL*1 A(N), B(N)
23          DO 10 I=1,N
24    10    A(I)=B(I)
25          RETURN
26          END


27          SUBROUTINE EQUAL(B,L,M,A,J,K)
     C***   EQUATE SUBSCRIPT TERMS
     C A(J THRU K)=B(L THRU M) RESPECTIVELY.
     C REQUIRED: K-J=M-L
28          LOGICAL*1 A(K), B(M)
```

```
29              N=M-L+1
30              DO 10 I=1,N
31              J1=J-1+I
32              L1=L-1+I
33      10      A(J1)=B(L1)
34              RETURN
35              END


36              SUBROUTINE FALSE (A,N)
        C*** A=FFF...F, N=BITS
37              LOGICAL*1 A(N)
38              DO 10 I=1,N
39      10      A(I) =.FALSE.
40              RETURN
41              END


42              SUBROUTINE FILMEM (M,L,N)
        C*** FILL FIRST L LOCATIONS IN MEMORY,M, FROM DATA CARDS
        C    5 OCTAL NUMBERS/CARD; FORMAT I10
        C   N=BITS /WORD
43              IMPLICIT INTEGER(O)
44              LOGICAL*1 L1(50),M(L,N)
45              DIMENSION O(50)
46              READ(5,79) (O(I),I=1,L)
47              DO 20 J=1,L
48              CALL OTOL(O(J),L1,N)
49              DO 20 I=1,N
50      20      M(J,I)=L1(I)
51              RETURN
52      79      FORMAT(5I10)
53              END


54              SUBROUTINE LODA(M,B,A,L,N)
        C*** MEM., M, LOCATION B IS LOADED IN REGISTER A
        C   L= WORDS/ MEM.    N= BITS/ WORD
55              LOGICAL*1 A(N), B(N), M(L,N)
56              CALL LOTD (B,J,N)
57              DO 10 I=1,N
58      10      A(I)=M(J,I)
59              RETURN
60              END


61              SUBROUTINE LODM(A,M,B,L,N)
        C*** REGISTER A IS LOADED IN MEM., M, LOCATION B
        C   L= WORDS/ MEM.    N= BITS/ WORD
62              LOGICAL*1 A(N), B(N), M(L,N)
63              CALL LOTD (B,J,N)
64              DO 10 I=1,N
65      10      M(J,I)=A(I)
66              RETURN
67              END


68              SUBROUTINE LOTD (L,O,N)
        C*** L= LOGICAL, O=DECIMAL, N=BITS/WORD
69              INTEGER O
70              LOGICAL*1 L(N)
        C     ACCOUNT  FOR SIGN
```

```
71              J =0
72              IF ( .NOT. L(1) ) GO TO 5
73              CALL TCOM (L,N )
74              J=1
     C    CONVERSION PROPER
75       5      O=0
76              DO 2 I=1, N,3
77              O=O*8
78              IF(L( I+2 ))O=O+1
79              IF(L(I+1)) O=O+2
80       2      IF(L(I)) O=O+4
81              IF (J.EQ. 1) O=-O
82              RETURN
83              END


84              SUBROUTINE LOTO (L,O,N)
     C*** L= LOGICAL, O=OCTAL, N=BITS/L
85              INTEGER O
86              LOGICAL*1 L(N)
     C    ACCOUNT  FOR SIGN
87              J =0
88              IF ( .NOT. L(1) ) GO TO 5
89              CALL TCOM (L,N )
90              J=1
     C    CONVERSION PROPER
91       5      O=0
92              DO 2 I=1, N,3
93              O=O*10
94              IF(L( I+2 ))O=O+1
95              IF(L(I+1)) O=O+2
96       2      IF(L(I)) O=O+4
97              IF (J.EQ. 1) O=-O
98              RETURN
99              END


100             SUBROUTINE OR (B,C,A,N)
     C***    C=A OR B. N= BITS
101             LOGICAL*1 A(N), B(N), C(N)
102             DO 10 I=1,N
103      10     A(I)=B(I).OR.C(I)
104             RETURN
105             END


106             SUBROUTINE OTOL (OO, L,N )
     C***OO=OCTAL; L= LOGICAL-T,F;
     C        N=BITS/L; BIT 1=SIGN
107             INTEGER O, OM, OO
108             LOGICAL*1 L(N)
109             O=OO
     C    CHECK FOR NEGATIVE
110             J=0
111             IF (O.GE. 0 ) GO TO 10
112             O=-O
113             J=1
     C    CONVERSION PROPER
114      10     CALL FALSE (L,N)
115             K=N-2
116             DO 8 I=1,K,3
```

```
117    2      OM = O-( O/10)*10
118           O =O/10
119    1      IF ( OM.EQ. 0) GO TO 8
120           GO TO (3,4,3,4,3,4,3 ), CM
121    3      L(N+1-I ) =.TRUE.
122    4      GO TO (6,5,5,6,6,5,5 ), OM
123    5      L(N-I ) = .TRUE.
124    6      GO TO  (8,8,8,7,7,7,7), CM
125    7      L( N-1-I  ) = .TRUE.
126    8      CONTINUE
127           IF (J.NE. 1) GO TO 11
128           CALL TCCM (L,N )
129    11     RETURN
130           END


131           SUBROUTINE PRTMEM (M,L,N)
       C***   PRINT FIRST L LOCATIONS IN MEMORY,M;
       C      N= BITS/WORD.
132           LOGICAL*1 M(L,N)
133           WRITE(6,89)
134           DO 20 J=1,L
135    20     WRITE (6,93) J, (M(J,I), I=1,N)
136           WRITE (6,85)
137           RETURN
138    85     FORMAT ('0'/ '0')
139    89     FORMAT      ( 20X,' MEMORY CONTENTS' /20X,
              $ '-------------------' /)
140    93     FORMAT (19X,I3, 60L2)
141           END


142           SUBROUTINE SL (A,B,N )
       C***   REGISTER A IS SHIFTED LEFT  1 BIT; RESULT IN B;
       C      N=BITS. F TO LSB, BIT N.
143           LOGICAL*1 A(N), B(N)
144           M= N-1
145           DO 20 I=1,M
146    20     B (I)   = A (I+1)
147           B(N) = .FALSE.
148           RETURN
149           END


150           SUBROUTINE SR (A,B,N )
       C***   REGISTER A IS SHIFTED RIGHT 1 BIT; RESULT IN B;
       C      N=BITS. F TO MSB, BIT 1.
151           LOGICAL*1 A(N), B(N)
152           M= N-1
153           DO 20 I=1,M
154    20     B (N-I+1 ) = A(N-I)
155           B(1) = .FALSE.
156           RETURN
157           END


158           SUBROUTINE TADD (X,Y,S,C,N)
       C***   TWO'S COMP, ADD. S= X+Y WHERE N= BITS. BIT #1=SIGN.
159           LOGICAL*1 C,CC,S(N), X(N),Y(N),P,Q,D
160           C=.FALSE.
161           DO 1 J=1,N
```

```
162          I = N+1-J
163          P =.NOT. X(I)
164          Q= .NOT. Y(I)
165          D =.NOT. C
166          CC =(X(I) .AND.C ) .OR. (X(I) .AND. Y(I) )
        $      .OR. (Y(I) .AND. C)
167          S(I) = (P.AND. Q .AND. C) .OR. (P.AND. Y(I)
        $      .AND.  D)
        $      .OR. (X (I).AND. Q .AND. D)  .OR.  (X(I)
        $      .AND. Y(I) .AND. C)
168    1     C =CC
169          RETURN
170          END


171          SUBROUTINE TCOM (A,N)
       C*** TWO'S COMPLEMENT; A=A'+1. N= BITS
172          LOGICAL*1 A(N), J(12)
173          CALL COM (A,A,N)
174          CALL FALSE (J,N)
175          J(N) =.TRUE.
176          CALL TADD (A,J,A,C,N)
177          RETURN
178          END


179          SUBROUTINE TESTEQ(A,B,C,N)
       C***   IF A=B, SINGLE BIT C=1
180          COMMON D(500),E(500)
181          LOGICAL*1 A(N),B(N),C,D,E
182          CALL COM(A,D,N)
183          CALL COM(B,E,N)
184          CALL AND(D,E,E,N)
185          CALL AND(A,B,D,N)
186          CALL OR(D,E,D,N)
187          CALL COM(D,D,N)
188          CALL TESTZE(D,C,N)
189          RETURN
190          END


191          SUBROUTINE TESTZE (A, B, N)
       C***   IF N BIT NUMBER A=0, SINGLE BIT B=1.
192          LOGICAL*1 A(N), B
193          B=.TRUE.
194          DO 10 I=1,N
195    10    IF(A(I)) B=.FALSE.
196          RETURN
197          END
```

SUMMARY OF PSEUDO-LINC I/O INSTRUCTIONS

Mnemonics	Machine Code	Meaning
REC n	70 n*	R Eset Call flip-flop n.
IOF	7100†	Turn Interrupt flip-flop O Ff.
ION	7200	Turn Interrupt flip-flop O N.
LAD	7300	Load Accumulator from Decoder.
WED n	73 n	WritE accumulator into Device n.
TAM	7400	Transfer data from the Accumulator to the Mask.
RED n	74 n	R Ead data from Device number n into accumulator.
SOH n	75 n	Skip On flag High from device n.
SIL n	77 n	Skip if Interrupt call flip-flop of device n is Low.

*Device numbers are allowed to have the following range: $01 \leq n \leq 77$.
†The code $7i00$ ($0 \leq i \leq 7$) indicates an instruction which does not require a device number.

INDEX

A

A (*see* Accumulator)
Accumulator, 45, 234
Adder:
    fast, 263-69
    full, 259
    half, 301
    parallel, 260
    serial, 260-62
    speedup, 278, 280-83
Addition, 24
Addressing modes:
    direct, 353
    immediate, 353
    indexed, 352-53
    indirect, 351-53
Add-to-memory, 413-14
Aiken, Howard, 6
Algebra:
    Boolean, 63-71
    laws, 68-69
Algorithm, 44, 52
Alphameric codes, 37-38
ALU (*see* Arithmetic and Logic Unit)
AND gate, 63-64
Arabic notation, 19

Architecture, 213, 348-92
Archival storage, 303
Arithmetic and Logic Unit, 258-302
Arithmetic operations, RTL, 217
Array processor, 388-91
ASCII, 38
Assembler, 55-57
Associative memory, 246-49
Asynchronous, 131, 379-83
Atanasoff, John, 6

B

Babbage, Charles, 5-6
Base, 19-20
BCD, 30
Binary adder, 259
Binary addition, 24
Binary codes (*see* Codes)
Binary multiplication, 27
Binary numbers, 20-23
Bit, 21
Boole, George, 63, 106
Boolean algebra, 63-71
Booth algorithm, 287-90
Borrow, 263

Branch and control instructions, 47, 49
Bug, 51
Bus structures, 355-58

C

Cache memory, 384
Calculator, 360
Canonical expansion, 71-72
Carry conditions, 264
Carry look-ahead, 264
Cathode ray tube, 227, 395
Character generator, 227-30
CLA, 265
Clocked flip-flops, 128
Clock pulse, 128
CM (*see* Control memory)
Codes:
    alphameric, 37-38
    ASCII, 38
    BCD, 30
    error-detecting, 35-37
    gray, 33-34, 137-38
    self-complementing, 32
    special, 37-39
    unit-distance, 33-35
    weighted, 29-32
Combinational circuits, 110
Comparator, 226-27, 294
Comparison division, 291-93
Compiler, 57-58
Complement (for subtraction), 24-27
Complementary approach, 80-83
Computers:
    analog, 11-12
    classification of, 11-12
    general-purpose, 11
    special-purpose, 11
Conditional operations, RTL, 220
Control memory, 340-45, 363-65
Control memory map, 375

Conversion:
    fractions, 23-24
    integers, 20-23
Converter, D/A, 228
Core Memory, 304-12
Cost, design criteria, 142-46
Counter:
    BCD, 126-28
    down, 129-31
    gray-code, 137-39
    modulo-5, 136-37
    ring, 135-36
    ripple, 129-33
    synthesis of, 122-33
    three-bit up, 123-26
CPU, 45
CROM (*see* Control memory)
Cross assemblers, 378
CRT (*see* Cathode ray tube)
Cutoff, transistor, 166
Cycle stealing, 404

D

D/A converter, 228
Data channels, 412-13
Data transfer methods:
    DMA, 396, 403-7
    interrupts, 396, 400-403
    programmed, 396-400
Davis, Ruth M., 16
Decoder, 100-101
    balanced, 315-20
    circuits, 312-20
    comparison (table), 320
    parallel, 312-14
    tree, 314-15
Delay, propagation, 159-60
De Morgan's theorem, 69-70
Demultiplexer, 102
Destructive readout (*see* Nondestructive readout)

D flip-flop, 140
Digital computer characteristics, 13-16
Digital computers, evolution, 4-7
Digital logic families, 207
Diode:
    characteristics, 152-55
    gates, 152-56
Diode-Transistor Logic (DTL), 169-73
Direct addressing, 353
Direct-coupled Transistor Logic
    (DCTL), 173-75
Direct memory access, (DMA), 396,
    403-7
Disk, 303
Division, 28, 290-99
    comparison, 291-93
    nonrestoration, 291, 297-99
    restoration, 291, 293-97
DMA, 396, 403-7
Don't-care, 80
Dual, 70

E

EBCDIC, 38
ECAP, 14, 57, 58
Eckert, J. Presper, 6
ECL, 182-85
Edge triggering, 116
EDSAC, 6
Emitter-Coupled Logic (ECL), 182-85
Emulation, 372
End-around carry, 26
ENIAC, 6
Error-detecting codes, 35-37
Evolution (heuristic), 134
Excess-3 code, 32
Excitation maps, memory, 117
Exclusive-OR, 72
Execution cycle, 235-38

F

Fan-out, 161-64
Ferrite core, 305
Fetch cycle, 235
Flag, 397-400
Flip-flop, 115-22
    D-type, 138-40
    dynamic, 199-203
    elementary, 188
    J-K, 126-28, 190-94
    master-slave, 189
    MOSFET, 199-203
    realization, 187-95
    S-R, 115-20, 187-90, 192-93
Formats, instruction, 349
FORTRAN, 57-58
    simulation, 241-53
Full adder, 259
Full subtractor, 263
Function of n variables, 72-75

G

Gates, 64
    AND, 154-55
    diode, 152-56
    MOSFET, 196-98
    NAND, 74, 86
    NOR, 74, 85-86
    OR, 154
    transistor, 165-87
    transmission, 201
General purpose computer, 11
Gray code, 33-34, 137-38

H

Half adder, 301
Hardware, 16

Heuristic design, 133-42
Hexadecimal numbers, 20
Hollerith, Herman, 4
Horizontal microprogramming, 363
Hysteresis loop, 305

I

IC, 203-8
ILLIAC IV, 358, 390
Immediate addressing, 353
Indexed addressing, 352-53
Index register, 352-55
Indirect addressing, 351-53
Induction, perfect, 67-68
Input-output instructions, 396
Instruction, single-address, 47
Instruction cycle, 235
Instruction formats, 349
Instruction register, 234
Instructions, LINC, 49
Instruction words, 348-55
Integrated circuit (IC), 203-8
    families, 206-8
Interface designs, 395
Interleaving, memory, 385-86
Interrupt:
    circuits, 396, 400-403
    multilevel, 408-12
    service routine, 402-3
Inverter, 66, 168
I/O (*see* Input-output instructions)
Iteration, 134

J

J-K flip-flop, 126-28

K

Karnaugh map, 75-83, 117

L

Languages, programming, 54-58
Large-scale integation (LSI), 203
Large-scale systems, 378-92
Latch, 188
Leading-edge triggering, 192
LINC (*see* Pseudo-LINC)
Literal, 85
Logic:
    blocks, 156
    diagrams, 83-85
    exclusive-OR, 91-95
    NAND-NOR, 85-91
    negative, 156-57
    positive, 154, 156-57
    transistor (*see under specific circuit
        name*)
    wired, 185-87
Logical design, 63-105
Logical model subroutines, 416-20
Logic operations, RTL, 217
LSI, 203

M

M (*see* Memory)
MA (*see* Memory address register)
Machine language, 47, 55
Machine organization, 45-47
Macrodefinition (MACRO), 57
Macroprogram, 359
Magnetic core memory, 304-12
Manchly, John W., 6
Map, 34
Map simplification, 75-83
Mark I computer, 6
Mask register, 408
Mass memory, 303
Maxterm, 72
MB (*see* Memory buffer register)
Medium-scale integration (MSI), 203

Memory, 233
  associative, 246-49
  bipolar, 325-27
  cache, 384, 386
  cell practical, 325
  coincidence selection, 321-25
  coincident-current, 307
  control, 340-45
  core, 304-12
  cycle time, 304
  disk, 303
  excitation maps, 115-17
  interleaving, 385-86
  linear selection, 320
  MOS, 328-30
  nondestructive, 303
  organization, 308
  PROM, 333-35
  RAM (*see* Memory, random access),
    304
  random-access, 304
  read-only, 304, 333-35
  ROM (*see* Memory, read-only)
  ROM applications, 335-39
  scratch pad, 386
  semiconductor, 320-32
  3-D, 307
  TU1, 330-31
  2-D, 310-11
  2 1/2-D, 311-12
  virtual, 384-86
  volatility, 303
Memory address register, 45, 234
Memory buffer register, 233
Memory excitation maps, 117
Memory operations, RTL, 220
Memory-increment, 413-14
Microcomputer, 359
Microinstructions, 340, 362-68
Microprocessor, 1, 358-78
Microprogram, 55, 359
Microprogramming:
  horizontal, 363

Microprogramming (*cont.*)
  vertical, 363
Minicomputer, 1, 4
Minterms, 71-72
Mnemonics, 48
Modeling, 16
Module synthesis, 134
Modulo, 24
MOS (*see* MOSFET)
MOSFET:
  circuits, 195-203
  complimenting, 198
  flip-flops, 199-203
  n-channel, 196
  p-channel, 196
MSI, 203
Multiple bit processing, 278
Multiplexer, 102-5
Multiplication, 27, 269-90
  Booth algorithms, 287-90
  combinational logic, 278
  fast, 277-90
  1's and 2's complement, 271-77
  positive numbers, 269
  sign and magnitude, 271
Multiprocessor, 388-91

N

NAND, 74
NAND-NOR logic, 85-91
Negative logic, 156-57
Noise considerations, 193
Nondestructive readout, 303
Nonrestoration division, 297-99
Nonvolatile memory, 303
NOR, 74
NOT operation, 65-66
Numbers:
  binary, 20-23
  hexadecimal, 20
  octal, 20-21
Number base conversion, 20-24

O

Octal numbers, 20-21
1's complement, 26
On-line, 395
Organization:
  machine, 45-47
  memory, 308
OR gate, 64-65
Overflow, 277, 299-301

P

Parity, 35-37, 95-97
PC (*see* Program counter)
Perfect induction, 67-68
Peripheral processor, 413
Pipeline, processor, 389, 391
Polling, 401, 407
Positive logic, 154, 156-57
Postulates, Boolean, 67
Problem-Oriented languages, 14
Processor, peripheral, 413
Production, 214
Product-of-sums, 72
Program, 44
Program counter, 45, 234
Programmed transfer, 396-400
Programming, 54-58
PROM, 304, 333-35
Propagation delay, 159-60
Pseudo-LINC, 47-54
  instructions, a brief list, 49
  I/O instructions, summary, 421
Pull-up, active, 180

Q

Quotient, 291, 295

R

Radix, 19
RAM (*see* Memory, random-access)
Read-only memories, 105
Redundant states, 114-15
Reflected code, 33-34
Register Transfer Logic (RTL), 214-20,
    238-40
  statements, limitations on, 239-40
  tables of operations, 215, 217, 219,
    220
Relative addressing, 55-56
Resistor-Transitor Logic (RTL), 175-77
Restoration division, 293-97
Ripple method, 129-33
ROM (*see* Memory, read-only)
Rotate operation, RTL, 219
RTL (*see* Register Transfer Logic *and*
    Resistor-Transistor Logic)

S

Saturation, transistor, 166
Scale, 358
Scale operations, RTL, 219
Selector channels, 413
Self-complementing codes, 32-33
Semiconductor memory (*see* Memory)
Sense voltage, 306
Sequential circuits, 110-46
Serial addition, 260-62
Shifting over zeros, 283-86
Shift operations, RTL, 219
Sign-and-magnitude multiplication,
    271
Sign bit, 26
Simplification:
  map method, 75-83
  other methods, 83
  theorems, 66-71

Simulation, 16, 241-53
  arithmetic model, 249-53
  logic model, 241-49
  tables of statements, 243, 250
Software, 44-58
S-R flip-flop, 115-20
SSI, 203
State, memory, 111
State assignment, 113-14
State diagram, 111-12
State generator, 230-32
State table, 113-15
Stealing, cycle, 404
Stilbitz, George, 6
Subroutines, logical model, 416-20
Subtraction, 24-27, 262-63
Sum-of-products, 72
Synchronous circuits, 110
Synthesis, circuit, 115-33

T

Table:
  state, 113
  truth, 64
Teletypewriter, 45, 47
T flip-flop, 124
Theorems, 66-71
  De Morgan's, 69-70
  elementary Boolean, 67
  main Boolean, 69
Timing diagram, 116, 223-24
Totem-pole output, 180
Transfer instructions, 47

Transistor:
  characteristics, 165-67
  cutoff region, 166
  gates, 165-87
  hybrid parameters, 168
  models, 165-67
  saturation region, 166
Transistor-Transistor Logic (TTL),
    177-82
Transmission gate, 201
Tristate logic, 331
Truth table, 64
TTL (*see* Transistor-Transistor Logic)
2's complement, 26

U

Uniprocessor, 387, 391
Unit-distance codes, 33-35

V

Veitch diagram, 75
Vertical microprogramming, 363
Virtual memory, 384-85
Volatile memory, 303
Voltage margin, 158
Von Neumann machine, 387

W

Wired logic, 185-87
Worst-case design, 161
Write, 326